T0220198

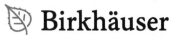

Samson Abramsky • Juha Kontinen
Jouko Väänänen • Heribert Vollmer
Editors

Dependence Logic

Theory and Applications

 Birkhäuser

Editors

Samson Abramsky
Department of Computer Science
University of Oxford
Oxford, UK

Juha Kontinen
Department of Mathematics and Statistics
University of Helsinki
Helsinki, Finland

Jouko Väänänen
Department of Mathematics & Statistics
University of Helsinki
Helsinki, Finland

Heribert Vollmer
Institut für Theoretische Informatik
Leibniz Universität Hannover
Hannover, Germany

ISBN 978-3-319-81123-9   ISBN 978-3-319-31803-5   (eBook)
DOI 10.1007/978-3-319-31803-5

Mathematics Subject Classification (2010): 03B15, 03B70, 03C80

Printed on acid-free paper

This book is published under the trade name Birkhäuser
The registered company is Springer International Publishing AG, CH

Preface

In February 2013, the editors of this volume organized a Dagstuhl Seminar on "Dependence Logic: Theory and Applications" at the Schloss Dagstuhl Leibniz Center for Informatics in Wadern, Germany. This event consisted of both invited and contributed talks by some of the approximately 40 participants. After the conclusion of the seminar, the organizers invited a number of speakers to write surveys that present the state-of-the-art knowledge in their area of expertise. The Dagstuhl Seminar was followed by an Academy Colloquium of the Koninklijke Nederlandse Akademie van Wetenschappen (KNAW) and a further Dagstuhl Seminar "Logics for Dependence and Independence" in June 2015. Also from these latter meetings, a few participants were invited to contribute to this volume.

All contributions were peer-reviewed by experts in the field and revised before they were included in this volume.

We thank the Directorate of Schloss Dagtuhl and the board of the KNAW for their support, the speakers of the seminars for making all three meetings a successful event, the referees and, above all, the contributors of this volume for their informative and well-written articles. We also thank Benjamin Levitt at the publisher's office for his support and guidance.

Oxford, UK — Samson Abramsky
Helsinki, Finland — Juha Kontinen
Helsinki, Finland — Jouko Väänänen
Hannover, Germany — Heribert Vollmer
June 2016

Contents

Contributors

A. Blass Mathematics Department, University of Michigan, Ann Arbor, MI, USA

I. Ciardelli ILLC, University of Amsterdam, Amsterdam, The Netherlands

J. Corander Department of Mathematics and Statistics, University of Helsinki, Helsinki, Finland

A. Durand IMJ-PRG, CNRS UMR 7586, Université Paris Diderot, Paris, France

P. Galliani School of Engineering and Informatics, University of Sussex, Falmer, Brighton, UK

E. Grädel Math. Foundations of Computer Science, RWTH Aachen University, Aachen, Germany

Å. Hirvonen Department of Mathematics and Statistics, University of Helsinki, Helsinki, Finland

W. Hodges Herons Brook, Sticklepath, Okehampton, Devon EX20 2PY, UK

J. Kontinen Department of Mathematics and Statistics, University of Helsinki, Helsinki, Finland

S. Link The University of Auckland, Auckland, New Zealand

H. Nyman Department of Mathematics and Statistics, Åbo Akademi University, Turku, Finland

E. Pacuit Department of Philosophy, University of Maryland, College Park, MD, USA

J. Pensar Department of Mathematics and Statistics, Åbo Akademi University, Turku, Finland

J. Väänänen Department of Mathematics and Statistics, University of Helsinki, Helsinki, Finland
Institute for Logic, Language and Computation, University of Amsterdam, Amsterdam, The Netherlands

H. Vollmer Institut für Theoretische Informatik, Leibniz Universität Hannover, Hannover, Germany

F. Yang Department of Philosophy and Religious Studies, Utrecht University, Utrecht, The Netherlands

Introduction

Samson Abramsky, Juha Kontinen, Jouko Väänänen, and Heribert Vollmer

Dependence logic was introduced by Jouko Väänänen in 2007. It extends first-order logic by new atomic dependence formulas (dependence atoms)

$$=(x_1, \ldots, x_n),\qquad\qquad(1)$$

the meaning of which is that the value of x_n is functionally determined by the values of $x_1, \ldots, x_{n-1}$. On the semantical side, dependence logic bases its semantics in the concept of a set X of assignments instead of, as is the case for usual first-order logic, a single assignment. Such sets are called *teams*. A team X is said to satisfy the dependence atom above if for any two assignments, if they agree on the variables $x_1, \ldots, x_{n-1}$ then they also agree on x_n. By viewing a team X as a database over attributes $x_1, \ldots, x_n$, dependence atoms correspond exactly to functional dependencies studied extensively in database theory. Dependence logic

S. Abramsky (✉)
Department of Computer Science, University of Oxford Wolfson Building, Parks Road, Oxford
OX1 3QD, United Kingdom
e-mail: samson.abramsky@cs.ox.ac.uk

J. Kontinen
Department of Mathematics and Statistics, University of Helsinki, P.O. Box 68,
00014 Helsinki, Finland
e-mail: juha.kontinen@helsinki.fi

J. Väänänen
Department of Mathematics and Statistics, University of Helsinki, Helsinki, Finland

Institute for Logic, Language and Computation, University of Amsterdam, Amsterdam,
The Netherlands
e-mail: jouko.vaananen@helsinki.fi

H. Vollmer
Institut für Theoretische Informatik, Leibniz Universität Hannover, Appelstr. 4,
30167 Hannover, Germany
e-mail: vollmer@thi.uni-hannover.de

© Springer International Publishing Switzerland 2016
S. Abramsky et al. (eds.), *Dependence Logic*, DOI 10.1007/978-3-319-31803-5_1

was introduced as a systematic extension of first-order logic by a means to explicitly talk about dependence among variables. Earlier attempts in this direction were the definition of *partially ordered quantifiers* by Henkin and the introduction of *Independence-friendly Logic* by Hintikka and Sandu.

In the past few years, the area of dependence logic has developed rapidly. One of the breakthroughs in the area was the introduction of *Independence Logic* that replaces the dependence atoms of dependence logic by independence atoms $\mathbf{x} \perp_{\mathbf{z}} \mathbf{y}$. The intuitive meaning of the atom $\mathbf{x} \perp_{\mathbf{z}} \mathbf{y}$ is that, for any fixed values of the variables $\mathbf{z}$, the variables $\mathbf{x}$ are independent of the variables $\mathbf{y}$ in the sense that knowing the value of $\mathbf{x}$ does not tell us anything about the value of $\mathbf{y}$. In databases, independence atoms correspond to embedded multivalued data dependencies. Furthermore, independence atoms and statistical conditional independence have also interesting connections.

More recently Galliani showed that independence atoms can be further analysed by the so-called inclusion $\mathbf{x} \subseteq \mathbf{y}$ and exclusion $\mathbf{x}|\mathbf{y}$ atoms. Both of these atoms have also been studied in database theory. The meaning of the atom $\mathbf{x} \subseteq \mathbf{y}$ is that all values of $\mathbf{x}$ in a team appear also as a value of $\mathbf{y}$, whereas the meaning of $\mathbf{x}|\mathbf{y}$ is that the values taken by $\mathbf{x}$ and $\mathbf{y}$ are distinct. Inclusion atoms have very interesting properties in the team semantics setting, e.g., they give rise to a variant of dependence logic that corresponds to the complexity class PTIME.

The idea of dependencies and independencies among variables has also been introduced into the context of propositional and (propositional) modal logic. This area has developed rapidly in the past few years, leading to many expressivity and complexity results.

This volume contains 11 articles discussing different aspects of logics for dependence and independence. Among these, there are articles addressing purely logical issues or computational aspects of dependence logic, but there are also several articles concerned with applications of dependence logic in various areas.

The chapter by A. Durand, J. Kontinen, and H. Vollmer gives a comprehensive survey of many propositional, modal, and first-order variants of dependence logic. It summarizes the state of the art regarding the expressive power as well as computational questions such as the complexity of the satisfiability or the model checking problems. The contribution by J. Väänänen discusses a theory of dependence developed by the German logician Kurt Grelling in an unpublished article from 1939 in the team semantics context. The chapter by P. Galliani contributes new results concerning expressive power of various variants of dependence logic with different sets of logical connectives and generalized dependence atoms. The contribution by E. Grädel studies the connections between inclusion logic and the least fixed-point logic. The chapter by W. Hodges discusses compositionality in the team semantics context. The topic of the survey of Å. Hirvonen are model-theoretic independence notions.

The notions of dependence and independence are central in many scientific areas, some of which are addressed in the remaining papers of this volume.

Inquisitive semantics is a new area aiming to give a logical account of information exchange as a process of requesting and providing information. The contribution by I. Ciardelli gives a comprehensive introduction to inquisitive logic, and also discusses the intimate connections between inquisitive logic and certain variants of dependence logic. The contribution by S. Link studies dependencies in databases by addressing the relationships between implication problems for fragments of statistical conditional independencies, embedded multivalued dependencies, and propositional logic. On the other hand, the article by H. Nyman, J. Pensar, and J. Corander reviews various Markovian models used to characterize dependencies and causality among variables in multivariate systems. The topic of the contribution by E. Pacuit and F. Yang is applications of dependence logic in social choice theory. The article defines an axiomatization of the famous Arrow's theorem using independence logic. The survey by A. Blass gives an introduction to the theory of secret-sharing pointing out connections to dependence and independence logic.

Expressivity and Complexity
of Dependence Logic

Arnaud Durand, Juha Kontinen, and Heribert Vollmer

1 Introduction

In this article we review recent results on expressivity and complexity of first-order, modal, and propositional dependence logic and some of its variants such as independence and inclusion logic. Dependence logic was introduced by Jouko Väänänen in [56]. On the syntactic side, it extends usual first-order logic by the so-called dependence atoms

$$=(x_1, \ldots, x_n),$$

the meaning of which is that the value of x_n is functionally determined by the values of $x_1, \ldots, x_{n-1}$. The semantics of dependence logic is defined using sets of assignments, teams, rather than single assignments as in first-order logic. Since the introduction of dependence logic in 2007, the area of team semantics has evolved into a general framework for logics in which various notions of dependence and independence can be formalized and studied. In this paper we mainly consider

A. Durand (✉)
IMJ-PRG, CNRS UMR 7586, Université Paris Diderot, Case 7012, 75205 Paris cedex 13, France
e-mail: durand@logique.jussieu.fr

J. Kontinen
Department of Mathematics and Statistics, University of Helsinki, P.O. Box 68,
00014 Helsinki, Finland
e-mail: juha.kontinen@helsinki.fi

H. Vollmer
Institut für Theoretische Informatik, Leibniz Universität Hannover, Appelstr. 4,
30167 Hannover, Germany
e-mail: vollmer@thi.uni-hannover.de

© Springer International Publishing Switzerland 2016

S. Abramsky et al. (eds.), *Dependence Logic*, DOI 10.1007/978-3-319-31803-5_2

5

variants of dependence logic arising by replacing/supplementing dependence atoms with further dependency notions, and we also study propositional and modal variants.

In Section 2 we review the basic definitions and results on first-order dependence logic and its variants (extensions and fragments). It is divided into three subsections of which the two first ones deal with results related to *expressive power* and *definability*. In particular, results charting the expressive power of certain natural syntactic fragments of dependence logic and its variants will be discussed in Section 2.4. Section 2.5 reviews results on the complexity of *satisfiability* and *model checking* in the (first-order) dependence logic context. In Section 3 we turn to modal and propositional versions of dependence logic. After introducing the basic notions and logics, we will again first touch expressivity questions and then turn to the complexity of algorithmic problems arising in this context, mostly the complexity of satisfiability and model checking. The paper concludes in Section 4 with a list of open questions.

2   First-order Dependence Logic

2.1   Team semantics

In this section we define the basics of the team semantics as presented in the monograph [56] by Väänänen. The origins of this definition go back to a paper by Wilfrid Hodges [35], in which he gave a Tarski-style semantics for Hintikka and Sandu's *independence-friendly logic* $\mathfrak{IF}$ [34]. Hodges originally used the term "trump semantics", somewhat reflecting the game-theoretic nature of the previously only known non-compositional semantics for $\mathfrak{IF}$.

Definition 1. Let $\mathcal{M}$ be a structure with domain M, and V a finite set of variables. Then

- A *team* X over $\mathcal{M}$ with domain $\mathrm{Dom}(X) = V$ is a finite set of assignments $s: V \to M$.
- For a tuple $\mathbf{v} = (v_1, \ldots, v_n)$, where $v_i \in V$, $X(\mathbf{v}) := \{s(\mathbf{v}) : s \in X\}$ is an n-ary relation of M, where $s(\mathbf{v}) := (s(v_1), \ldots, s(v_n))$.
- For $W \subseteq V$, $X \upharpoonright W$ denotes the team obtained by restricting all assignments of X to W.
- The set of free variables of a formula ϕ is defined analogously as in first-order logic, and is denoted by $\mathrm{Fr}(\phi)$. In particular, all non-first-order atoms considered in this article (see Definition 3) are treated as atomic formulas, and hence all variable occurrences in them are considered to be free.

With the above notions defined, we are now ready to present the semantics of dependence logic. In this article we consider two variants of the semantics called the *strict* and the *lax semantics* introduced in [16]. The original semantics given in [56] is a combination of these variants (with the lax disjunction and the strict existential quantifier). For any logic, e.g., dependence logic, whose formulas have the downwards closure property of Proposition 4, the two variants of the semantics are easily seen to be equivalent. On the other hand, for inclusion and independence logic the semantics are not equivalent [16]. A serious disadvantage of the strict semantics is the failure of the locality property in the case of inclusion and independence logic (see Proposition 1).

We will first define the lax version of the team semantics for first-order formulas in negation normal form. For an assignment s, $\mathcal{M} \models_s \alpha$ below refers to satisfaction in first-order logic. We denote by $s[m/v]$ the assignment such that $s[m/v](x) = m$ if $x = v$, and $s[m/v](x) = s(x)$ otherwise.

Definition 2 (Lax Semantics). Let $\mathcal{M}$ be a structure, X a team over $\mathcal{M}$, and ϕ a formula such that $\mathrm{Fr}(\phi) \subseteq \mathrm{Dom}(X)$. Then X *satisfies* ϕ in $\mathcal{M}$, $\mathcal{M} \models_X \phi$, if

lit: For a first-order literal α, $\mathcal{M} \models_X \alpha$ if and only if for all $s \in X$, $\mathcal{M} \models_s \alpha$.

$\vee$: $\mathcal{M} \models_X \psi \vee \theta$ if and only if there are Y and Z such that $Y \cup Z = X$, $\mathcal{M} \models_Y \psi$ and $\mathcal{M} \models_Z \theta$.

$\wedge$: $\mathcal{M} \models_X \psi \wedge \theta$ if and only if $\mathcal{M} \models_X \psi$ and $\mathcal{M} \models_X \theta$.

$\exists$: $\mathcal{M} \models_X \exists v \psi$ if and only if there exists a function $F : X \to \mathscr{P}(M) \setminus \{\emptyset\}$ such that $\mathcal{M} \models_{X[F/v]} \psi$, where $X[F/v] = \{s[m/v] : s \in X, m \in F(s)\}$.

$\forall$: $\mathcal{M} \models_X \forall v \psi$ if and only if $\mathcal{M} \models_{X[M/v]} \psi$, where $X[M/v] = \{s[m/v] : s \in X, m \in M\}$.

A sentence is a formula without free variables. A sentence ϕ is *true* in $\mathcal{M}$ (abbreviated $\mathcal{M} \models \phi$) if $\mathcal{M} \models_{\{\emptyset\}} \phi$. Sentences ϕ and ϕ' are *equivalent*, $\phi \equiv \phi'$, if for all models $\mathcal{M}$, $\mathcal{M} \models \phi \Leftrightarrow \mathcal{M} \models \phi'$.

In the strict semantics, the semantic rule for disjunction is replaced by

$\mathcal{M} \models_X \psi \vee \theta$ if and only if, there are Y and Z such that $Y \cap Z = \emptyset$, $Y \cup Z = X$, $\mathcal{M} \models_Y \psi$ and $\mathcal{M} \models_Z \theta$,

and the semantic rule for existential quantifier by

$\mathcal{M} \models_X \exists v \psi$ if and only if, there exists a function $F : X \to \mathscr{P}(M) \setminus \{\emptyset\}$ such that $|F(s)| = 1$ for all $s \in X$, and $\mathcal{M} \models_{X[F/v]} \psi$.

It is worth noting that functions quantified in the strict semantics version of the existential quantifier correspond exactly to functions $F : X \to M$. Hence the notation $X[F/v]$ can be naturally extended to cover also functions $F : X \to M$.

The meaning of first-order formulas is invariant under the choice between the strict and the lax semantics. Furthermore, first-order formulas have the following *flatness property*:

Theorem 1 (Flatness). *Let $\mathcal{M}$ be a structure and X a team of $\mathcal{M}$. Then for a first-order formula ϕ the following are equivalent:*

1. $\mathcal{M} \models_X \phi$,
2. *For all $s \in X$, $\mathcal{M} \models_s \phi$.*

It is worth noting that in [56] also a general notion of flatness of a formula is defined by replacing the second item above by "For all $s \in X$, $\mathcal{M} \models_{\{s\}} \phi$".

Next we will give the semantic clauses for the non-first-order atoms and connectives considered in this paper. We begin with the new atomic formulas:

Definition 3. • Let $\mathbf{x}$ be a tuple of variables and let y be another variable. Then $=(\mathbf{x}, y)$ is a *dependence atom*, with the following semantic rule:

> $\mathcal{M} \models_X =(\mathbf{x}, y)$ if and only if for all $s, s' \in X$, if $s(\mathbf{x}) = s'(\mathbf{x})$, then $s(y) = s'(y)$.

• Let $\mathbf{x}$, $\mathbf{y}$, and $\mathbf{z}$ be tuples of variables (not necessarily of the same length). Then $\mathbf{y} \perp_{\mathbf{x}} \mathbf{z}$ is a *conditional independence atom*, with the semantic rule

> $\mathcal{M} \models_X \mathbf{y} \perp_{\mathbf{x}} \mathbf{z}$ if and only if for all $s, s' \in X$ such that $s(\mathbf{x}) = s'(\mathbf{x})$, there exists a $s'' \in X$ such that $s''(\mathbf{xyz}) = s(\mathbf{xy})s'(\mathbf{z})$.

Furthermore, we will write $\mathbf{x} \perp \mathbf{y}$ as a shorthand for $\mathbf{x} \perp_\emptyset \mathbf{y}$, and call it a *pure independence atom*.

• Let $\mathbf{x}$ and $\mathbf{y}$ be two tuples of variables of the same length. Then $\mathbf{x} \subseteq \mathbf{y}$ is an *inclusion atom*, with the semantic rule

> $\mathcal{M} \models_X \mathbf{x} \subseteq \mathbf{y}$ if and only if $X(\mathbf{x}) \subseteq X(\mathbf{y})$.

• Let $\mathbf{x}$ and $\mathbf{y}$ be two tuples of variables of the same length. Then $\mathbf{x} \mid \mathbf{y}$ is an *exclusion atom*, with the semantic rule

> $\mathcal{M} \models_X \mathbf{x} \mid \mathbf{y}$ if and only if $X(\mathbf{x}) \cap X(\mathbf{y}) = \emptyset$.

We denote the set of all dependence atoms by $=(\ldots)$. Analogously, all independence, inclusion and exclusion atoms are denoted by $\perp_c$, $\subseteq$, and $\mid$, respectively. For a collection $\mathcal{C} \subseteq \{=(\ldots), \perp_c, \subseteq, \mid\}$, we write $\mathcal{FO}(\mathcal{C})$ (omitting the set parenthesis of $\mathcal{C}$) for the logic obtained by adding all atoms listed in $\mathcal{C}$ to the syntax of first-order logic. Independence atoms (or independence logic) were first considered in [21], and inclusion atoms go back to [16]. In our notation, dependence logic, independence logic, and inclusion logic are denoted by $\mathcal{FO}(=(\ldots))$, $\mathcal{FO}(\perp_c)$, and $\mathcal{FO}(\subseteq)$, respectively. We also use the notation $\mathcal{D}$ as a shortcut for $\mathcal{FO}(=(\ldots))$. The fragment of independence logic containing only pure independence atoms in denoted $\mathcal{FO}(\perp)$.

Under the lax semantics, all of the above logics satisfy the following locality property [16]:

Proposition 1. *Let* $\phi \in \mathcal{FO}(=(\ldots), \perp_c, \subseteq, |)$. *Then for all models* $\mathcal{M}$ *and teams* X,

$$\mathcal{M} \models_X \phi \Leftrightarrow \mathcal{M} \models_{X \restriction Fr(\phi)} \phi.$$

On the other hand, under the strict semantics Proposition 1 fails for inclusion and independence logic [16].

The aforementioned atoms are particular instances of a general notion of *generalized dependence atom* [44]. The semantics of a generalized dependence atom A_Q is determined (essentially) by a class Q of structures and teams over which the atomic formula $A_Q(\mathbf{x}_1, \ldots, \mathbf{x}_n)$ is satisfied (see [44] for details; we will consider $\mathcal{FO}$-definable generalized dependence atoms as depicted in Table 1).

Next we will define connectives and quantifiers that will also be discussed in the next section. One of the most natural extensions of dependence logic is obtained by the classical negation ($\sim$) with the usual interpretation:

$$\mathcal{M} \models_X \sim \phi \text{ iff } \mathcal{M} \not\models_X \phi.$$

This extension was introduced in [57], and the logic obtained was called *Team Logic* ($\mathcal{TL}$). The classical disjunction $\varoslash$ (also sometimes referred to as intuitionistic disjunction) has also been considered especially in the modal team semantics context, see Section 3. The connective $\varoslash$ has the expected interpretation

$$\mathcal{M} \models_X \phi \varoslash \psi \text{ iff } \mathcal{M} \models_X \phi \text{ or } \mathcal{M} \models_X \psi.$$

In [1] two new connectives called the *intuitionistic* ($\rightarrow$) and the *linear implication* ($\multimap$) were introduced giving rise to an extension of dependence logic called BID:

$$\mathcal{M} \models_X \phi \rightarrow \psi \quad \text{iff} \quad \text{for all } Y \subseteq X, \text{ if } \mathcal{M} \models_Y \phi \text{ then } \mathcal{M} \models_Y \psi.$$

$$\mathcal{M} \models_X \phi \multimap \psi \quad \text{iff} \quad \text{for all } Y, \text{ if } \mathcal{M} \models_Y \phi \text{ then } \mathcal{M} \models_{X \cup Y} \psi.$$

Quantifiers, other than the familiar $\exists$ and $\forall$, have also been studied in the team semantics setting [12, 13]. From the complexity theoretic point of view, the following majority quantifier introduced in [7] is interesting:

$$\mathcal{M} \models_X \mathsf{M}x\phi(x) \text{ iff for at least } |M|^{|X|}/2 \text{ many functions } F\colon X \to M, \text{ we have}$$

$$\mathcal{M} \models_{X[F/x]} \phi(x).$$

2.2 Normal forms

In order to analyse the expressive power of dependence logic and to compare it with other formalisms, it is useful to obtain normal forms such as this one proved in [56].

Theorem 2. *Every dependence logic sentence is equivalent to some sentence of the form:*

$$\phi := \forall \mathbf{x} \exists \mathbf{y} \, (\bigwedge_{i \in I} =\!(\mathbf{x_i}, y_i) \wedge \theta)$$

where $I \subseteq \mathbb{N}$, $\mathbf{x_i}$ is a subsequence of $\mathbf{x}$, and y_i is a member of $\mathbf{y}$.

Such a result is an analogue of Skolem normal form for first-order logic. It separates clearly the functional dependencies introduced between subsets of variables from the regular part of the formula. It also makes intuitively clear that to be translated into an extension of first-order logic one would need second-order quantification to express these dependencies between variables. Refinements of such a normalization result are at the heart of various characterizations of dependence-like logics and their fragments. For example, the analogue of Theorem 2 for independence logic (with dependence atoms replaced by independence atoms) was shown in [23]. Furthermore, a prenex normal form theorem for formulas of $\mathcal{FO}(=(\ldots), \bot_c, \subseteq)$ was shown for the strict and the lax semantics in [18] and [26], respectively.

2.3 Expressive Power

In this section we review results on the expressive power of the variants of dependence logic of the previous subsections.

As it turns out, the expressive power of sentences of dependence logic corresponds to that of existential second-order logic [56], and hence to the complexity class non-deterministic polynomial time (NP) via the well-known theorem of Fagin [14]. In the following, we will not distinguish in notation between a logic and the classes of models defined by its sentences, and we will use the equality symbol to denote that logics are equivalent for sentences, and we will use equality for logics and complexity classes in the same vein.

Theorem 3. $\mathcal{D} = \mathsf{NP} = \mathcal{ESO}$

The direction $\mathcal{ESO} \leq \mathcal{D}$ is proved by utilizing the fact that every $\mathcal{ESO}$-sentence can be transformed to the so-called Skolem normal form. On the other hand, the direction $\mathcal{D} \leq \mathcal{ESO}$ is proved by essentially simulating the team semantics of dependence logic in $\mathcal{ESO}$ with an extra relation symbol interpreting the team.

An interesting consequence of the team semantics of dependence logic is that Theorem 3 does not immediately settle the question also for open formulas. In fact, all $\mathcal{D}$-formulas have the following *Downwards Closure* property:

Theorem 4 (Downwards Closure). *Let ϕ be a $\mathcal{D}$-formula. Then for all structures $\mathcal{M}$ and teams X, if $\mathcal{M} \models_X \phi$ and $Y \subseteq X$, then $\mathcal{M} \models_Y \phi$.*

It was shown in [39] that the open formulas of dependence logic can define exactly the downward closed properties of teams expressible in $\mathcal{ESO}$ (again with an extra relation symbol for the team). Furthermore, already dependence atoms combined with disjunction give rise to NP-complete decision problems [37]. Define the formulas ϕ_1 and ϕ_2 as follows:

- $\phi_1 := \; =(x, y) \vee =(u, v)$,
- $\phi_2 := \; =(x, y) \vee =(u, v) \vee =(u, v)$.

Then the question of deciding whether a finite team X satisfies ϕ_1 is NL-complete, and for ϕ_2, already NP-complete.

As one might expect, the expressive power of dependence logic with the classical negation ($\mathcal{TL}$) increases to full second-order logic, and hence to the complexity class Polynomial Hierarchy (PH).

Theorem 5. $\mathcal{TL} = \mathcal{SO} = $ PH

This result is already shown in [56], but a direct translation of $\mathcal{SO}$ sentences into $\mathcal{TL}$-sentences was later given in [48]. Furthermore, in [38] it was shown that any property of teams definable in second-order logic can be expressed in team logic. It is worth noting that, for example, in general $\mathcal{TL}$-formulas are not closed downwards, e.g., the formula

$$\sim \; =(x) \tag{1}$$

expresses that x has at least two distinct values.

Interestingly, the two new connectives (implications) introduced in [1] preserve downwards closure when added to dependence logic. It was observed in [1] that any sentence of BID-logic can be translated into second-order logic. In fact, by the result of [62], already the intuitionistic implication alone increases the expressive power of dependence logic to full second-order logic.

Theorem 6. $\mathcal{D}(\rightarrow) = \mathcal{SO} = $ PH

This result utilizes the universal quantification implicit in the semantic rule of the intuitionistic implication. On the other hand, in [7, 8] the extension $\mathcal{D}(M)$ of dependence logic by the majority quantifier M was defined and studied. The main result of that paper is stated as follows:

Theorem 7. $\mathcal{D}(M) = $ CH.

Above CH refers to the complexity class the counting hierarchy CH $\supseteq$ PH. Theorems 7 and 5 imply that, for sentences, $\mathcal{D}(M)$ is at least as expressive as $\mathcal{TL}$ over finite structures. On the other hand, this result does not extend to open formulas since $\mathcal{D}(M)$-formulas have the downward closure property unlike $\mathcal{TL}$-formulas (see, e.g., formula (1)).

The aforementioned results show that dependence logic and its extensions allow us to logically characterize NP and some of its super classes. In [11] the question whether PTIME corresponds to a natural fragment of dependence logic was considered. For $\mathcal{ESO}$ (also $\mathcal{SO}$) it is known that the so-called Horn fragment $\mathcal{SO}\exists$-Horn of $\mathcal{ESO}$ captures **PTIME** over successor structures [19]. In [11] a fragment $\mathcal{D}^*$-Horn equivalent to $\mathcal{SO}\exists$-Horn was identified. The formulas of $\mathcal{D}^*$-Horn have the form

$$\forall\mathbf{x}\exists\mathbf{y}(\bigwedge_i = (\mathbf{z}_i, y_i) \wedge \bigwedge_j C_j),$$

where $\mathbf{z}_i$ is subsequence of $\mathbf{x}$, the clauses C_j (i.e. disjunctions of $\mathcal{FO}$-literals) are assumed to satisfy a certain Horn condition, and the existentially quantified variables y_i are only allowed to appear in certain identity atoms of C_j (see [11] for the exact definition). The main result of [11] shows that

Theorem 8. *Over finite successor structures, $\mathcal{D}^*$-Horn = $\mathcal{SO}\exists$-Horn.*

Theorem 8 implies that

$$\mathcal{D}^*\text{-Horn} = \textbf{PTIME}$$

over finite successor structures. In the article [11] the expressive power of open formulas of $\mathcal{D}^*$-Horn is also characterized.

All of the results discussed in this section use the original semantics of dependence logic. It is easy to check that the results hold also for both variants of the semantics. Next we will consider the expressive power of inclusion logic. It turns out that the expressive power of inclusion logic is not invariant under the choice between the strict and the lax semantics.

The expressive power of inclusion logic under the lax semantics was studied in [17]. The main result of that paper shows that

Theorem 9. *Over the lax semantics,*

$$\mathcal{FO}(\subseteq) = \text{GFP}^+,$$

where GFP^+ is the so-called Positive Greatest Fixed Point Logic. It is known that over finite structures GFP^+ is equi-expressive with Least Fixed Point Logic (LFP), and furthermore for ordered finite structures LFP = **PTIME** by the famous result of Immerman [36] and Vardi [60]. Therefore, it follows that

$$\mathcal{FO}(\subseteq) = \textbf{PTIME}$$

over ordered finite structures. In drastic contract with Theorem 9, it was observed in [18] that, over the strict semantics, inclusion logic is equivalent to $\mathcal{ESO}$ and hence captures **NP**.

Theorem 10. *Over the strict semantics,*

$$\mathcal{FO}(\subseteq) = \mathcal{ESO}.$$

This result is based on a simulation of dependence atoms in a dependence logic sentence (in the $\forall^*\exists^*$-normal form) by certain inclusion logic formulas. This simulation is not possible in general but only over teams that are generated by evaluating a $\forall^*\exists^*$-block of quantifiers with the strict semantics.

2.4 Refining the correspondence with $\mathcal{ESO}$

In this part we investigate how the correspondence between existential second-order logic and $\mathcal{FO}(\mathscr{C})$ for subsets $\mathscr{C}$ of dependence-like atoms can be refined. In particular we examine what is the effect of bounding the number of variables and the so-called arity of atoms which roughly concerns the number of distinct variables involved in them.

By relating fragments of $\mathcal{FO}(\mathscr{C})$ to fragments of existential second-order logic, one may hope to obtain separation results in dependence logics through hierarchy theorems in complexity or to give evidence that such results would have non-trivial consequences in complexity theory. In either way, this provides interesting insight on the expressive power of these logics.

Let us first define the notion of arity of an atom.

Definition 1. Let $k \in \mathbb{N}$.

- A dependence atom $=(\mathbf{x}, y)$ is of arity k if the length of $\mathbf{x}$ is k.
- An independence atom $\mathbf{y} \perp_{\mathbf{x}} \mathbf{z}$ is of arity k if $\mathbf{xyz}$ contains $k+1$ distinct variables.
- An inclusion atom $\mathbf{x} \subseteq \mathbf{y}$ is of arity k if the length of $\mathbf{x}$ and $\mathbf{y}$ is k.

We now define the corresponding fragments of $\mathcal{FO}(\mathscr{C})$ and existential second-order logic.

Definition 2. Let $\mathscr{C}$ be a subset of $\{=(\ldots), \perp_c, \subseteq, |\}$. Let $k \in \mathbb{N}$. Then:

- $\mathcal{FO}(\mathscr{C})(k\text{−ary})$ is the class of sentences of $\mathcal{FO}(\mathscr{C})$ in which all atoms of $\mathscr{C}$ are of arity bounded by k.
- $\mathcal{FO}(\mathscr{C})(k\forall)$ is the class of sentences of $\mathcal{FO}(\mathscr{C})$ in which every variable is quantified exactly once and at most k universal quantifiers occur.
- For convenience, we set by $\mathcal{D}(k\text{−ary})$ the class $\mathcal{FO}(=(\ldots))(k\text{−ary})$ and by $\mathcal{D}(k\text{−ary})$ the class $\mathcal{FO}(=(\ldots))(k\forall)$.

Definition 3. Let $k \in \mathbb{N}$.

- $\mathcal{ESO}(k\text{−ary})$ is the class of $\mathcal{ESO}$-sentences

$$\bullet \quad \exists X_1 \ldots \exists X_n \psi,$$

in which the relation symbols X_i are at most k-ary and ψ is a first-order formula.

- $\mathcal{ESO}_f(k-\text{ary})$ is the class of $\mathcal{ESO}$-sentences

$$\exists f_1 \ldots \exists f_n \psi,$$

in which the function symbols f_i are at most k-ary and ψ is a first-order formula.
- $\mathcal{ESO}_f(m\forall)$ is the class of $\mathcal{ESO}$-sentences in Skolem normal form

$$\exists f_1 \ldots \exists f_n \forall x_1 \ldots \forall x_r \psi,$$

where $r \leq m$.

Such fragments of $\mathcal{ESO}$ have been widely studied and in particular their relationship with complexity classes. Roughly speaking, controlling the number of first-order variables in existential second-order logic amounts to control the polynomial degree in non-deterministic polynomial-time computations. To be more precise, it is known (see[22]) that, for $k \geq 1$:

$$\mathcal{ESO}_f(k\text{-ary}, k\forall) = \mathcal{ESO}_f(k\forall) = \mathsf{NTIME}_{\mathsf{RAM}}(n^k).$$

where $\mathsf{NTIME}_{\mathsf{RAM}}(n^k)$ is the class of problems decidable by a non-deterministic random access machine in time $O(n^k)$. We can now relate the expressive power of these various fragments.

First, reusing of variables is a key issue in team semantics. It turns out that the following result is true (see [6]).

Proposition 2. *Any sentence of dependence logic is logically equivalent to a sentence in which at most one variable is universally quantified (possibly several times).*

For what concerns dependence logic, the correspondence with existential second-order logic for the fragments with bounded arity and bounded number of universal variables is as follows [6]:

Theorem 11. *For all integers $k \geq 1$,*

- $\mathcal{D}(k-\text{ary}) = \mathcal{ESO}_f(k-\text{ary})$,
- $\mathcal{D}(k\forall) \leq \mathcal{ESO}_f(k\forall) \leq \mathcal{D}(2k\forall)$.

Roughly speaking, dependence logic can be seen as existential second-order logic with functions (dependence atoms) but "without proper names" for these functions. Hence composition of functions, that can be done freely in existential second-order logic, can be simulated only by using intermediate variables in dependence logic. So, as long as, only the arity of dependence atoms is fixed, one can obtain an exact correspondence between the fragments (as stated in Theorem 11). By contrast an exact correspondence between the fragments $\mathcal{D}(k\forall)$ and $\mathcal{ESO}_f(k\forall)$ seems unlikely (see the second item of Theorem 11). However, it is possible to establish an exact correspondence between $\mathcal{D}(k\forall)$ and some syntactically restricted fragment of $\mathcal{ESO}_f(k\forall)$ (see [6]).

Because inclusion and independence logic do not have the downward closure property, the situation is drastically different depending on whether the lax or the strict semantics are used (see [18, 24–26]).

Theorem 12. *Let $k \geq 1$. The following holds in the lax semantics:*

- $\mathcal{FO}(\subseteq)(k-\text{ary}) < \mathcal{FO}(\subseteq)(k+1-\text{ary})$,
- $\mathcal{FO}(\subseteq)(k-\text{ary}) \leq \mathcal{ESO}_f(k-\text{ary}) = \mathcal{FO}(\perp_c)(k-\text{ary})$,
- $\mathcal{FO}(\perp)(2\forall) = \mathcal{FO}(\perp)$,
- $\mathcal{FO}(\subseteq)(1\forall) = \mathcal{FO}(\subseteq)$.

For the strict semantics, the following results are true:

- $\mathcal{FO}(\subseteq)(k\forall) = \mathcal{ESO}_f(k\forall) = \mathsf{NTIME}_{\mathrm{RAM}}(n^k)$,
- $\mathcal{FO}(\perp_c)(k\forall) \leq \mathcal{ESO}_f((k+1)\forall)$,
- $\mathcal{ESO}_f(k\forall) \leq \mathcal{FO}(\perp_c)(2k\forall)$.

The results above imply that there is an infinite expressivity hierarchy for $\mathcal{D}(k\forall)$, $\mathcal{FO}(\subseteq)(k\forall)$, and $\mathcal{FO}(\perp_c)(k\forall)$. Indeed, it is well known (by a slight adaptation of classical non-deterministic time hierarchy [5]) that, for any integer $k \geq 1$, $\mathsf{NTIME}_{\mathrm{RAM}}(n^k)$ is strictly included in $\mathsf{NTIME}_{\mathrm{RAM}}(n^{k+1})$. Hence, for example, $\mathcal{D}(k\forall)$ is strictly less expressive than $\mathcal{D}((k+1)\forall)$.

Similarly, one might ask whether there is a strict hierarchy based on arity for $\mathcal{D}$ or $\mathcal{FO}(\perp_c)$. For example, is $\mathcal{D}(k+1-\text{ary})$ strictly more expressive than $\mathcal{D}(k-\text{ary})$ for all (or some) $k \geq 1$. Such a hierarchy would imply the existence of a similar hierarchy for $\mathcal{ESO}_f(k-\text{ary})$ which is a long-standing open question (for empty signature, this is known as the *Spectrum Arity Hierarchy Conjecture* [15]).

Finally, let us examine the situation when exclusion atoms are allowed in the syntax. It turns out that none of the approach above helps to control the arity correspondence between the corresponding fragments. By introducing mainly two new concepts, namely inclusion quantifier (an adaptation of the idea of quantifier relativization applied to inclusion atoms) and term value preserving disjunction, the following result is obtained in [51]:

Theorem 13. *For all integers $k \geq 1$, $\mathcal{FO}(\subseteq, |)(k-\text{ary}) = \mathcal{ESO}(k-\text{ary})$.*

Note that in this result the correspondence is with the relational fragment of existential second-order logic: no quantification on functions is allowed.

2.5 Satisfiability and Model Checking

In this section we briefly review results on satisfiability and model checking in the first-order dependence logic context.

We begin by recalling these problems for a logic $\mathscr{L}$:

- The *Satisfiability Problem* SAT[$\mathscr{L}$] is defined as

$$\mathrm{SAT}[\mathscr{L}] := \{\phi \in \mathscr{L} \mid \text{there is a structure } \mathcal{M} \text{ such that } \mathcal{M} \models \phi\}.$$

- The *Model-Checking Problem* is defined as

$$MC[\mathcal{L}] := \{(\phi, \mathbb{M}) \mid \mathbb{M} \models \phi\}.$$

Since, for sentences, dependence logic is equivalent to $\mathcal{ESO}$ the classical results [4, 55] on the Decision Problem (Entscheidungsproblem) of $\mathcal{FO}$ imply that SAT[$\mathcal{D}$] is undecidable (Π_1^0-complete). In the case of first-order logic, these negative results spurred an extensive investigation of decidable fragments of $\mathcal{FO}$. Henkin [33] was the first to consider fragments of first-order logic with a fixed number of variables. Satisfiability for the fragments with three or more variables is easily seen to be undecidable, but for two variables, it is decidable (see, e.g., [3]). The two-variable fragment of $\mathcal{FO}$ is denoted by $\mathcal{FO}^2$.

In [40], the two-variable fragment $\mathcal{D}^2$ (i.e. the sentences of $\mathcal{D}$ in which only variables x and y appear) of dependence logic was studied and the following result was obtained:

Theorem 14. SAT[$\mathcal{D}^2$] *is* NEXPTIME-*complete.*

The complexity of the problem remains the same also for finite satisfiability, although there are sentences of $\mathcal{D}^2$ having only infinite models (the so-called infinity axioms). The proof of Theorem 14 is based on a polynomially bounded translation of sentences of $\mathcal{D}^2$ to $\Sigma_1^1(\mathcal{FOC}^2)$- sentences, i.e., sentences of the form

$$\exists R_1 \ldots \exists R_m \phi,$$

where ϕ is a $\mathcal{FOC}^2$-sentence. The logic $\mathcal{FOC}^2$ extends $\mathcal{FO}^2$ by counting quantifiers $\exists^{\geq i} x \phi(x)$ expressing that $\phi(x)$ is satisfied by at least i distinct elements. Theorem 14 then follows from the fact that satisfiability is NEXPTIME-complete for $\mathcal{FOC}^2$ [50]. It is worth noting that since $\mathcal{D}^2$ is not closed under classical negation, Theorem 14 does not solve the complexity of the validity problem for $\mathcal{D}^2$. In fact, this problem is still open.

Recently Theorem 14 has been generalized to logics of the form $\mathcal{FO}^2(\mathcal{A})$ that replace dependence atoms of $\mathcal{D}^2$ by a certain collections $\mathcal{A}$ of generalized dependence atoms [41]. The result therein shows that SAT[$\mathcal{FO}^2(\mathcal{A})$] is NEXPTIME-complete if the atoms in $\mathcal{A}$ are themselves $\Sigma_1^1(\mathcal{FOC}^2)$-definable. This result can be used to show, e.g., that the satisfiability and finite satisfiability problems of the two-variable fragments of inclusion, exclusion, and independence logic are all NEXPTIME-complete.

The complexity of model checking of dependence logic and its variants has been studied in [20]. The paper introduces a general model-checking game for logics with team semantics, and by analyzing the corresponding games, shows several results on the complexity of model checking in the team semantics framework.

Theorem 15. *The model-checking problem for* $\mathcal{D}$ *is* NEXPTIME-*complete.*

In [20] it is also showed that NEXPTIME is an upper bound for the complexity of model checking for any variant $\mathcal{FO}(\mathscr{C})$ of $\mathcal{D}$ such that the atoms in $\mathscr{C}$ are PTIME-computable.

3 Propositional and Modal Dependence Logic

This section will be devoted to the study of team-based logics over Kripke structures, such as modal dependence logic and, as a special case thereof, propositional dependence logic. Again, we will survey results on expressiveness of the most important logics and on the complexity of the satisfiability and model-checking problems. We will first turn to modal logics and later mention briefly some results for the purely propositional case.   .

3.1 Preliminaries

We start by introducing team semantics for usual modal logic. The central semantic concept here is the *Kripke model*, which is a tuple $K = (W, R, \pi)$ where W is a nonempty set of worlds, $R \subseteq W \times W$, and $\pi: P \rightarrow 2^W$, where P is a set of propositional variables. The idea here is that π determines which variables hold (are set to *true*) in each world. So every world carries a propositional assignment. In analogy to first-order logic, we want to express dependencies among the values of certain variables; so we have to evaluate formulas relative to a set of assignments, hence in our case a set of worlds. A *team* of a model K is thus defined to be a set $T \subseteq W$. The central basic concept underlying Väänänen's modal dependence logic and all its variants, that modal formulas are evaluated not in a world but in a *team*, is made precise in the definitions to follow below.

Before that, we would like to point out that another formalism, *independence-friendly modal logic*, has also been considered. In analogy to Hintikka and Sandu's independence-friendly logic $\mathcal{IF}$ [34], the so-called slash modalities are introduced. Consider the example formula $\square_1(\lozenge_2/\square_1)p$. It is evaluated like $\square\lozenge p$, but now the witness for $\lozenge$ has to be chosen independently of the witnesses for $\square$; hence the formula states, when evaluated at w, that there is a world u accessible from all worlds v that are accessible from w, a kind of "confluence property". There are several competing formalisms for modal independence-friendly logic, cf., e.g., [54].

Väänänen [58] when introducing modal dependence logic $\mathcal{MDL}$ made a step analogous to the introduction of dependencies in first-order dependence logic by extending the logical language with atoms $=(p_1, \ldots, p_n)$ to express dependencies among propositional variables. *Caveat:* In modal dependencies we can express dependencies among variables, not among worlds, as one might first expect when thinking of the standard translation of modal logic into first-order logic.

In the following, we will define the syntax of the modal logics we consider by grammars in extended Backus-Naur form (EBNF).

Definition 4 (Syntax of $\mathcal{ML}$).

$$\varphi ::= p \mid \neg p \mid (\varphi \wedge \varphi) \mid (\varphi \vee \varphi) \mid \Diamond \varphi \mid \Box \varphi,$$

where p is an atomic proposition.

Definition 5 (Semantics of $\mathcal{ML}$). Let $K = (W, R, \pi)$ be a Kripke model, let $T \subseteq W$ be a team, and let φ be an $\mathcal{ML}$-formula. We define when $K, T \models \varphi$ holds inductively:

- If $\varphi = p$, then $K, T \models \varphi$ if and only if $T \subseteq \pi(p)$.
- If $\varphi = \neg p$, then $K, T \models \varphi$ if and only if $T \cap \pi(p) = \emptyset$.
- If $\varphi = \psi \vee \chi$ for some formulas ψ and χ, then $K, T \models \varphi$ if and only if $T = T_1 \cup T_2$ with $K, T_1 \models \psi$ and $K, T_2 \models \chi$.
- If $\varphi = \psi \wedge \chi$ for some formulas ψ and χ, then $K, T \models \varphi$ if and only if $K, T \models \psi$ and $K, T \models \chi$.
- If $\varphi = \Diamond \psi$ for some formula ψ, then $K, T \models \varphi$ if and only if there is some team T' of K such that $K, T' \models \psi$, for each $w \in T$, there is some $w' \in T'$ with $(w, w') \in R$, and for each $w' \in T'$, there is some $w \in T$ with $(w, w') \in R$.
- If $\varphi = \Box \psi$ for some formula ψ, then $K, T \models \varphi$ if and only if $K, T' \models \psi$, where T' is the set $\{w' \in W \mid (w, w') \in R \text{ for some } w \in T\}$.

Team semantics for $\mathcal{ML}$ shares the so-called *flatness property*, see also Theorem 1:

Lemma 1 (Flatness of $\mathcal{ML}$). *For all K, T, φ, $K, T \models \varphi$ if and only if for all $w \in T$, we have $K, w \models \varphi$.*

So there is no essential semantic effect of team semantics compared to the usual semantics for modal logic. This changes when enriching the language with dependence atoms.

Definition 6 (Syntax of $\mathcal{MDL}$).

$$\varphi ::= p \mid \neg p \mid =(\{p, \} p) \mid (\varphi \wedge \varphi) \mid (\varphi \vee \varphi) \mid \Diamond \varphi \mid \Box \varphi,$$

where p is an atomic proposition.

Before defining the semantics of dependence atoms, we introduce a useful shorthand notation.

Definition 7. Let $\mathbf{p} = (p_1, \ldots, p_n)$ be a sequence of atomic propositions and w, w' be worlds of a Kripke model $K = (W, R, \pi)$. Then w and w' are equivalent under π over $\mathbf{p}$, denoted by $w \equiv_{\pi, \mathbf{p}} w'$, if the following holds:

$$\pi(w) \cap \{p_1, \ldots, p_n\} = \pi(w') \cap \{p_1, \ldots, p_n\}.$$

Definition 8 (Semantics of $\mathcal{MDL}$). We extend Definition 5 by the following clause:

- $K, T \models\ =(\mathbf{p}, q)$ if and only if for all $w_1, w_2 \in T$: if $w_1 \equiv_{\pi,\mathbf{p}} w_2$, then $w_1 \equiv_{\pi,q} w_2$.

Theorem 16 (Downwards Closure). *Let ϕ be a $\mathcal{MDL}$-formula. Then for all Kripke model K and teams T, if $K, T \models \phi$ and $T' \subseteq T$, then $K, T' \models \phi$.*

As in the first-order case, also the independence atom has been introduced into modal logic, thus leading to modal independence logic [42].

Definition 9 (Syntax of $\mathcal{MIL}$).

$$\varphi ::= p \mid \neg p \mid \{p\} p \perp_{\{p\}} \{p\} p \mid (\varphi \wedge \varphi) \mid (\varphi \vee \varphi) \mid \Diamond\varphi \mid \Box\varphi,$$

where p is an atomic proposition.

Definition 10 (Semantics of $\mathcal{MIL}$). We extend Definition 5 by the following clause, where $\mathbf{p}, \mathbf{q}, \mathbf{r}$ are sequences of atomic propositions:

- $K, T \models \mathbf{p} \perp_{\mathbf{r}} \mathbf{q}$ if and only if for all $w_1, w_2 \in T$ such that $w_1 \equiv_{\pi,\mathbf{r}} w_2$ there exists $w_3 \in T$ such that $w_1 \equiv_{\pi,\mathbf{r}} w_3$, $w_1 \equiv_{\pi,\mathbf{p}} w_3$, and $w_2 \equiv_{\pi,\mathbf{q}} w_3$.

We note that one motivation behind the introduction of $\mathcal{MIL}$ was that it can be used to express security of cryptographic protocols, see [42].

At this point, where we have extended basic modal logic by a *dependence* as well as an *independence* atom, we would like to point out that there is a general way to introduce such so-called *generalized dependence atoms*, expressing further $\mathcal{FO}$-definable *properties on teams*. To make this precise, consider an $\mathcal{FO}$-formula φ. We say that φ defines the atom D if

$$K, T \models D(p_1, \ldots, p_n) \iff \mathfrak{A} \models \varphi,$$

where the structure $\mathfrak{A}$ has universe T and unary relations $A_{p_i}^{\mathfrak{A}}$ with $w \in A_{p_i}^{\mathfrak{A}}$ iff $p_i \in \pi(w)$.

$\mathcal{FO}$-definitions of dependence, independence, and some further team properties can be found in Table 1. There, boldface symbols denote sequences of propositional variables.

A further possible extension of the so far considered modal logics is by introducing additional propositional connectives. In Section 2.1, different forms of implication and negation have been defined. Here we just define $\mathcal{MTL}$, *modal team logic*, to extend $\mathcal{ML}$ by a second type of negation, denoted by $\sim$ and interpreted just as classical negation.

Definition 11 (Syntax of $\mathcal{MTL}$).

$$\varphi ::= p \mid \neg p \mid \sim\varphi \mid (\varphi \wedge \varphi) \mid (\varphi \vee \varphi) \mid \Diamond\varphi \mid \Box\varphi,$$

where p is a propositional variable.

Table 1 Definitions of some generalized dependence atoms

Notation	Atom	$\mathcal{FO}$-defining formula
$=(\mathbf{p}, q)$	Dependence	$\forall w \forall w' \left(\bigwedge\limits_{i=1}^{n} \left(A_{p_i}(w) \leftrightarrow A_{p_i}(w')\right) \right.$ $\left. \rightarrow \left(A_q(w) \leftrightarrow A_q(w')\right) \right)$
$\mathbf{p} \perp_{\mathbf{r}} \mathbf{q}$	Independence	$\forall w \forall w' \left(\bigwedge\limits_{i=1}^{n} \left(A_{r_i}(w) \leftrightarrow A_{r_i}(w')\right) \right.$ $\rightarrow \exists w'' \left(\bigwedge\limits_{i=1}^{n} \left(A_{r_i}(w'') \leftrightarrow A_{r_i}(w)\right) \right.$ $\wedge \bigwedge\limits_{i=1}^{m} \left(A_{p_i}(w'') \leftrightarrow A_{p_i}(w)\right)$ $\left. \left. \wedge \bigwedge\limits_{i=1}^{s} \left(A_{q_i}(w'') \leftrightarrow A_{q_i}(w')\right) \right) \right)$
$\mathbf{p} \subseteq \mathbf{q}$	Inclusion	$\forall w \exists w' \bigwedge\limits_{i=1}^{n} \left(A_{p_i}(w) \leftrightarrow A_{q_i}(w')\right)$
$\mathbf{p} \mid \mathbf{q}$	Exclusion	$\forall w \forall w' \bigvee\limits_{i=1}^{n} \left(A_{p_i}(w) \leftrightarrow \neg A_{q_i}(w')\right)$
NE	Non-Emptiness	$\exists w \top$

Definition 12 (Semantics of $\mathcal{MTL}$). We extend Definition 5 by the following clause:

- If $\varphi = \sim\!\psi$ for some formula ψ, then $K, T \models \varphi$ if and only if $K, T \not\models \psi$.

We note that usually (see [47]), $\mathcal{MTL}$ also contains dependence atoms; however, since these atoms can be expressed in $\mathcal{MTL}$ we omit them in the syntax. To make clear what the power of classical negation in this context is, we discuss the definition of dependence as well as some additional propositional connectives in $\mathcal{MTL}$.

First, the classical disjunction $\varovee$ is readily expressed in $\mathcal{MTL}$: $\varphi \varovee \psi$ is logically equivalent to $\sim\!(\sim\!\varphi \wedge \sim\!\psi)$. Next we note that, analogously to the first-order case [1], the atom $=(p_1, \ldots, p_n)$ is logically equivalent with

$$\left(\bigwedge_{1 \leq i \leq n-1} =(p_i) \right) \rightarrow =(p_n),$$

where $\rightarrow$ is the modal version of the intuitionistic implication with the semantics

$$K, T \models \varphi \rightarrow \psi \text{ iff for all } T' \subseteq T: \text{if } K, T' \models \varphi \text{ then } K, T' \models \psi.$$

The connective $\rightarrow$ has a short logically equivalent definition in $\mathcal{MTL}$ (see [47]), hence so does the atom $=(p_1, \ldots, p_n)$. The intuitionistic implication has been studied in the modal team semantics context in [63].

We want to mention one final extension of modal dependence logic in this section, the so-called extended modal dependence logic, $\mathcal{EMDL}$. It allows $\mathcal{ML}$-formulas instead of atoms inside the dependence atom [10].

Definition 13 (Syntax of $\mathcal{EMDL}$).

$$\varphi ::= p \mid \neg p \mid =(\{\psi,\} \psi) \mid (\varphi \wedge \varphi) \mid (\varphi \vee \varphi) \mid \Diamond\varphi \mid \Box\varphi,$$

where p is an atomic proposition and ψ is an $\mathcal{ML}$-formula.

Definition 14 (Semantics of $\mathcal{EMDL}$). We extend Definition 5 by the following clause:

- $K, T \models =(\psi_1, \ldots, \psi_n)$ if for all $w_1, w_2 \in T$, if $K, \{w_1\} \models \psi_i \Leftrightarrow K, \{w_2\} \models \psi_i$ for $1 \leq i \leq n-1$, then $K, \{w_1\} \models \psi_n \Leftrightarrow K, \{w_2\} \models \psi_n$.

The interest in $\mathcal{EMDL}$ stems mainly from the fact, that it allows us to formulate some basic temporal dependencies. We give only one very simple example: The formula

$$=(\Diamond p, \Diamond^2 p, \ldots, \Diamond^n p, p)$$

expresses that "truth of p at this moment only depends on the truth of p in the previous n time steps" (on frame classes where the relation R denotes a backwards-oriented time-relation). So in a sense, $\mathcal{EMDL}$ can be seen as a *basic temporal dependence logic*.

We would like to point out that $\mathcal{EMDL}$ shares the *downwards closure property* of $\mathcal{MDL}$, but analogously to the first-order case, neither $\mathcal{MIL}$ nor $\mathcal{MIL}$ has this property.

3.2 Expressivity

The first results on expressive power of modal team-based logics are due to Sevenster [53]:

Theorem 17. *1. $\mathcal{MDL}$ is strictly more expressive than $\mathcal{ML}$.*
2. On singleton teams of evaluation, $\mathcal{MDL}$ is as expressive as $\mathcal{ML}$.

While the first result simply follows from the fact that $\mathcal{ML}$ is closed under union (of teams) but $\mathcal{MDL}$ is not, the second result requires an interesting proof that we would like to sketch. Given an $\mathcal{MDL}$-formula φ, we first use existentially quantified (Boolean) Skolem functions to replace dependence atoms. Next, we replace the existential quantifier by a big classical disjunction $\otimes$ over all possibilities for such functions. The result now follows since over singleton teams, the interpretations of the connectives $\otimes$ and $\vee$ agree.

This proof sketch points out the importance of classical disjunction. If we add $\otimes$ to $\mathcal{ML}$ or $\mathcal{MDL}$ we obtain the same expressive power, namely that of $\mathcal{EMDL}$ [10, 29]:

Theorem 18. $\mathcal{ML} < \mathcal{MDL} < \mathcal{ML}(\oslash) \equiv \mathcal{MDL}(\oslash) \equiv \mathcal{EMDL}$.

Concerning modal independence logic, the following is known [42]:

Theorem 19. *1.* $\mathcal{ML} < \mathcal{MDL} < \mathcal{MIL}$.
2. On singleton teams of evaluation, $\mathcal{MIL}$ is as expressive as $\mathcal{ML}$.

Again, the first result follows from simple closure properties. We will sketch the proof of the second result, since it will lead us to an important topic. Given an $\mathcal{MIL}$-formula φ, it is clear that on singleton teams it captures a property of Kripke models that

- is invariant under modal bisimulation (more on that in the next paragraph),
- only depends on the worlds that can be reached in a number of steps bounded by the modal depth of the formula.

These observations allow us to construct an $\mathcal{ML}$-formula that describes (by a big disjunction) all possibilities for satisfying Kripke models.

The just given proof can be extended to show that on singleton teams, $\mathcal{ML}$ extended by any $\mathcal{FO}$-definable dependence atoms (see Table 1) is as expressive as $\mathcal{ML}$.

The ideas used in the just sketched proof go back to a fundamental result by Johan van Benthem [59], characterizing exactly the properties of Kripke structures that modal logic $\mathcal{ML}$ can define in terms of the so-called bisimilarity. His results can be generalized to modal team logic, as we describe next.

Definition 15 (k-bisimulation). Let $K_1 = (W_1, R_1, \pi_1)$ and $K_2 = (W_2, R_2, \pi_2)$ be Kripke models. We define inductively what it means for worlds $w_1 \in W_1$ and $w_2 \in W_2$ to be k-bisimilar, for some $k \in \mathbb{N}$, written as $(K_1, w_1) \rightleftharpoons_k (K_2, w_2)$.

- $(K_1, w_1) \rightleftharpoons_0 (K_2, w_2)$ holds if for each propositional variable p, we have that $K_1, w_1 \models p$ if and only if $K_2, w_2 \models p$.
- $(K_1, w_1) \rightleftharpoons_{k+1} (K_2, w_2)$ holds if the following three conditions are satisfied:

  1. $(K_1, w_1) \rightleftharpoons_0 (K_2, w_2)$,
  2. for each successor w_1' of w_1 in K_1, there is a successor w_2' of w_2 in K_2 such that $(K_1, w_1') \rightleftharpoons_k (K_2, w_2')$ (*forward* condition),
  3. for each successor w_2' of w_2 in K_2, there is a successor w_1' of w_1 in K_1 such that $(K_1, w_1') \rightleftharpoons_k (K_2, w_2')$ (*backward* condition).

Full bisimulation is defined analogously as follows:

Definition 16 (full bisimulation). Let $K_1 = (W_1, R_1, \pi_1), K_2 = (W_2, R_2, \pi_2)$ be Kripke models and let w_1 and w_2 be worlds of K_1 and K_2. Then (K_1, w_1) and (K_2, w_2) are *bisimilar*, written as $(K_1, w_1) \rightleftharpoons (K_2, w_2)$, if there is a relation $Z \subseteq W_1 \times W_2$ such that $(w_1, w_2) \in Z$, and Z fulfils the following closure property:

- $(K_1, w_1) \rightleftharpoons_0 (K_2, w_2)$, for all $(w_1, w_2) \in Z$,
- for each successor w_1' of w_1 in K_1, there is a successor w_2' of w_2 in K_2 with $(w_1', w_2') \in Z$ (forward condition),

- for each successor w_2' of w_2 in K_2, there is a successor w_1' of w_1 in K_1 with $(w_1', w_2') \in Z$ (backward condition).

The famous theorem by van Benthem characterizing the expressive power of modal logics can now be stated as follows:

A *property of pointed models* is a class of pairs (K, w), where K is a Kripke model and w a world of K. For a formula φ we say that φ *expresses* the property $\{(K, w) \mid K, w \models \varphi\}$ (under the usual Kripke semantics). A team property is bisimulation-invariant if it is closed under bisimulation.

Theorem 20 (van Benthem's Theorem [59]). *Let P be a property of pointed Kripke structures. There is an $\mathcal{ML}$-formula which expresses P if and only if there is a first-order formula which expresses P and P is bisimulation-invariant.*

The result of van Benthem has been transferred into the field of modal logics with team semantics, as we want to explain next. First, the notion of bisimulation can very naturally be lifted to teams.

Definition 17 (team bisimulation). Let $K_1 = (W_1, R_1, \pi_1)$, $K_2 = (W_2, R_2, \pi_2)$ be Kripke models, let T_1 and T_2 be teams of K_1 and K_2. Then (K_1, T_1) and (K_2, T_2) are k-bisimilar, written as $K_1, T_1 \leftrightarrows_k K_2, T_2$ if the following holds:

- for each $w_1 \in T_1$, there is some $w_2 \in T_2$ such that $(K_1, w_1) \leftrightarrows_k (K_2, w_2)$,
- for each $w_2 \in T_2$, there is some $w_1 \in T_1$ such that $(K_1, w_1) \leftrightarrows_k (K_2, w_2)$.

Analogously, we say that (K_1, T_1) and (K_2, T_2) are (fully) bisimilar, written as $K_1, T_1 \leftrightarrows K_2, T_2$ if the following holds:

- for each $w_1 \in T_1$, there is some $w_2 \in T_2$ such that $(K_1, w_1) \leftrightarrows (K_2, w_2)$,
- for each $w_2 \in T_2$, there is some $w_1 \in T_1$ such that $(K_1, w_1) \leftrightarrows (K_2, w_2)$.

Now we can characterize the expressive power of some of our modal logics. The expressive power of a logic here is defined to be the set of properties expressible in it. More precisely, a *team property* is a class of pairs (K, T), where K is a Kripke model and T a team of K. For a formula φ we say that φ *expresses* the property $\{(K, T) \mid K, T \models \varphi\}$. A team property is bisimulation-invariant if it is closed under bisimulation.

The following characterizations of the expressive power of $\mathcal{EMDL}$ and $\mathcal{MIL}$ were obtained in [29, 43]:

Theorem 21. *Let P be a team property.*

1. *There is an $\mathcal{EMDL}$-formula which expresses P if and only if P is invariant under k-bisimulation for some k and downwards-closed.*
2. *There is an $\mathcal{MIL}$-formula which expresses P if and only if P is invariant unter k-bisimulation for some k.*

While the just given result already completely settles the question of expressivity of $\mathcal{EMDL}$ and $\mathcal{MIL}$ in terms of bisimulation, the following complete analogue of van Benthem's Theorem was finally obtained in [43]:

Theorem 22. *Let P be a team property. There is an* $\mathcal{MTL}$*-formula which expresses P if and only if there is a first-order formula which expresses P and P is bisimulation-invariant.*

It is worth to note that all our extensions of modal logic by further connectives or generalized dependence atoms define only team properties that are bisimulation-invariant. Hence the just given theorem implies that modal team logic $\mathcal{MTL}$ gives an upper bound with respect to expressivity for all these logics. In particular, modal logic with team semantics and classical negation is sufficient to express all FO-definable generalized dependence atoms. This observation can be strengthened as follows [43]:

Let $\mathsf{ML}^{\mathsf{FO}}$ denote the extension of $\mathcal{ML}$ by all generalized dependence atoms D that are $\mathcal{FO}$-definable without identity, in the extended setting, i.e., dependence atoms are applied not only to propositions but $\mathcal{ML}$-formulae.

Theorem 23. $\mathsf{ML}^{\mathsf{FO}}$ *is equally expressive as* $\mathcal{MTL}$.

3.3 Complexity

The starting point for complexity studies of modal logic is a paper by Ladner from 1977 [45] in which he proved the following theorem:

Theorem 24. *Satisfiability for modal logic* $\mathcal{ML}$ *is* **PSPACE**-*complete.*

His result easily carries over from the usual modal semantics to team semantics. The upper bound follows from Ladner's so-called witness algorithm, which is not so important for us here. The lower bound is given by a reduction from the standard **PSPACE**-complete problem QBF, the evaluation problem for quantified Boolean formulas, where alternations of modalities are used to force a satisfying Kripke model to imitate the evaluation tree of the given formula [2].

In a very clever way this was extended by Sevenster [53] as follows:

Theorem 25. *Satisfiability for* $\mathcal{MDL}$ *is* **NEXPTIME**-*complete.*

To prove the upper bound, one has to express the dependencies by Boolean Skolem functions, similar as in the proof of Theorem 17. Then we can use nondeterminism to guess those functions in exponential time, and check satisfiability. For the lower bound, Sevenster presents a reduction from Dependence-QBF, a variant of the above-mentioned QBF extended by dependencies among variables [49], a problem which is **NEXPTIME**-complete. This reduction is essentially Ladner's reduction extended by dependence atoms.

Lohmann and Vollmer [46] extended Sevenster's result by determining the complexity of every fragment of $\mathcal{MDL}$, given by any subset of the modalities $\Box$ and $\Diamond$, by restricting the allowed propositional connectives to any subset of $\{\wedge, \vee, \neg, \top, \bot, \oslash\}$ and by considering fragments with and without dependence atoms. Their results can be summarized as in Table 2. An entry "+" in the table

Table 2 Complexity of satisfiability for fragments of modal dependence logic

□	◇	∧	∨	¬	T	⊥	=()	⊘	Complexity
+	+	+	*	+	*	*	+	*	NEXPTIME
+	+	+	+	+	*	*	−	*	PSPACE
+	+	+	+	−	*	+	*	*	PSPACE
+	+	+	−	+	*	*	−	+	Σ_2^p
+	+	+	−	−	*	+	*	+	Σ_2^p
+	+	+	−	+	*	*	−	−	coNP
+	+	+	−	−	*	+	*	−	coNP
+	−	+	+	+	*	*	*	*	NP
−	+	+	+	+	*	*	*	*	NP
+	−	+	−	+	*	*	*	+	NP
−	+	+	−	+	*	*	*	+	NP
+	−	+	−	+	*	*	*	−	PTIME
−	+	+	−	+	*	*	*	−	PTIME
+	−	+	*	−	*	*	*	*	PTIME
−	+	+	*	−	*	*	*	*	PTIME
*	*	−	*	*	*	*	*	*	PTIME
*	*	*	*	−	*	−	*	*	trivial
−	−	+	+	+	*	*	*	*	NP
−	−	+	*	+	*	*	*	+	NP
−	−	*	−	*	*	*	*	−	PTIME
−	−	*	*	−	*	*	*	*	PTIME

means that the syntactic element is allowed, "−" means it is forbidden, and "*" means that the complexity does not depend on whether the element is present or not. All rows in the table denote completeness results for the respective complexity class under polynomial-time many-one reductions, except those for **PTIME**.

In a similar way, the complexity of the model-checking problem for fragments of modal dependence logic has been classified in [9]. For some fragments, model checking is **NP**-complete, for others it is solvable in polynomial time.

Theorem 26. *Model checking for* $\mathcal{MDL}$ *is* **NP**-*complete*.

We want to turn to the explanation of one special case in Table 2, the complexity of Poor Man's Logic. In the context of modal logic, *poor man's formulas* are just formulas that do not contain ∨. Hemaspaandra [31, 32] showed that poor man's modal logic is **PSPACE**-complete over the class of Kripke structures in which every world has at most two successors. The proof is again by a reduction from QBF, where we express the QBF-tree by alternations of modalities. Important now is that without disjunction, we cannot express the tree-structure. In fact, satisfiability for poor man's modal logic over K (the class of all Kripke structures) is only **coNP**-complete. The requirement that in every model each world has at most two successors is essential for the complexity. Lohmann and Vollmer [46] showed:

Theorem 27. *Poor man's modal dependence logic, i.e., the fragment of modal dependence logic allowing only the propositional connectives* $\wedge$ *and* $\neg$ *(besides the two modalites and dependence atoms), is* NEXPTIME-*complete.*

The proof relies as before on a reduction from Dependence-QBF. We express the QBF-tree by alternations of modalities. As in the case of Hemaspaandra's proof we cannot enforce tree-structure of the Kripke model without disjunction. In this case, however, we can use dependence atoms to ensure that everything in the model that does not belong to the tree is essentially nothing else than a copy of a subtree; hence, in difference to Hemaspaandra, we do not need the requirement about the number of successors of a world.

Next, we turn to modal independence logic. The following has been proven in [42]:

Theorem 28. *Satisfiability for* $\mathcal{MIL}$ *is* NEXPTIME-*complete.*

While the lower bound follows trivially from the complexity result for $\mathcal{MDL}$, the upper bound is proven by an embedding of $\mathcal{MIL}$ into the Gödel-Kalmár-Schütte fragment of all $\mathcal{FO}$-sentences with prefix $\exists^* \forall^2 \exists^*$ (without function symbols, without equality) in a satisfiability preserving way. This fragment is decidable in NEXPTIME [3]. This proof thus is different from the one given by Sevenster (and it is not clear how to extend his ideas to include the independence atom), but it yields Sevenster's result as a corollary.

The embedding into the $\mathcal{FO}$-fragment is even possible in a much more general context [42]:

Theorem 29. *Satisfiability for* $\mathcal{ML}$ *extended by dependence atoms that can be defined in* $\exists^* \forall^2 \exists^*$ *is in* NEXPTIME.

For a large class of generalized dependence atoms, satisfiability can thus be placed into the class NEXPTIME, while for a few particular atoms, we have completeness—above we already mentioned this for dependence and independence, and in [30], satisfiability for modal inclusion logic $\mathcal{MINCL}$, i.e., $\mathcal{ML}$ extended by the inclusion atom (see Table 1), was shown to be NEXPTIME-complete as well.

Also the model-checking problem for modal logic extended by generalized dependence atoms was studied in [42].

Theorem 30. *Model checking for* $\mathcal{ML}$ *extended by* $\mathcal{FO}$-*definable dependence atoms is in* NP.

Recall that by Theorem 26 model checking for $\mathcal{MDL}$ (and also for $\mathcal{MIL}$) is NP-complete.

3.4   Propositional Logic

Propositional logic with team semantics is nothing else than modal logic with team semantics but without connections between the team members (worlds). The syntax is the following:

Definition 18 (Syntax of $\mathcal{PL}$).

$$\varphi ::= p \mid \neg p \mid (\varphi \wedge \varphi) \mid (\varphi \vee \varphi), \quad \text{where } p \in \Phi.$$

The semantics is exactly the same as for the modal case. Propositional logic can now be extended by different generalized dependence atoms (such as those given in Table 1) as well as different propositional connectives (intuitionistic implication, classical disjunction, etc.) The expressive power of some of these logics (i.e. the classes of teams that are definable) has been studied by Yang [63]; in particular, propositional dependence logic ($\mathcal{PL}$ plus dependence atoms) can define all nonempty downwards-closed team properties, and propositional team logic ($\mathcal{PL}$ plus classical negation) can define all team properties.

Concerning complexity questions, the following results are known:

Theorem 31. *1. The satisfiability and the model-checking problems for propositional dependence logic and propositional independence logic are* NP-*complete [9, 27, 46].*
2. *The satisfiability problem for propositional inclusion logic is* EXPTIME-*complete [30], while its model-checking problem is in* PTIME *(Lauri Hella, personal communication).*
3. *The model-checking problem for propositional dependence logic extended by classical negation is* PSPACE-*complete [47].*
4. *The satisfiability problem for propositional inclusion logic extended by classical negation as well as for propositional independence logic extended by classical negation is complete for the class* AEXPTIME(poly) *of all problems solvable by alternating Turing machines in exponential time making only a polynomial number of alternations; the model-checking problem for both logics is* PSPACE-*complete [30].*

Besides the above-summarized results, also the validity problem for propositional logic and some of its extensions have been studied. It is worth noting that, unlike for classical logics closed under the classical negation, the satisfiability problem and the validity problem for most of the logics discussed in this article are not dual to each other. Partial results for axiomatizability have also been obtained [27, 52, 61, 64].

4 Conclusion

1. A number of complexity questions for fragments of first-order dependence logic or variants remain unsettled. The complexity of the validity problem for $\mathcal{D}^2$ is one example. More generally, one might ask what decidability results for first-order logic, e.g., for formula classes defined by restricted quantifier prefixes, transfer into the context of team-based logics.
2. While many expressivity results of modal and propositional logic have been stated in this survey and this issue is more or less settled in the first-order case, it has to be mentioned that some very basic cases in modal logic still remain unsettled. In particular, what is the expressive power of propositional or modal independence logic? It is worth noting that very recently a version of Theorem 21 for (extended) modal inclusion logic has been shown in [28].
3. In Section 3.2 we stated equal expressivity for many dialects of modal logics. However, we did not touch the topic of succinctness. As an example, while on singleton teams of evaluation, $\mathcal{MIL}$ equally expressive than $\mathcal{ML}$, it can be proven that it is exponentially more succinct [42]. Also, in Theorem 18 we stated that $\mathcal{EMDL}$ and $\mathcal{ML}$ plus classical disjunction have the same expressive power. In [29] it was shown that any translation from $\mathcal{EMDL}$ to $\mathcal{ML}$ with disjunction necessarily leads to an exponential blow-up in formula size. In most other cases, the question of succinctness remains unsettled so far.
4. The question of axiomatizability has not been covered in detail in this survey. We mention that axiomatizability of some sublogics of $\mathcal{MIL}$ has been studied, e.g., in [63] and [52], but remains open for many logics. Related to this is the complexity of the tautology problem. In particular it remains open if we can axiomatize $\mathcal{MIL}$.
5. While we mentioned a number of complexity results on modal dependence logic and some of its extensions, this issue remains unsettled for full $\mathcal{MIL}$. In particular, what is the complexity of satisfiability and validity of $\mathcal{MIL}$?

Acknowledgements The authors thank the anonymous referee for corrections and valuable comments. The second author was supported by grants 292767, 275241, and 264917 of the Academy of Finland.

References

1. Abramsky, S., Väänänen, J.: From IF to BI: a tale of dependence and separation. Synthese **167**(2, Knowledge, Rationality & Action), 207–230 (2009). doi:10.1007/s11229-008-9415-6. http://dx.doi.org/10.1007/s11229-008-9415-6
2. Blackburn, P., de Rijke, M., Venema, Y.: Modal logics. In: Cambridge Tracts in Theoretical Computer Science, vol. 53. Cambridge University Press, Cambridge (2001)
3. Börger, E., Grädel, E., Gurevich, Y.: The Classical Decision Problem. Perspectives in Mathematical Logic. Springer, Berlin (1997)
4. Church, A.: A note on the Entscheidungsproblem. J. Symb. Log. **1**(1), 40–41 (1936)

5. Cook, S.A.: A hierarchy for nondeterministic time complexity. In: Conference Record, Fourth Annual ACM Symposium on Theory of Computing, pp. 187–192. ACM, New York (1972)
6. Durand, A., Kontinen, J.: Hierarchies in dependence logic. ACM Trans. Comput. Log. **13**(4), 1–21 (2012)
7. Durand, A., Ebbing, J., Kontinen, J., Vollmer, H.: Dependence logic with a majority quantifier. In: Chakraborty, S., Kumar, A. (eds.) FSTTCS, LIPIcs, vol. 13, pp. 252–263. Schloss Dagstuhl - Leibniz-Zentrum fuer Informatik, Germany (2011)
8. Durand, A., Ebbing, J., Kontinen, J., Vollmer, H.: Dependence logic with a majority quantifier. J. Log. Lang. Inf. **24**(3), 289–305 (2015). doi:10.1007/s10849-015-9218-3. http://dx.doi.org/10.1007/s10849-015-9218-3
9. Ebbing, J., Lohmann, P.: Complexity of model checking for modal dependence logic. In: SOFSEM 2012: Theory and Practice of Computer Science. Lecture Notes in Computer Science, vol. 7147, pp. 226–237. Springer, Berlin, Heidelberg (2012)
10. Ebbing, J., Hella, L., Meier, A., Müller, J.S., Virtema, J., Vollmer, H.: Extended modal dependence logic. In: WoLLIC. Lecture Notes in Computer Science, vol. 8071, pp. 126–137. Springer, Berlin, Heidelberg (2013)
11. Ebbing, J., Kontinen, J., Müller, J., Vollmer, H.: A fragment of dependence logic capturing polynomial time. Log. Methods Comput. Sci. **10**(3) (2014). doi:10.2168/LMCS-10(3:3)2014. http://dx.doi.org/10.2168/LMCS-10(3:3)2014
12. Engström, F.: Generalized quantifiers in dependence logic. J. Log. Lang. Inf. **21**, 299–324 (2012). http://dx.doi.org/10.1007/s10849-012-9162-4. 10.1007/s10849-012-9162-4
13. Engström, F., Kontinen, J.: Characterizing quantifier extensions of dependence logic. J. Symb. Log. **78**, 0–9 (2013)
14. Fagin, R.: Generalized first-order spectra and polynomial-time recognizable sets. In: Complexity of Computation. Proceedings of SIAM-AMS Symposium in Applied Mathematics, New York, 1973, pp. 43–73. SIAM-AMS Proceedings, vol. VII. American Mathematical Society, Providence, RI (1974)
15. Fagin, R.: Finite-model theory - a personal perspective. Theor. Comput. Sci. **116**(1&2), 3–31 (1993)
16. Galliani, P.: Inclusion and exclusion dependencies in team semantics - on some logics of imperfect information. Ann. Pure Appl. Log. **163**(1), 68–84 (2012)
17. Galliani, P., Hella, L.: Inclusion logic and fixed point logic. In: Rocca, S.R.D. (ed.) Computer Science Logic 2013 (CSL 2013). Leibniz International Proceedings in Informatics (LIPIcs), vol. 23, pp. 281–295. Schloss Dagstuhl–Leibniz-Zentrum fuer Informatik, Dagstuhl, Germany (2013). doi:http://dx.doi.org/10.4230/LIPIcs.CSL.2013.281. http://drops.dagstuhl.de/opus/volltexte/2013/4203
18. Galliani, P., Hannula, M., Kontinen, J.: Hierarchies in independence logic. In: Rocca, S.R.D. (ed.) Computer Science Logic 2013 (CSL 2013). Leibniz International Proceedings in Informatics (LIPIcs), vol. 23, pp. 263–280. Schloss Dagstuhl–Leibniz-Zentrum fuer Informatik, Dagstuhl, Germany (2013). doi:http://dx.doi.org/10.4230/LIPIcs.CSL.2013.263. http://drops.dagstuhl.de/opus/volltexte/2013/4202
19. Grädel, E.: Capturing complexity classes by fragments of second-order logic. Theor. Comput. Sci. **101**(1), 35–57 (1992)
20. Grädel, E.: Model-checking games for logics of imperfect information. Theor. Comput. Sci. **493**, 2–14 (2013). doi:10.1016/j.tcs.2012.10.033. http://dx.doi.org/10.1016/j.tcs.2012.10.033
21. Grädel, E., Väänänen, J.A.: Dependence and independence. Stud. Logica **101**(2), 399–410 (2013)
22. Grandjean, E., Olive, F.: Graph properties checkable in linear time in the number of vertices. J. Comput. Syst. Sci. **68**(3), 546–597 (2004)
23. Hannula, M.: Axiomatizing first-order consequences in independence logic. Ann. Pure Appl. Logic **166**(1), 61–91 (2015). doi:10.1016/j.apal.2014.09.002. http://dx.doi.org/10.1016/j.apal.2014.09.002

24. Hannula, M.: Hierarchies in inclusion logic with lax semantics. In: Banerjee, M., Krishna, S.N. (eds.) Proceedings of Logic and Its Applications - 6th Indian Conference, ICLA 2015, Mumbai, India, January 8–10, 2015. Lecture Notes in Computer Science, vol. 8923, pp. 100–118. Springer, Berlin (2015). doi:10.1007/978-3-662-45824-2_7

25. Hannula, M.: On variants of dependence logic: axiomatizability and expressiveness. Ph.D. thesis, University of Helsinki (2015)

26. Hannula, M., Kontinen, J.: Hierarchies in independence and inclusion logic with strict semantics. J. Log. Comput. 25(3), 879–897 (2015). doi:10.1093/logcom/exu057. http://dx.doi.org/10.1093/logcom/exu057

27. Hannula, M., Kontinen, J., Virtema, J., Vollmer, H.: Complexity of propositional independence and inclusion logic. In: Mathematical Foundations of Computer Science 2015 - 40th International Symposium, MFCS 2015, Proceedings, Part I. Lecture Notes in Computer Science, vol. 9234, pp. 269–280. Springer, Berlin (2015)

28. Hella, L., Stumpf, J.: The expressive power of modal logic with inclusion atoms. In: Esparza, J., Tronci, E. (eds.) Proceedings Sixth International Symposium on Games, Automata, Logics and Formal Verification, GandALF 2015, Genoa, Italy, 21–22nd September 2015, EPTCS, vol. 193, pp. 129–143 (2015). doi:10.4204/EPTCS.193.10

29. Hella, L., Luosto, K., Sano, K., Virtema, J.: The expressive power of modal dependence logic. In: Advances in Modal Logic 10, Invited and Contributed Papers from the Tenth Conference on "Advances in Modal Logic," Held in Groningen, The Netherlands, August 5–8, 2014, pp. 294–312. College Publications, London (2014). http://www.aiml.net/volumes/volume10/Hella-Luosto-Sano-Virtema.pdf

30. Hella, L., Kuusisto, A., Meier, A., Vollmer, H.: Modal inclusion logic: being lax is simpler than being strict. In: Italiano, G.F., Pighizzini, G., Sannella, D. (eds.) Mathematical Foundations of Computer Science 2015 - 40th International Symposium, MFCS 2015, Milan, Italy, August 24–28, 2015, Proceedings, Part I. Lecture Notes in Computer Science, vol. 9234, pp. 281–292. Springer, Berlin (2015)

31. Hemaspaandra, E.: The complexity of poor man's logic. J. Log. Comput. 11(4), 609–622 (2001). Corrected version: [32]

32. Hemaspaandra, E.: The complexity of poor man's logic. CoRR cs.LO/9911014v2 (2005). http://arxiv.org/abs/cs/9911014v2

33. Henkin, L.: Logical Systems Containing Only a Finite Number of Symbols. Presses De l'Université De Montréal, Montreal (1967)

34. Hintikka, J., Sandu, G.: Informational independence as a semantical phenomenon. In: Logic, Methodology and Philosophy of Science, VIII (Moscow, 1987). Studies in Logic and the Foundations of Mathematics, vol. 126, pp. 571–589. North-Holland, Amsterdam (1989). doi:10.1016/S0049-237X(08)70066-1. http://dx.doi.org/10.1016/S0049-237X(08)70066-1

35. Hodges, W.: Compositional semantics for a language of imperfect information. Log. J. IGPL 5, 539–563 (1997)

36. Immerman, N.: Relational queries computable in polynomial time. Inf. Control. 68 (1-3), 86–104 (1986). doi:10.1016/S0019-9958(86)80029-8. http://dx.doi.org/10.1016/S0019-9958(86)80029-8

37. Kontinen, J.: Coherence and computational complexity of quantifier-free dependence logic formulas. Stud. Logica 101(2), 267–291 (2013)

38. Kontinen, J., Nurmi, V.: Team logic and second-order logic. Fundamenta Informaticae 106 (2–4), 259–272 (2011)

39. Kontinen, J., Väänänen, J.: On definability in dependence logic. J. Log. Lang. Inf. 18(3), 317–332 (2009)

40. Kontinen, J., Kuusisto, A., Lohmann, P., Virtema, J.: Complexity of two-variable dependence logic and if-logic. Inf. Comput. 239, 237–253 (2014). doi:10.1016/j.ic.2014.08.004. http://dx.doi.org/10.1016/j.ic.2014.08.004

41. Kontinen, J., Kuusisto, A., Virtema, J.: Decidable fragments of logics based on team semantics. CoRR abs/1410.5037 (2014). http://arxiv.org/abs/1410.5037

42. Kontinen, J., Müller, J., Schnoor, H., Vollmer, H.: Modal independence logic. In: Advances in Modal Logic 10, Invited and Contributed Papers from the Tenth Conference on "Advances in Modal Logic," Held in Groningen, The Netherlands, August 5–8, 2014, pp. 353–372. College Publications, London (2014). http://www.aiml.net/volumes/volume10/Kontinen-Mueller-Schnoor-Vollmer.pdf

43. Kontinen, J., Müller, J.S., Schnoor, H., Vollmer, H.: A Van Benthem Theorem for Modal Team Semantics. In: 24th EACSL Annual Conference on Computer Science Logic (CSL 2015). Leibniz International Proceedings in Informatics (LIPIcs), vol. 41, pp. 277–291. Schloss Dagstuhl–Leibniz-Zentrum fuer Informatik, Dagstuhl, Germany (2015)

44. Kuusisto, A.: A double team semantics for generalized quantifiers. J. Log. Lang. Inf. **24**(2), 149–191 (2015). doi:10.1007/s10849-015-9217-4. http://dx.doi.org/10.1007/s10849-015-9217-4

45. Ladner, R.E.: The computational complexity of provability in systems of modal propositional logic. SIAM J. Comput. **6**(3), 467–480 (1977)

46. Lohmann, P., Vollmer, H.: Complexity results for modal dependence logic. Stud. Logica **101**(2), 343–366 (2013)

47. Müller, J.S.: Satisfiability and model checking in team based logics. Ph.D. thesis, Leibniz Universität Hannover (2014)

48. Nurmi, V.: Dependence logic: Investigations into higher-order semantics defined on teams. Ph.D. thesis, University of Helsinki (2009)

49. Peterson, G., Reif, J., Azhar, S.: Lower bounds for multiplayer noncooperative games of incomplete information. Comput. Math. Appl. **41**(7-8), 957–992 (2001). doi:10.1016/S0898-1221(00)00333-3. http://www.sciencedirect.com/science/article/B6TYJ-43P387P-19/2/cd72ba7ccb5f3c2a2b65eb3c45aa2ca7

50. Pratt-Hartmann, I.: Complexity of the two-variable fragment with counting quantifiers. J. Log. Lang. Inf. **14**, 369–395 (2005). doi:10.1007/s10849-005-5791-1. http://portal.acm.org/citation.cfm?id=1080942.1080949

51. Rönnholm, R.: Capturing k-ary existential second order logic with k-ary inclusion-exclusion logic. CoRR **abs/1502.05632** (2015). http://arxiv.org/abs/1502.05632

52. Sano, K., Virtema, J.: Axiomatizing propositional dependence logics. In: Kreutzer, S. (ed.) 24th EACSL Annual Conference on Computer Science Logic, CSL 2015, September 7–10, 2015, Berlin, Germany. LIPIcs, vol. 41, pp. 292–307. Schloss Dagstuhl - Leibniz-Zentrum fuer Informatik (2015). doi:10.4230/LIPIcs.CSL.2015.292

53. Sevenster, M.: Model-theoretic and computational properties of modal dependence logic. J. Log. Comput. **19**(6), 1157–1173 (2009). doi:10.1093/logcom/exn102. http://logcom.oxfordjournals.org/cgi/content/abstract/exn102v1

54. Tulenheimo, T., Sevenster, M.: On modal logic, IF logic, and IF modal logic. In: Advances in Modal Logic 6, Papers from the Sixth Conference on "Advances in Modal Logic," Held in Noosa, Queensland, Australia, on 25–28 September 2006, pp. 481–501. College Publications, London (2006). http://www.aiml.net/volumes/volume6/Tulenheimo-Sevenster.ps

55. Turing, A.: On computable numbers, with an application to the entscheidungsproblem. Proc. Lond. Math. Soc. Ser. 2 **42**, 230–265 (1936)

56. Väänänen, J.: Dependence Logic. London Mathematical Society Student Texts, vol. 70. Cambridge University Press, Cambridge (2007). doi:10.1017/CBO9780511611193. http://dx.doi.org/10.1017/CBO9780511611193

57. Väänänen, J.: Team Logic. In: Johan van Benthem Benedikt Löowe, D.G. (ed.) Interactive Logic. Texts in Logic and Games, vol. 1, pp. 281–302. Amsterdam University Press, Amsterdam (2007)

58. Väänänen, J.: Modal dependence logic. In: Apt, K.R., van Rooij, R. (eds.) New Perspectives on Games and Interaction. Texts in Logic and Games, vol. 4, pp. 237–254. Amsterdam University Press, Amsterdam (2008)

59. van Benthem, J.: Modal Logic and Classical Logic. Bibliopolis, Berkeley, CA (1985)

60. Vardi, M.Y.: The complexity of relational query languages (extended abstract). In: Lewis, H.R., Simons, B.B., Burkhard, W.A., Landweber, L.H. (eds.) Proceedings of the 14th Annual ACM Symposium on Theory of Computing, May 5–7, 1982, San Francisco, California, USA, pp. 137–146. ACM, New York (1982). doi:10.1145/800070.802186
61. Virtema, J.: Complexity of validity for propositional dependence logics. In: Proceedings Fifth International Symposium on Games, Automata, Logics and Formal Verification, GandALF 2014, Verona, Italy, September 10–12, 2014. Electronic Proceedings in Theoretical Computer Science, vol. 161, pp. 18–31 (2014). doi:10.4204/EPTCS.161.5. http://dx.doi.org/10.4204/EPTCS.161.5
62. Yang, F.: Expressing second-order sentences in intuitionistic dependence logic. Stud. Logica **101**(2), 323–342 (2013)
63. Yang, F.: On extensions and variants of dependence logic – a study of intuitionistic connectives in the team semantics setting. Ph.D. thesis, University of Helsinki (2014)
64. Yang, F., Väänänen, J.: Propositional logics of dependence. Annals of Pure and Applied Logic 167, 557–589 (2016)

Grelling on Dependence

Jouko Väänänen

Abstract Kurt Grelling wrote a paper in 1939 presenting various concepts of dependence. The paper remained unpublished, but deserves to be read today. Many of the ideas of the paper have been subsequently reinvented but one concept, which we call G-dependence, is still genuinely new, and that is the main topic of this paper. We isolate some basic properties of G-dependence and pose the question of finding simple axioms for it.

1 Introduction

In 1939 [9] the mathematician and logician Kurt Grelling[1] developed, in co-operation with Paul Oppenheim[2], a completely abstract theory of dependence and independence with apparently no connection to algebra or probability theory, although a study of dependence was emerging in these fields too. His starting point seems to have been the so-called Gestalt Theory [19], but his prime example was the earthly way in which commercial price depends on supply and demand. This paper will present an overview of Grelling's theory.

The concepts of linear and algebraic dependence have a long history. They were known early in the 19th century in the methodology of solving systems of equations. Likewise, the concept of independence of events was at the heart of probability

---

[1]Kurt Grelling was born in Berlin in 1886. He studied in Göttingen with David Hilbert as his supervisor, graduating in 1910 with a thesis on set theory. Already in 1908 he published a paper with L. Nelson on Russell's paradox, in which they introduced what is today known as the Grelling-Nelson paradox. He was subsequently influential in philosophy in the so-called Berlin Circle. He perished in Auschwitz in 1942.

[2]Paul Oppenheim was born in Frankfurt am Main in 1885, emigrated from Germany to Belgium in 1933, and then to USA in 1939. He contributed to philosophy, especially Gestalt Theory, and to philosophy of science, working with Grelling and also with Carl Gustav Hempel (1905–1997).

J. Väänänen (✉)
Department of Mathematics and Statistics, University of Helsinki, Helsinki, Finland

Institute for Logic, Language and Computation, University of Amsterdam, Amsterdam, The Netherlands
e-mail: jouko.vaananen@helsinki.fi

© Springer International Publishing Switzerland 2016
S. Abramsky et al. (eds.), *Dependence Logic*, DOI 10.1007/978-3-319-31803-5_3

theory from the moment of its conception. In the thirties van den Waerden [21] and Whitney [22] pointed out similarities in the properties of linear and algebraic dependence and suggested an abstract concept which covers both, without assuming any background algebraic operations. This concept is nowadays known as the concept of a matroid or a pregeometry. Also in the thirties, Kolmogoroff [16] gave an axiomatic basis for probability theory, including an exact formulation of independence of events, with events being essentially just abstract sets.

Apart from algebra, probability theory and logic, the concepts of dependence and independence were, of course, used throughout the centuries of development of experimental science. In physics Galileo argued already in the 16th century that the time of descent of two falling bodies is dependent on the height from where they are dropped but independent of their weight. In the 18th century in philosophy Hume developed the theory of causality, continued by Mill in the 19th century; and causality *is* intimately connected with dependence: an effect depends in a natural sense on its cause, and on the other hand, what is independent of the effect cannot be its cause either. In biology Mendel argued in 1866 that what are now called alleles of genes of plants are totally dependent on the alleles of the parents, but alleles of different genes are inherited independently of each other. When we come to the early 19th century the concept of independence was widely used in physics, chemistry, biology, statistics, and other fields. In quantum physics the concept of independence of events received notoriety from the EPR-phenomenon [5], according to which two, however, distant entangled particles apparently depend on each other in an instantaneous cause-effect sense, although they are in the sense of classical physics totally independent of each other. The entanglement phenomenon was proved with a mathematical argument in 1964 by Bell [2], and later also experimentally demonstrated.

In an unrelated development in computer science in the seventies Codd [4] introduced the concept of functional dependence in a relational database. This is, as we argue below, essentially equivalent to Grelling's concept of dependence. Many variations of functional dependence were introduced in database theory including the concept of independence (under the name 'embedded multivalued dependence' [6]). This concept is a special case of the probability theoretic concept of independence.

We use *team semantics* [20] as a universal approach covering all the individual cases of dependence.

Notation. If A is a set, $\mathscr{P}(A)$ is the power-set of A. We use x, y, z, and so on for finite set. Then x_i, y_j, z_k, and so on refer to individual elements of x, y, z, respectively. If x and y are sets, then xy is a shorthand notation for the union $x \cup y$. If $x \subseteq I$ and $s : I \longrightarrow M$ is a function, then $s \upharpoonright x$ is the restriction of s to the domain x.

2 Dependence

Grelling's interest in dependence arose from Gestalt theory. Gestalt theory goes back to the essay *On 'Gestalt Qualities'* by the philosopher Christian von Ehrenfels, published in 1890. This theory started in philosophy and then moved to psychology. An example of a Gestalt concept is the phenomenon that our perception "fills" the missing parts in Figure 1. The missing parts are in a sense *functionally dependent* on—or in the closure of—the visible parts. Instead of a visual image we could consider a piece of music or poetry. Grelling started to develop a mathematical theory of such dependence.

Grelling states, in his own notation, a definition equivalent to the following:

Definition 1 (Grelling [9]). Let I be a set of functions g of the same variable x (or more generally, a finite string of variables). The closure $\mathrm{cl}(\varphi)$ of $\varphi \subseteq I$ consists of those functions f in I for which the following holds: for any x_1 and x_2, if each function g in φ satisfies $g(x_1) = g(x_2)$, then $f(x_1) = f(x_2)$. We say that ψ *depends (only) on* φ if $\psi \subseteq \mathrm{cl}(\varphi)$.

Typically g would be a magnitude with x as a parameter. A typical parameter would be time. We can think of Grelling's functions as columns of a table of data and his arguments x as indexes of the rows of the table (see Figure 2), although Grelling does not refer explicitly to a table of data as his model.

The closure operation cl of Definition 1 satisfies, as Grelling observes, the axioms

C1 $x \subseteq \mathrm{cl}(x)$.
C2 If $x \subseteq y$, then $\mathrm{cl}(x) \subseteq \mathrm{cl}(y)$.
C3 $\mathrm{cl}(\mathrm{cl}(x)) \subseteq \mathrm{cl}(x)$.

Fig. 1 A gestalt phenomenon.

Fig. 2 Grelling's setup

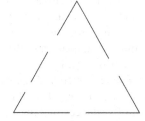

x	f_1	f_2	$\cdots$	f_m
a_1	$f_1(a_1)$	$f_2(a_1)$	$\cdots$	$f_m(a_1)$
$\vdots$	$\vdots$	$\vdots$		$\vdots$
a_k	$f_1(a_k)$	$f_2(a_k)$	$\cdots$	$f_m(a_k)$
$\vdots$	$\vdots$	$\vdots$		$\vdots$

Fig. 3 A closure operation.

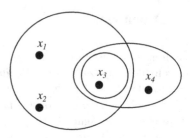

Any function $\mathscr{P}(I) \longrightarrow \mathscr{P}(I)$ as above is nowadays called a *closure operation* (Fig. 3). If I is a vector space, then $\text{cl}(x)$ can be taken to be the *linear span* $[x]$ of x. Now dependence means the same as linear dependence. Likewise, in an algebraically closed field we can let $\text{cl}(x)$ be the algebraic closure of x, and then dependence means algebraic dependence. If we have a topology on I, we can let $\text{cl}(x)$ be the topological closure of x, and dependence means being in the set or on the boundary of the set. For an example from model theory, suppose $\mathscr{M}$ is an infinite first order structure. An element a of M is *algebraic over* $x \subseteq M$ if there is a first order formula (with identity) $\phi(y, z_1, \ldots, z_m)$ and elements $b_1, \ldots, b_m$ of M such that the set

$$A = \{a \in M : \mathscr{M} \models \phi(a, b_1, \ldots, b_m)\}$$

is finite and $a \in A$. We get a closure operation cl on M be letting $\text{cl}(x)$ be the set of elements of M which are algebraic over x. If G is a finite graph we can let $\text{cl}(x)$ be the set of points which are on a closed walk (a sequence of vertices starting and ending at the same vertex, with each two consecutive vertices in the sequence connected by an edge in the graph) including x. In this case dependence on x means being within a closed walk from some elements of x. If I is the set of attributes (fields) of a database, then $\text{cl}(x)$ can be taken to consist of those attributes y_i for which the attributes in x *functionally determine* y_i, i.e., if we take any two tuples from the database and they agree about the attributes in x they also agree about the attribute y_i. Dependence in this case means the same as functional dependence in database theory. A classical closure operation in logic is logical consequence: Suppose I is the set of all first order sentences in a given vocabulary. Let $\text{cl}(x)$ be the set of sentences that logically follow from the sentences of x. Then cl is a closure operation on I. Dependence in this structure means the same as logical consequence. For an example from recursion theory, suppose I is uncountable the set of all subsets of $\mathbb{N}$. For a (typically finite) subset x of I let $\text{cl}(x)$ consist of those (countably many, if x is countable) subsets of $\mathbb{N}$ which can be computed with some finite number of oracles from x. In this case dependence means being computable from. Similarly we can let $\text{cl}(x)$ be the constructible closure $L(x) \cap I$ of x in the sense of Gödel's notion of constructibility in set theory. In a sense the most general form of a closure operation is the following: If F is any set of n-ary functions, for various $n \in \mathbb{N}$, on a set I, we can take $\text{cl}(x)$ to be the closure of $x \subseteq I$ under all the functions in F. Then

y depends in this sense on x, if the elements of y can be obtained from elements of x by repeated applications of functions in F.

Rather than defining dependence on the basis of a closure operation we can start with something that can be called a 'dependence relation' and define a closure operation from it. Indeed, in [18] a binary relation $x \Rightarrow y$ ("y depends (only) on x") on $\mathscr{P}(I)$ is called a *dependence relation*, if the following Armstrong Axioms ([1]) are satisfied:

D1 $x \Rightarrow x$
D2   If $x \Rightarrow yz$, then $xu \Rightarrow y$
D3   If $x \Rightarrow y$ and $y \Rightarrow z$, then $x \Rightarrow z$
D4   If $x \Rightarrow y$ and $x \Rightarrow z$, then $x \Rightarrow yz$

The equivalence

$$x \Rightarrow y \text{ iff } y \subseteq \mathrm{cl}(x)$$

ties closure operations and dependence relations together. If the closure operation satisfies (C1)–(C3), then the dependence relation satisfies (D1)–(D4), and for finite I the converse holds, too.

We now present Grelling's definition in the framework of team semantics (equivalently, relational databases), which makes it easier to compare the concept with similar concepts in logic, computer science, algebra, model theory, and statistics. In [20] the concept of a *team*, and *team semantics*, was introduced, based on the concept of a *trump* in [13, 14]. The idea of team semantics is to use *sets* of assignments, rather than *single* assignments, to define the meaning of logical formulas.

Definition 2 ([20]). Suppose I is a finite set of variables and M is a set. Any function $s : I \longrightarrow M$ is called an *assignment*. Any set of assignments is called a *team* with domain I. A function $\langle s_j : j \in J \rangle$ with assignments $s_j : I \longrightarrow M$ as values is called a *multiteam*.

The difference between teams and multiteams is that the latter allows repetition of the same assignment, giving rise to the concept of the *probability* of an assignment (exploited in [15]). If $X = \langle s_j : j \in J \rangle$ is a multiteam, then $\{s_j : j \in J\}$ is a team and we write $s \in X$ whenever $s = s_j$ for some $j \in J$. Teams can be represented as a table, see Figure 4. In this respect they are essentially relational databases.

Fig. 4 A team.

s	x_1	x_2	$\cdots$	x_m
s_1	$s_1(x_1)$	$s_1(x_2)$	$\cdots$	$s_1(x_m)$
$\vdots$	$\vdots$	$\vdots$		$\vdots$
s_k	$s_1(a_k)$	$s_2(a_k)$	$\cdots$	$s_k(x_m)$
$\vdots$	$\vdots$	$\vdots$		$\vdots$

Grelling	Team semantics	Team semantics notation
(none)	team	X
function f	variable f	variable x
argument value x	assignment x	assignment s
$f(x)$	$x(f)$	$s(x)$

Fig. 5 Grelling's notation and team semantics compared.

Grelling's setup is actually the setup of team semantics using multiteams. What he calls a function is in team semantics a variable. This difference in vocabulary is familiar from probability theory: a random variable is actually a function. Also in mathematics, if $y = f(x)$, then y is at the same time a variable and a function. What Grelling calls an argument is in team semantics an assignment. In a sense, team semantics thinks of the n-tuple

$$(f_1(x), \ldots, f_n(x)),$$

where $I = \{f_1, \ldots, f_n\}$, as an assignment. Finally, the value $f(x)$ is written in team semantics as $x(f)$, or using the notation of team semantics, as $s(x)$ (see Figure 5).

Definition 3 (Team closure operation). Suppose X is a team or a multiteam with domain I. We obtain a closure operation by letting $cl_X(x)$ consist of those $y \in I$ for which

$$\forall s, s' \in X(s \restriction x = s' \restriction x \longrightarrow s \restriction y = s' \restriction y). \tag{E}$$

We call such a closure operation cl_X a *team closure operation*. Dependence of y (only) on x means in this closure operation functional dependence, i.e., the existence of a function f on $^{<\omega}M$ such that

$$\forall s \in X(s \restriction y = f(s \restriction x)).$$

We denote this associated dependence relation by $\Rightarrow_X$.

The concept Grelling calls *dependence*[3] is exactly that of Definition 3. If M is a field, I is a set of vectors in a vector space V over M, the assignments s of a team are interpreted as basis vectors of V, and we interpret the columns of a team X as coefficients which expresses any i in the domain of X as a linear combination of the vectors $s \in X$, then $cl(x)$ in the sense of team semantics means the same as linear span and dependence means the same as linear dependence, provided we demand in Definition 3 that the function f is linear.

---

[3]He uses the symbol 'Equidep(y, x)' for the dependence of y on x.

Fig. 6 A team.

s	x_1	x_2	x_3	x_4
s_1	1	0	1	1
s_2	2	1	1	1
s_3	0	2	1	1
s_4	0	2	1	0

Grelling points out that constancy is a special case of dependence. This can occur if $cl_X(\emptyset) \neq \emptyset$. In team semantics $i \in cl_X(\emptyset)$, i.e., i is constant, means

$$\forall s, s' \in X(s(i) = s'(i)).$$

In the team of Figure 6 we have $cl_X(\emptyset) = \{x_3\}$, i.e., x_3 is constant. Another trivializing case, as Grelling points out, is the case that $s \restriction y$ is different for each $s \in X$. Then $cl(y) = I$ and every $i \in I$ is dependent on y. Such a y is called a *key* in database theory. In the team of Figure 6 the set $\{x_1, x_4\}$ is a key. By means of a key one can identify every entry in the database.

The team closure (or dependence) operations are sufficiently general to cover all cases of closure (or dependence) operations. The following result is essentially due to Armstrong [1]:

Theorem 1 ([1]). *For every dependence relation $\Rightarrow$ there is a team X such that $\Rightarrow$ is the team dependence relation $\Rightarrow_X$.*

Proof. For a start, let us pick some x, y such that $x \not\Rightarrow y$. Let $Z_{x,y}$ be the set of variables z_i such that $x \Rightarrow z_i$, and let $V_{x,y}$ be the set of remaining variables. Thus $y \cap V_{x,y} \neq \emptyset$. Suppose $a_{x,y}$ and $b_{x,y}$ are two arbitrary distinct elements (e.g. 0 and 1). Let $X_{x,y} = \{s_0, s_1\}$ where $s_0(v) = a_{x,y}$ for $v \in Z_{x,y} \cup V_{x,y}$, $s_1(v) = a_{x,y}$ for $v \in Z_{x,y}$ and $s_1(v) = b_{x,y}$ for $v \in V_{x,y}$. Obviously, $x \not\Rightarrow_{X_{x,y}} y$. Let us finally let X be the union of all $X_{x,y}$, where $x, y \subseteq I$ are such that $x \not\Rightarrow y$. To make this union more coherent, let us assume all the $a_{x,y}$ and $b_{x,y}$ are different for different x, y. We call this the Disjointness Assumption. Suppose now $u \Rightarrow v$. We show $u \Rightarrow_X v$. Suppose for this end $s, s' \in X$ such that $s \restriction u = s' \restriction u$. By the Disjointness Assumption $s, s' \in X_{x,y}$ for some x, y such that $x \not\Rightarrow y$. Since $s \restriction u = s' \restriction u$, we must have $u \subseteq Z_{x,y}$. Since $u \Rightarrow v$, we have also $v \subseteq Z_{x,y}$. Thus $s \restriction v = s' \restriction v$, i.e., $u \Rightarrow_X v$. Suppose then $u \not\Rightarrow v$. In this case there are $s, s' \in X_{u,v} \subseteq X$ such that $s \restriction u = s' \restriction u$ but $s \restriction v \neq s' \restriction v$, i.e., $u \Rightarrow_X v$. $\square$

Corollary 1. *For every closure operation* cl *there is a team X such that* cl *is the team closure operation* cl_X.

Note that if I is finite, the team X constructed in the above theorem is also finite. On the other hand, if I is the set of all complex numbers and cl is the algebraic closure, then the team X has continuum size.

In summary, there are two alternative approaches to dependence: the closure operation approach and the dependence relation approach. Both approaches can be subsumed under the team semantics approach. In algebra the closure operation approach seems the most natural. In computer science the dependence relation approach seems the most appropriate. But all in all, in a finite domain I, there is just one theory of dependence, governed by the rules (C1)–(C3), or equivalently by the rules (D1)–(D4).

3 Interdependence and mutual dependence

Grelling discusses at length *interdependence* and *mutual dependence*, concepts derived from the basic concept of dependence. These are cases where several variables all depend on each other as if they were bound together by an equation. Indeed, if in a vector space

$$\mathbf{x} - 2\mathbf{y} + 3\mathbf{z} = \mathbf{0}, \tag{1}$$

then $\mathbf{x}$ depends on $\mathbf{y}$ and $\mathbf{z}$ but equally $\mathbf{y}$ depends on $\mathbf{x}$ and $\mathbf{z}$, and $\mathbf{z}$ depends on $\mathbf{x}$ and $\mathbf{y}$. This is an example of interdependence. The law of Boyle

$$\frac{pV}{T} = \text{constant}$$

on the relationship between pressure (p), volume (V), and temperature (T) of a gas is used as an example in [10]. Also Hintikka [12] emphasizes the importance of interdependence from the point of view of applications to science.

Definition 4. Suppose cl is a closure operation and $x \in \mathscr{P}(I)$. We say that x is *interdependent* if $\text{cl}(x \setminus \{x_i\}) = x$ for all $x_i \in x$, or equivalently, x_i depends on $x \setminus \{x_i\}$ for each $x_i \in x$.

In an interdependent set every element is dependent on a subset of the remaining elements. In the team of Figure 6 the set $\{x_1, x_2\}$ is interdependent. In a vector space any set of vectors, which forms a linearly dependent set although every proper subset is linearly independent, is interdependent. If G is a finite graph any set of elements, which reside on a single closed walk, is an interdependent set.

A particularly strong form of interdependence is, what Grelling calls *interquidependence*: $\text{cl}(y) = x$ for all proper non-empty subsets y of x. In this case any individual element of x has all of x as its closure. Another closely related form of the concept of 'being dependent on each other' is the following:

Definition 5. Suppose cl is a closure operation on a set I. Two subsets x and y of I are said to be *mutually dependent* if x is dependent on y and y is dependent on x, i.e., $\text{cl}(x) = \text{cl}(y)$ (Figure 7).

The mutual dependence of x and y, which is of course a symmetrical property, means in team semantics that if we know the values of x we can figure out the values of y and vice versa. In Figure 6 the sets $\{x_1\}$ and $\{x_2\}$ are mutually dependent.

Fig. 7 x and y are mutually
dependent.

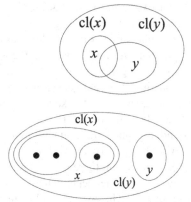

Fig. 8 x and y are mutually
dependent but $x \cup y$ is not
interdependent.

Obviously, the singleton sets $\{x_i\}$ and $\{x_j\}$ are mutually dependent if and only if $\{x_1, x_2\}$ is interdependent. For non-singleton sets this is not true: $x \cup y$ can be interdependent without x and y being mutually dependent (see Figure 8).

As with interdependence, mutual dependence is a concept derived from dependence, so to understand mutual dependence it is sufficient to develop an understanding of dependence itself.

4 G-dependence

Grelling also considers a variant of dependence which is new even today. The variant is based on the idea that if x depends on y, then this is often not only a one-way relationship between x and y. It is often the case, and indeed meant to be the case, that being able to compute x from y gives also a clue in the other direction: information about x limits what y can be. Grelling writes[4]:

> "The notion 'Equidep' [of dependence] is being based on the statement (E) [of Definition 3], i.e., on the assumption that equality of the values of y is implied by the equality of the corresponding values of the other variables [x] involved (that was the very reason for the choice of the symbol). Now it seems to be equally evident that a variable which is said to depend upon other variables must vary with them. In order to explain this new notion let us consider a method often employed by scientists in testing the dependence of one phenomenon on other phenomena. Suppose we have a certain phenomenon a and want to test its dependence upon a group of phenomena: b, c, d. Then we often proceed in the following way: first we keep b and c constant and let d alone vary; then, if a varies also, we infer that d is one of the phenomena upon which a is depending. Suppose we do the same thing with c and find that a does not vary when c alone among the group b, c, d has been made to vary. In that case we would say that a does not depend upon c, etc." [9]

---

[4]We have changed names and symbols from the original in order to be consistent with the current paper.

For example, to say that the time of descent depends on the height of the drop seems to imply that changing the height leads to a change in the time of descent. To say that whether it rains depends on whether the wind is from the west seems to imply that a turn of the wind to the west would bring about a change of weather. To say that our genes depend on the genes of our parents seems to suggest that if our parents had different genes, we would have different genes as well. Whenever we type a password on the keyboard, the sound the keys being pressed make depends obviously on the password because it is the typing of the password that produces the sound. But since different keys make different sounds, anyone who hears the sounds may with appropriate instruments be able to guess the password [11].

In conclusion, there is an element in the way we use the word "depends" which suggests that the dependence is sometimes meant to be more 'active' than in Definition 1. The context of closure operations seems too general to formulate such a concept, but in the context of team semantics Grelling proposes the following:

Definition 6 (G-dependence). Suppose X is a team or a multiteam with domain I. We say that $y \subseteq I$ is *G-dependent*[5] (or *G-depends*) on $x \subseteq I$ if the following holds for all $s, s' \in X$: If $s(x_i) \neq s'(x_i)$ for exactly one $x_i \in x$, then $s(y_j) \neq s'(y_j)$ for at least one $y_j \in y$.

In other words, if a unique element of x changes value in the team X, so does some element of y. As mentioned above, the point of assuming a difference $s(x_i) \neq s'(x_i)$ for *exactly* one x_i is that several differences might cancel out the difference $s(y_j) \neq s'(y_j)$. For example, y_1 may be G-dependent on $\{x_1, x_2\}$ and if exactly one of x_1 and x_2 changes, then y_1 changes, but if both x_1 and x_2 change, then y_1 might stay the same as the changes in y_1 implied by the change in x_1 might be cancelled by the change in x_2.

Changing exactly one factor at a time is sometimes called *ceteris paribus* meaning "other things being equal". G-dependence of y on x can be described also as a change in x having the effect of a change in y, ceteris paribus. The use of 'ceteris paribus' is sometimes criticized as it may not be clear what the 'other things' are that are assumed equal. In the concept of G-dependence the 'other things' that are equal are explicitly mentioned.

Example 1. Let us assume a company has a database with attributes *salary*, *demand-level*, *performance-level*. The company wants to maintain the constraint that salary depends on demand- and performance-levels. It is possible that demand-level is increased, performance-level is lowered, and the salary stays the same. Therefore, a change in the pair

(demand-level, performance-level)

---

[5]Grelling calls this "varequidependence." We use "G-dependence", G for Grelling.

need not imply a change in the salary. But the company wants to maintain the policy that if exactly one of the two levels changes then the salary should change as well. So here salary is G-dependent on demand- and performance-levels. To just say that salary depends on demand- and performance-levels does not (according to our technical definition of dependence) exclude, e.g., the unintended possibility that everyone has the same salary.

Example 2. A hotel advertises that the room price depends on various factors, such as the floor, the size, the wing, and extras. Some factors push the price up, other factors lower the price. To really honour the statement that the price depends on such factors the hotel may want to make sure that the price G-depends on them.

Grelling observes the following basic properties of G-dependence:

Lemma 1 ([9]).

1. *x always G-depends on itself.*
2. *Every x G-depends on Ø.*
3. *If x is constant (i.e. dependent on Ø), then every y is G-dependent on x.*
4. *If x has only different values, then x G-depends on any y.*
5. *If $|x| = 1$, then y G-depends on x if and only if x depends on y.*

Proof. (1): Suppose $s, s' \in X$ such that for exactly one $x_i \in x$ we have $s(x_i) \neq s'(x_i)$. Then a fortiori for some $x_i \in X$ we have $s(x_i) \neq s'(x_i)$.

(2): To prove that x is G-dependent on Ø, we start with $s, s' \in X$ and we have to prove an implication of the form $\varphi \longrightarrow \psi$ where φ is false since Ø has no elements. Thus the implication is true for trivial reasons.

(3): This claim is proved as claim (2).

(4): To prove that x is G-dependent on y, we start with $s, s' \in X$ and we have to prove an implication of the form $\varphi \longrightarrow \psi$ where ψ is true by assumption. Thus the implication is true for trivial reasons.

(5): Suppose $x = \{x_0\}$. Let us first assume y is G-dependent on x. To prove that x is dependent on y, assume $s, s' \in X$ such that $s \upharpoonright y = s' \upharpoonright y$. If $s(x_0) \neq s'(x_0)$, then by G-dependence, $s(y_j) \neq s'(y_j)$ for some $y_j \in y$, contrary to $s \upharpoonright y = s' \upharpoonright y$. Let us then assume that x is dependent on y and prove that y is G-dependent on x. To that end, suppose $s, s' \in X$ such that $s(x_0) \neq s'(x_0)$. If there is no $y_j \in y$ such that $s(y_j) \neq s'(y_j)$, then $s \upharpoonright y = s' \upharpoonright y$, and by x being dependent on y we obtain $s(x_0) = s'(x_0)$, a contradiction. □

The following is a consequence of Lemma 1 item 5:

Corollary 2. *Dependence is definable from G-dependence in the following sense:* $\{y_1, \ldots, y_n\}$ *depends on x if and only if each y_i G-depends on x.*

We shall prove below a sequence of structural properties of G-dependence.

Lemma 2 (Monotonicity). *If x G-depends on y, and $x \subseteq x'$, then x' G-depends on y.*

Proof. Suppose $s, s' \in X$ such that there is exactly one $y_i \in y$ such that $s(y_i) \neq s'(y_i)$. Then $s(x_j) \neq s'(x_j)$ for some $x_j \in x'$, hence $s(x_j) \neq s'(x_j)$ for some $x_j \in x'$. $\square$

Lemma 3 (Addition Rule). *Suppose y G-depends on x, and y' G-depends on x', then yy' G-depends on xx'.*

Proof. Suppose $s, s' \in X$ such that there is exactly one $x_i \in xx'$ such that $s(x_i) \neq s'(x_i)$. We show $s(y_j) \neq s'(y_j)$ for some $y_j \in yy'$.

Case 1: $x_i \in x$. Then there is exactly one $x_i \in x$ such that $s(x_i) \neq s'(x_i)$. We obtain $s(y_j) \neq s'(y_j)$ for some $y_j \in y$ and hence for some $y_j \in yy'$.

Case 2: $x_i \in x'$. Then there is exactly one $x_i \in x'$ such that $s(x_i) \neq s'(x_i)$. We obtain $s(y_j) \neq s'(y_j)$ for some $y_j \in y'$ and hence for some $y_j \in yy'$. $\square$

Lemma 4 (Thinning Rule). *Suppose z depends on y, and y G-depends on xz. Then y G-depends on x.*

Proof. Suppose $s, s' \in X$ such that there is exactly one $x_i \in x$ such that $s(x_i) \neq s'(x_i)$. We show $s(y_j) \neq s'(y_j)$ for some $y_j \in y$. Assume otherwise, i.e., $s(y_j) = s'(y_j)$ for all $y_j \in y$. Then, since z depends on y, we have $s(z_k) = s'(z_j)$ for all $z_k \in z$. Hence there is exactly one $x_i \in xz$ such that $s(x_i) \neq s'(x_i)$. Since y G-depends on xz, $s(y_j) \neq s'(y_j)$ for some $y_j \in y$, a contradiction. $\square$

Lemma 5 (Exchange Rule). *Suppose $x \cap y$ depends on z, x G-depends on yz, and y G-depends on xz. Then z G-depends on xy.*

Proof. Suppose $s, s' \in X$ such that there is exactly one $x_i \in xy$ such that $s(x_i) \neq s'(x_i)$. We show $s(z_j) \neq s'(z_j)$ for some $z_j \in z$. Assume otherwise, i.e., $s(z_j) = s'(z_j)$ for all $z_j \in z$.

Case 1: $x_i \in x$. Then there is exactly one $x_i \in xz$ such that $s(x_i) \neq s'(x_i)$. We obtain $s(y_j) \neq s'(y_j)$ for some $y_j \in y$. If y_j is x_i we contradict the assumption that $x \cap y$ depends on z. Suppose then y_j is not x_i. But then $s(y_j) = s'(y_j)$, since there is exactly one $x_i \in xy$ such that $s(x_i) \neq s'(x_i)$. This contradiction shows that $s(z_j) \neq s'(z_j)$ for some $z_j \in z$.

Case 2: $x_i \in y$. Then there is exactly one $x_i \in yz$ such that $s(x_i) \neq s'(x_i)$. We obtain $s(x_j) \neq s'(x_j)$ for some $x_j \in x$. If x_j is x_i we contradict the assumption that $x \cap y$ depends on z. Suppose then x_j is not x_i. But then $s(x_j) = s'(x_j)$, since there is exactly one $x_i \in yz$ such that $s(x_i) \neq s'(x_i)$. This contradiction shows that $s(z_j) \neq s'(z_j)$ for some $z_j \in z$. $\square$

Lemma 6 (Singleton Transitivity Rule). *Suppose $|u| = 1$, $x \cap y$ depends on z, and*

(1) z G-depends on uy,
(2) z G-depends on ux,

Then z G-depends on xy.

Proof. Let $u = \{u_0\}$. Suppose $s, s' \in X$ such that there is exactly one $x_i \in xy$ such that $s(x_i) \neq s'(x_i)$. We show $s(z_j) \neq s'(z_j)$ for some $z_j \in z$. Assume otherwise, i.e., $s(z_j) = s'(z_j)$ for all $z_j \in z$. Hence $s \upharpoonright x \cap y = s' \upharpoonright x \cap y$.

Case 1: $x_i \in x$. If there is exactly one $x_i \in ux$ such that $s(x_i) \neq s'(x_i)$, then by (2) $s(z_j) \neq s'(z_j)$ for some $z_j \in z$, a contradiction. Hence u_0 is not x_i and $s(u_0) \neq s'(u_0)$. But now there is exactly one $u_0 \in uy$ such that $s(u_i) \neq s'(u_i)$, whence by (1) $s(z_j) \neq s'(z_j)$ for some $z_j \in z$, a contradiction.

Case 2: $x_i \in y$. If there is exactly one $y_i \in uy$ such that $s(y_i) \neq s'(y_i)$, then by (2) $s(z_j) \neq s'(z_j)$ for some $z_j \in z$, a contradiction. Hence u_0 is not y_i and $s(u_0) \neq s'(u_0)$. But now there is exactly one $u_0 \in ux$ such that $s(u_i) \neq s'(u_i)$, whence by (1) $s(z_j) \neq s'(z_j)$ for some $z_j \in z$, a contradiction. $\square$

Lemma 7 (Transitivity Rule). *If u G-depends on xy, $x \cap y$ depends on z, and z G-depends on xu and on yu, then z G-depends on xy.*

Proof. Suppose $s, s' \in X$ such that there is exactly one $x_i \in xy$ such that $s(x_i) \neq s'(x_i)$. We show $s(z_j) \neq s'(z_j)$ for some $z_j \in z$. Assume otherwise, i.e., $s(z_j) = s'(z_j)$ for all $z_j \in z$. Since u G-depends on xy, there is $u_j \in u$ such that $s(u_j) \neq s'(u_j)$. If $x_i \in x \setminus y$, then this contradicts the assumption that z G-depends on yu. If $x_i \in y \setminus x$, then this contradicts the assumption that z G-depends on xu. If $x_i \in x \cap y$, then this contradicts the assumption that $x \cap y$ depends on z. $\square$

Lemma 8 ([9]). *If x is interdependent, every y is G-dependent on x.*

Proof. If $x = \emptyset$, then the claim follows from Lemma 1 (2). If $|x| = 1$, then by interdependence, x is constant (i.e. dependent on $\emptyset$). Hence y is G-dependent on x by Lemma 1 (3). Suppose then $|x| > 1$. Suppose $s, s' \in X$ such that for exactly one $x_j \in x$ we have $s(x_j) \neq s(x_j)$, but still $s \upharpoonright y \neq s' \upharpoonright y$. Let $x_k \in x$ be different from x_j. Thus $s(x_k) = s(x_k)$. As x_j depends on x_k, $s(x_j) \neq s(x_j)$, a contradiction. $\square$

Definition 7. x is *G-interdependent* if every $x_i \in x$ is G-dependent on $x \setminus \{x_i\}$.

Theorem 2 ([9]). *Suppose $|x| > 1$. Then x is interdependent if and only if x is G-interdependent.*

Proof. Suppose first x is interdependent. To prove that x is G-interdependent, let $x_i \in x$. We show that $\{x_i\}$ is G-dependent on $x \setminus \{x_i\}$, which we know is non-empty. Suppose $s, s' \in X$ such that for exactly one $x_j \in x$ we have $s(x_j) \neq s(x_j)$. Since x_j is dependent on $x \setminus \{x_i\}$, we cannot have

$$s \upharpoonright (x \setminus \{x_i\}) = s' \upharpoonright (x \setminus \{x_i\}). \tag{2}$$

Thus there must be $x_k \in x \setminus \{x_i\}$ such that $s(x_k) \neq s'(x_k)$, contrary to the uniqueness of x_j. To conclude the proof that x_i is G-dependent on $x \setminus \{x_i\}$ we have to prove an implication $\varphi \longrightarrow \psi$, where φ has just been shown to be false. Thus the claim follows. Suppose then x is G-interdependent. To prove that x is interdependent, let $x_i \in x$. We show that $\{x_i\}$ is dependent on $x \setminus \{x_i\}$, which we again know is non-empty. Suppose $s, s' \in X$ such that (2) holds. We claim that $s(x_i) = s'(x_i)$. Suppose not. Then x_i is the one and only element of x for which s and s' disagree. Let $x_k \in x \setminus \{x_i\}$. By G-interdependence x_k is G-dependent on $x \setminus \{x_k\}$. Since there is exactly one $x_i \in x \setminus \{x_k\}$ where s and s' disagree, G-dependence implies $s(x_k) \neq s'(x_k)$, a contradiction. $\square$

The strongest and final variant of dependence Grelling introduces is the following concept:

Definition 8. Suppose X is a (multi)team. A subset x of I is said to be *strongly dependent* on a subset y of I in X, if x is dependent and G-dependent on y in X.

By Corollary 2 strong dependence is expressible in terms of G-dependence alone. In strong dependence we incorporate both aspects of dependence discussed above: isolating not only the y that x is dependent on in the sense that if we are given y, we can figure out what x is, but also making sure that everything in y is relevant. In other words, if the values of y are kept the same, the values of x stay the same, but the moment one factor in y is changes, immediately also x changes. There is a strong sense of x and y being dependent on each other, and to be sure, mutual dependence implies strong dependence. However, strong dependence is not equivalent to mutual dependence and is not necessarily even symmetric. In the team of Figure 9 y strongly depends on $x_1 x_2$ but $x_1 x_2$ does not (strongly) depend on y.

But strong dependence is symmetric on singletons:

Lemma 9 ([9]). *The following conditions are equivalent:*

1. $\{x_1\}$ *is strongly dependent on* $\{x_2\}$.
2. $\{x_2\}$ *is strongly dependent on* $\{x_1\}$.
3. $\{x_1\}$ *and* $\{x_2\}$ *are mutually dependent.*
4. $\{x_1, x_2\}$ *is interdependent.*

Proof. (1) implies (2): We prove first that $\{x_2\}$ is dependent on $\{x_1\}$. Suppose $s(x_1) = s'(x_1)$. If $s(x_2) \neq s'(x_2)$, then as $\{x_1\}$ is G-dependent on $\{x_2\}$, $s(x_1) \neq s'(x_1)$, a contradiction. Hence $s(x_2) = s'(x_2)$. Let us then prove that $\{x_2\}$ is G-dependent on $\{x_1\}$. Suppose $s(x_1) \neq s'(x_1)$. If $s(x_2) = s'(x_2)$, then as $\{x_1\}$ is dependent on $\{x_2\}$, $s(x_1) = s'(x_1)$, a contradiction. Hence $s(x_2) \neq s'(x_2)$.
(2) implies (3), and (3) implies (4) by definition.
(4) implies (1): We need only prove that $\{x_1\}$ is G-dependent on $\{x_2\}$. Suppose $s(x_2) \neq s'(x_2)$. If $s(x_1) = s'(x_1)$, then as $\{x_2\}$ is dependent on $\{x_1\}$, $s(x_2) = s'(x_2)$, a contradiction. Hence $s(x_1) = s'(x_1)$. $\square$

Fig. 9 Strong dependence is not symmetric.

	x_1	x_2	y
s_1	0	0	0
s_2	1	1	0

Grelling mentions the following Lemma as being of particular importance in science, for it implies that if $\{x_1\}$ is mutually dependent on y and one succeeds in keeping constant all the elements of y except x_2, then a strict correlation should be observed between x_1 and x_2.

Lemma 10 ([9]). *If $\{x_1\}$ is strongly dependent on y, and $y \setminus \{x_2\}$ is constant in X, then $\forall s, s' \in X(s(x_1) = s'(x_1) \leftrightarrow s(x_2) = s'(x_2))$.*

Proof. The direction "$\longrightarrow$" is a consequence of x_1 being G-dependent on y. The direction "$\leftarrow$" is a consequence of x_1 being dependent on y. □

We have derived various properties of G-dependence by appealing directly to the definition of its semantics. Are there simple rules which would give a mechanical method for deriving all the true properties of G-dependence?

Proposition 1. *Suppose $\Sigma \cup \{\varphi\}$ is a finite set of statements of the form "u is G-dependent on v" for various finite sets u and v of variables. There is an effective method for deciding whether every team which satisfies Σ also satisfies φ.*

Proof. We use the standard technique of reduction to the Bernays-Schönfinkel-Ramsey class (see, e.g., [3, Section 6.2.2]). Let $x = \{x_1, \ldots, x_n\}$ be all the finitely many variables occurring in $\Sigma \cup \{\varphi\}$. Let $x' = \{x'_1, \ldots, x'_n\}$ be a set of new variables such that $W \cap W' = \emptyset$. We take a new n-place relation symbol R. For any $u, v \subseteq W$ and $\psi \in \Sigma \cup \{\varphi\}$ of the form

$$\text{"} v \text{ is G-dependent on } u \text{"},$$

let ψ^* be the formula

$$\left(\bigvee_{x_i \in u} (\neg x_i = x'_i \wedge \bigwedge_{x_j \in u \setminus \{x_i\}} x_j = x'_j) \right) \longrightarrow \bigvee_{y_i \in v} \neg y_i = y'_i.$$

Let θ be the first order sentence

$$[\forall x_1 \ldots \forall x_n \forall x'_1 \ldots x'_n((R(x_1, \ldots, x_n) \wedge R(x'_1, \ldots, x'_n)) \longrightarrow \bigwedge_{\psi \in \Sigma} \psi^*)] \longrightarrow$$

$$\forall x_1, \ldots \forall x_n \forall x'_1 \ldots \forall x'_n((R(x_1, \ldots, x_n) \wedge R(x'_1, \ldots, x'_n)) \longrightarrow \varphi^*).$$

It is clear that every team which satisfies Σ also satisfies φ if and only if θ is a valid first order sentence. Since θ is equivalent to a universal-existential sentence in a relational vocabulary, its validity is decidable by the decidability of the Bernays-Schönfinkel-Ramsey class. □

In view of the above decidability result it seems reasonable to ask the following question:

Open Problem 3. *Find a (simple) complete set Γ of rules for G-dependence statements such that if $\Sigma \cup \{\varphi\}$ is a set of statements of the form "u is G-dependent on v" for various finite sets u and v of variables, then φ follows from Σ by the rules Γ if and only if every team which satisfies Σ also satisfies φ.*

By Proposition 1 such complete sets Γ must exist but what is needed is a set of *simple* rules, such as the Armstrong Axioms (D1)–(D4) above.

In summary, G-dependence captures important aspects of phenomena associated with dependence and one not brought about by (functional) dependence. It seems to be more relevant for the methodology of scientific research than for database theory. Still, G-dependence has a clear intuitive meaning also in the database context and can conceivably have applications there. While functional dependence has a simple complete axiomatization ([1]) essentially based on (D1)–(D4), no such Completeness Theorem is known for G-dependence.

5   G-dependence logic

In dependence logic [20] relations "x depends on y" are incorporated into first order logic by introducing a new type of atomic formula. If x and y are finite sequences of variables, then

$$=(y, x)$$

is an atomic formula of dependence logic, with the intuitive meaning "x depends on y". The semantics of $=(y, x)$ is defined by means of condition (E) of Definition 3 as follows: Suppose X is a (multi)team. Then

$$X \models =(y, x)$$

if and only if

$$\forall s, s' \in X(s \restriction y = s' \restriction y \longrightarrow s \restriction x = s' \restriction x).$$

The semantics of other atomic formulas as well as logical operations $\wedge, \vee, \exists, \forall$ is defined as in [20]. Let us denote the resulting extension of first order logic by $\mathscr{D}$.

In the same way we can add relations "x G-depends on y" to first order logic by introducing a new type of atomic formula

$$\backsim(y, x)$$

with the intuitive meaning "x G-depends on y". The semantics of $\backsim(y, x)$ is defined as follows: Suppose X is a (multi)team. Then

$$X \models \backsim(y, x)$$

if and only if for all $s, s' \in X$ the following holds: if $s(y_i) \neq s'(y_i)$ for exactly one $y_i \in y$, then $s(x_j) \neq s'(x_j)$ for at least one $x_j \in x$.

Let us denote the resulting logic by $\mathscr{G}$. The following simple properties of $\mathscr{G}$ are easy to prove:

Lemma 11. *G-dependence logic $\mathscr{G}$ is local[6], has the empty team property[7], and is downwards closed[8].*

Since dependence is definable in terms of G-dependence (Corollary 2), $\mathscr{D}$ is a sublogic of $\mathscr{G}$. It is easy to adapt the proof that formulas of $\mathscr{D}$ are definable (in non-empty teams) in existential second order logic ([20, Theorem 68]) to a proof that formulas of $\mathscr{G}$ are definable in existential second order logic. Since formulas of $\mathscr{G}$ are downward closed, and the expressive power of formulas $\varphi(x_1, \ldots, x_n)$ of $\mathscr{D}$ is on non-empty teams the same as the expressive power of sentences $\Phi(R)$ of existential second order logic with a predicate R (for the team) occurring negatively only ([17]), we obtain

Proposition 2. *G-dependence logic $\mathscr{G}$ and dependence logic $\mathscr{D}$ have the same expressive power (of formulas).*

It follows that the atom $\backsimeq(x, y)$ is expressible (in non-empty teams) by a formula $\varphi(x, y)$ of $\mathscr{D}$. How complicated is the simplest such formula $\varphi(x, y)$? The proof of Proposition 2 gives some formula $\varphi(x, y)$ but the best we can say, by using general facts abut $\mathscr{D}$, is that it is universal existential. Let us look at some other logics based on team semantics. *Independence logic*, based on the *independence* atoms $x \perp y$, was introduced in [8]. *Inclusion logic*, based on the *inclusion* atoms $x \subseteq y$, was introduced in [7]. Although the inclusion atom is definable in independence logic, we consider the extension $FO(\subseteq, \perp)$ of first order logic obtained by adding the atoms $x \subseteq y$ and the atoms $x \perp y$. The expressive power of formulas of this logic (even of independence logic alone) is the same as that of existential second order logic ([7]).

Because of Proposition 2 the validity problem for formulas of G-dependence logic $\mathscr{G}$ is non-arithmetical ([20, Theorem 120]). However, we point out that for general soft reasons, applicable to almost any axiomatization question in team semantics, a fragment of $\mathscr{G}$ can be axiomatized. Let $\vee_B$ be the *Boolean disjunction*, that is, a team satisfies $\varphi \vee_B \psi$ if and only if it satisfies φ or ψ. Let $\exists_1$ be the following version of the existential quantifier: a team X satisfies $\exists_1 x \varphi$ in a model if and only if there is an element a in the domain of the model such that the team, obtained by changing the value of $s(x)$ to a for each $s \in X$, satisfies φ (this is known in inquisitive logic as the *inquisitive existential quantifier*). Let $\mathscr{G}^-$ be the fragment of $\mathscr{G}$ obtained by closing the set of atomic formulas of $\mathscr{G}$ under $\wedge, \vee_B, \exists_1$ and $\forall$. Note that $\mathscr{G}^-$ still contains ordinary first order logic (and the G-dependence atoms).

---

[6]The truth of a formula depends only on the restriction of the team to the variables that are free in the formula.

[7]The empty team satisfies every formula.

[8]If a team satisfies a formula, then every subteam satisfies it.

Theorem 4. *Validity of formulas of the fragment $\mathscr{G}^-$ of $\mathscr{G}$ is recursively axiomatizable.*

Proof. We assume, for simplicity, that the non-logical vocabulary of $\mathscr{G}^-$ consists of one binary relation symbol S only. Let R^m be a new m-ary relation symbol for each m. It is straightforward to define a translation $\varphi(x_1, \ldots, x_n) \mapsto \tau_{\varphi(x_1, \ldots, x_n)}(R^n)$ from $\mathscr{G}^-$ to the Π_1^1-part of second order logic such that the following holds: A non-empty team X with domain $\{x_1, \ldots, x_n\}$ satisfies $\varphi(x_1, \ldots, x_n)$ in a model $\mathscr{M}$ if and only if $(\mathscr{M}, \mathrm{rel}(X))$ satisfies $\tau_{\varphi(x_1, \ldots, x_n)}(R^n)$, where $\mathrm{rel}(X)$ refers to X as a relation (rather than as a team). We use $\mathbf{x}$ to denote $x_1, \ldots, x_n$ for various n, and similarly $\mathbf{x}'$.

1. $\tau_{x_i = x_j}(R^m) = \forall \mathbf{x}(R^m(x_1, \ldots, x_m) \longrightarrow x_i = x_j)$, whenever $1 \leq i \leq m$ and $1 \leq j \leq m$.

2. $\tau_{\neg x_i = x_j}(R^m) = \forall \mathbf{x}(R^m(x_1, \ldots, x_m) \longrightarrow \neg x_i = x_j)$, whenever $1 \leq i \leq m$ and $1 \leq j \leq m$.

3. $\tau_{S(x_i, x_j)}(R^m) = \forall x(R^m(x_1, \ldots, x_m) \longrightarrow S(x_i, x_j))$, whenever $1 \leq i \leq m$ and $1 \leq j \leq m$.

4. $\tau_{\neg S(x_i, x_j)}(R^m) = \forall \mathbf{x}(R^m(x_1, \ldots, x_m) \longrightarrow \neg S(x_i, x_j))$, whenever $1 \leq i \leq m$ and $1 \leq j \leq m$.

5. $\tau_{\leftrightarrows(u,v)}(R^m) = \forall \mathbf{x} \forall \mathbf{x}'((R^m(x_1, \ldots, x_m) \wedge R^m(x_1', \ldots, x_m')) \longrightarrow (\bigvee_{x_i \in u}(\neg x_i = x_i' \wedge \bigwedge_{x_j \in u \setminus \{x_i\}} x_j = x_j')) \longrightarrow \bigvee_{y_i \in v} \neg y_i = y_i')$, whenever $u \subseteq \{1, \ldots, m\}$ and $v \subseteq \{1, \ldots, m\}$.

6. $\tau_{\varphi \wedge \psi}(R^m) = \tau_\varphi(R^m) \wedge \tau_\psi(R^m)$

7. $\tau_{\varphi \vee_B \psi}(R^m) = \tau_\varphi(R^M) \vee \tau_\psi(R^m)$

8. $\tau_{\exists_1 x_{i_0} \varphi(x_{i_0}, x_{i_1}, \ldots, x_{i_n})}(R^m) = \exists x_{i_0}' \forall R^{m+1}(\forall x_1 \ldots \forall x_{m+1}(R^{m+1}(x_1, \ldots, x_{m+1}) \leftrightarrow (R(x_1, \ldots, x_{i_0 - 1}, x_{i_0 + 1}, \ldots, x_{m+1}) \wedge x_{i_0} = x_{i_0}')) \longrightarrow \tau_{\varphi(x_{i_0}, \ldots, x_{i_n})}(R^{m+1}))$, whenever $\{i_1, \ldots, i_n\} \subseteq \{1, \ldots, m\}$ and $i_0 \in \{1, \ldots, m+1\}$

9. $\tau_{\forall x_{i_0} \varphi(x_{i_0}, x_{i_1}, \ldots, x_{i_n})}(R^m) = \forall R^{m+1}(\forall x_1 \ldots \forall x_{m+1}(R^{m+1}(x_1, \ldots, x_{m+1}) \leftrightarrow R(x_1, \ldots, x_{i_0 - 1}, x_{i_0 + 1}, \ldots, x_{m+1})) \longrightarrow \tau_{\varphi(x_{i_0}, \ldots, x_{i_n})}(R^{m+1}))$, whenever $\{i_1, \ldots, i_n\} \subseteq \{1, \ldots, m\}$ and $i_0 \in \{1, \ldots, m+1\}$.

We used here the fact that Π_1^1-formulas of second order logic are closed—up to logical equivalence—under disjunction, conjunction, and first order existential quantification. Now, a formula $\varphi(x_1, \ldots, x_n)$ of G-dependence logic is valid if and only if the Π_1^1-sentence $\tau_{\varphi(x_1, \ldots, x_n)}(R^n)$ of second order logic is valid. The latter concept is recursively axiomatizable by Gödel's Completeness Theorem. $\qquad \square$

Are there interesting formulas in $\mathscr{G}^-$ apart from the atomic formulas $\leftrightarrows(x, y)$ (and first order formulas)? We do not know. However, $\mathscr{G}'$ is not decidable, as it contains first order logic.

Open Problem 5. *Give a simple axiomatization for $\mathscr{G}^-$.*

6 Conclusion

We end with a quote from Grelling [9]. It is easy to agree with this even today: "The definitions which I have proposed here are nothing but attempts to solve the problem of dependence. Most of these concepts might not be applicable yet to the practical course of science. However I firmly believe and hope that further developments of these investigations will finally prove to be fairly useful for all sorts of scientists."

Note: When this paper was going to press, Tapani Hyttinen showed that G-dependence can be completely subsumed by ordinary dependence: $\backsim(y_1 \ldots y_n, x)$ is logically equivalent to $=(xy_2 \ldots y_n, y_1) \wedge =(xy_1y_3 \ldots y_n, y_2) \wedge \ldots \wedge =(xy_1 \ldots y_{n-1}, y_n)$.

Acknowledgements The author would like to thank the Simons Foundation for a fellowship and Isaac Newton Institute for Mathematical Sciences for its hospitality during the programme Mathematical, Foundational and Computational Aspects of the Higher Infinite supported by EPSRC Grant Number EP/K032208/.

References

1. Armstrong, W.W.: Dependency structures of data base relationships. IFIP Congress, 580–583 (1974)
2. Bell, J.: On the Einstein Podolsky Rosen paradox. Physics **1**, 195–200 (1964)
3. Börger, E., Grädel, E., Gurevich, Y.: The Classical Decision Problem. Universitext. Springer, Berlin (2001). Reprint of the 1997 original
4. Codd, E.F.: Further normalization of the data base relational model. IBM Research Report, San Jose, California, RJ909 (1971)
5. Einstein, A., Podolsky, B., Rosen, N.: Can quantum-mechanical description of physical reality be considered complete? Phys. Rev. **47**, 777–780 (1935)
6. Fagin, R.: Multivalued dependencies and a new normal form for relational databases. ACM Trans. Database Syst. **2**(3), 262–278 (1977)
7. Galliani, P.: Inclusion and exclusion dependencies in team semantics—on some logics of imperfect information. Ann. Pure Appl. Logic **163**(1), 68–84 (2012)
8. Grädel, E., Väänänen, J.: Dependence and independence. Stud. Logica **101**(2), 399–410 (2013)
9. Grelling, K.: A logical theory of dependence (included in [19]) (1939)
10. Grelling, K., Oppenheim, P.: Logical analysis of 'gestalt' as 'functional whole' (included in [19]) (1939)
11. Hanspach, M., Goetz, M.: On covert acoustical mesh networks in air. J. Commun. **8**(11), 758–767 (2013)
12. Hintikka, J.: Hyperclassical logic (a.k.a. IF logic) and its implications for logical theory. Bull. Symb. Log. **8**(3), 404–423 (2002)
13. Hodges, W.: Compositional semantics for a language of imperfect information. Log. J. IGPL **5**(4), 539–563 (electronic) (1997)
14. Hodges, W.: Some strange quantifiers. In: Structures in Logic and Computer Science. Lecture Notes in Computer Science, pp. 51–65. Springer, Berlin (1997)
15. Hyttinen, G., Paolini T., Väänänen, J.: Quantum team logic and bell's inequalities. Rev. Symb. Log. **8**(4), 722–742 (2015)

16. Kolmogoroff, A.: Grundbegriffe der Wahrscheinlichkeitsrechnung. Ergebnisse der Math. **2**(3), IV + 62 S (1933)
17. Kontinen, J., Väänänen, J.: On definability in dependence logic. J. Log. Lang. Inf. **18**(3), 317–332 (2009). Erratum: ibid. **20**(1), 133–134 (2011)
18. Paolini, G., Väänänen, J.: Dependence logic in pregeometries and omega-stable theories. J. Symb. Log. **81**(1), 32–55 (2016)
19. Smith, B. (ed.): Foundations of Gestalt Theory. Philosophia, Munich and Vienna (1988)
20. Väänänen, J.: Dependence Logic. London Mathematical Society Student Texts, vol. 70. Cambridge University Press, Cambridge (2007)
21. van der Waerden, B.L.: Moderne Algebra. Unter Benutzung von Vorlesungen von *E. Artin* und *E. Noether*. Bd. I. (Die Grundlehren der mathematischen Wissenschaften in Einzeldarstellungen mit besonderer Berücksichtigung der Anwendungsgebiete Bd. 23). VIII + 243 S. Berlin, J. Springer, Berlin (1930)
22. Whitney, H.: On the abstract properties of linear dependence. Am. J. Math. **57**, 509–533 (1935)

On Strongly First-Order Dependencies

Pietro Galliani

Abstract We prove that the expressive power of first-order logic with team semantics plus contradictory negation does not rise beyond that of first-order logic (with respect to sentences), and that the totality atoms of arity $k+1$ are not definable in terms of the totality atoms of arity k. We furthermore prove that all first-order nullary and unary dependencies are strongly first-order, in the sense that they do not increase the expressive power of first-order logic if added to it.

1 Introduction

In the last few years, team semantics [14, 18] has proved itself to be a very powerful theoretical framework for the study of dependency notions and their interaction; and, furthermore, some intriguing potential applications of team semantics in the areas of belief representation [7, 10], social choice and physics [2], and database theory [16] have been noticed.

As a natural generalization of Tarski semantics to the case of multiple assignments, team semantics allows to extend first-order logic in novel ways, in particular by adding to it *dependency atoms* that specify complex patterns of dependence and independence between variables; and much of the research in the area so far has been dedicated to the comparison of the logics thus obtained.

Many of these logics are much stronger than first-order logic itself – for instance, dependence logic is as expressive as the existential fragment of second-order logic [18], and inclusion logic is as expressive as greatest fixed point logic [9] – but this needs not be the case. Indeed, as shown in [8], many nontrivial dependency notions, such as the negations of functional dependence, inclusion, exclusion, and conditional independence, are *strongly first-order* in the sense that they do not increase the expressive power of first-order logic if added to it. The *totality atoms*, which assert that a certain tuple of variables takes all possible values in a team, are an especially interesting example of a strongly first-order dependency, and in this work we will study them in some depth.

P. Galliani (✉)
School of Engineering and Informatics, University of Sussex, Falmer, Brighton BN1 9QJ, UK
e-mail: pgallian@gmail.com

S. Abramsky et al. (eds.), *Dependence Logic*, DOI 10.1007/978-3-319-31803-5_4

It is important to emphasize here that these strongly first-order dependencies, despite not increasing the expressive power of first-order logic sentences, cannot be disposed of: even though every sentence containing them (but not other, stronger dependencies) has the same truth conditions of some first-order sentence, the satisfaction conditions of *formulas* containing them are not in general equivalent to the satisfaction conditions of any first-order formula with respect to team semantics.

The study of team semantics (and, in particular, of strongly first-order dependencies) can thus be seen as an attempt to investigate the nature of the *boundary* between first- and second-order logic; and, from a more practical point of view, dependencies which are strongly first-order are eminently treatable in that they do not increase the complexity of the logic.

The purpose of this work is to further investigate the properties of strongly first-order dependencies and – more in general – of team semantics-based extensions of first-order logic whose expressive power is no greater than that of first-order logic proper. In Section 3 we will investigate the effect of adding the contradictory negation operator to extensions of first-order logic by strongly first-order operator; then in Section 4 we will develop a *hierarchy theorem* for totality atoms, and in Sections 5 and 6 we will study dependency atoms of arity 0 or 1.

2 Preliminaries

In this section we will briefly recall some fundamental definitions, as well as some results that we will need to use later in this work.

Definition 1 (Team). Let $\mathfrak{M}$ be a first-order model with domain M and let V be a set of variables. A *team* X over $\mathfrak{M}$ with domain $\mathbf{Dom}(X) = V$ is a set of assignments $s : V \to M$.

Given such a team X and a tuple $\mathbf{v}$ of variables in $\mathbf{Dom}(X)$, we write $X(\mathbf{v})$ for the relation $\{s(\mathbf{v}) : s \in X\}$; and given a first-order formula θ, we write $(X \upharpoonright \theta)$ for the team $\{s \in X : \mathfrak{M} \models_s \theta\}$ obtained by taking only the assignments of X which satisfy θ (according to Tarski semantics).

For the purposes of this work, we will only consider the so-called *lax* version of team semantics, and we will only work with formulas in negation normal form:

Definition 2. Let $\mathfrak{M}$ be a first-order model, let X be a team over it, and let $\phi(\mathbf{v})$ be a first-order formula in negation normal form and with free variables in $\mathbf{v} \subseteq \mathbf{Dom}(X)$. We say that X *satisfies* $\phi(\mathbf{v})$ in $\mathfrak{M}$, and we write $M \models_X \phi(\mathbf{v})$, if and only if this can be deduced from the following rules:

TS-lit: For all first-order literals α, $\mathfrak{M} \models_X \alpha$ if and only if for all $s \in X$, $\mathfrak{M} \models_s \alpha$ according to Tarski semantics;

TS-$\vee$: $\mathfrak{M} \models_X \psi \vee \theta$ if and only if there exist $Y, Z \subseteq X$ such that $X = Y \cup Z$, $\mathfrak{M} \models_Y \psi$ and $\mathfrak{M} \models_Z \theta$;

TS-$\wedge$: $\mathfrak{M} \models_X \psi \wedge \theta$ if and only if $\mathfrak{M} \models_X \psi$ and $\mathfrak{M} \models_X \theta$;

TS-∃: $\mathfrak{M} \models_X \exists v \psi$ if and only if there exists a function $F : X \to \mathscr{P}(M) \backslash \{\emptyset\}$ such that, for $Y = X[F/v] = \{s[m/v] : m \in F(s)\}$, we have that $\mathfrak{M} \models_Y \psi$;

TS-∀: $\mathfrak{M} \models_X \forall v \psi$ if and only if $M \models_{X[M/v]} \psi$, where $X[M/v] = \{s[m/v] : s \in X, m \in M\}$.

A sentence ϕ is said to be *true* in a model $\mathfrak{M}$ if and only if $\mathfrak{M} \models_{\{\emptyset\}} \phi$; and in this case, we write $\mathfrak{M} \models \phi$.

The next result shows that, in the case of first-order logic, team semantics may indeed be reduced to Tarski semantics:

Proposition 1 ([18]). *For all first-order formulas ϕ, all models $\mathfrak{M}$ and all teams X, $\mathfrak{M} \models_X \phi$ if and only if for all $s \in X$ we have that $\mathfrak{M} \models_s \phi$ according to Tarski semantics. In particular, for all first-order sentences ϕ we have that $\mathfrak{M} \models_{\{\emptyset\}} \phi$ if and only if $\mathfrak{M} \models \phi$ according to Tarski semantics.*

However, team semantics allows us to extend first-order logic in novel ways, for instance, by operators such as the *intuitionistic implication* [1]

TS-intimp: $\mathfrak{M} \models_X \phi \to \psi$ if and only if for all $Y \subseteq X$, $\mathfrak{M} \models_Y \phi \Rightarrow \mathfrak{M} \models_Y \psi$,

the *contradictory negation* [19]

TS-∼: $\mathfrak{M} \models_X {\sim} \phi$ if and only if $\mathfrak{M} \not\models_X \phi$,

the *classical disjunction* [18]

TS-⊔: $\mathfrak{M} \models_X \phi \sqcup \psi$ if and only if $\mathfrak{M} \models_X \phi$ or $\mathfrak{M} \models_X \psi$,

or the *possibility operator* [8]

TS-◇: $\mathfrak{M} \models_X \Diamond \phi$ iff there exists a $Y \subseteq X$, $Y \neq \emptyset$ s.t. $\mathfrak{M} \models_Y \phi$

or by means of novel atoms corresponding to notions of constancy and functional dependence [18]

TS-con: $\mathfrak{M} \models_X =(\mathbf{v})$ iff for all $s, s' \in X$, $s(\mathbf{v}) = s'(\mathbf{v})$;

TS-fdep: $\mathfrak{M} \models_X =(\mathbf{v}, \mathbf{w})$ iff for all $s, s' \in X$, $s(\mathbf{v}) = s'(\mathbf{v}) \Rightarrow s(\mathbf{w}) = s'(\mathbf{w})$,

inclusion dependence [6]

TS-inc: $\mathfrak{M} \models_X \mathbf{v} \subseteq \mathbf{w}$ iff $X(\mathbf{v}) \subseteq X(\mathbf{w})$

(conditional) independence [12]

TS-ind: $\mathfrak{M} \models_X \mathbf{v} \perp_{\mathbf{u}} \mathbf{w}$ iff for all $s, s' \in X$ with $s(\mathbf{u}) = s'(\mathbf{u})$ there exists a $s'' \in X$ with $s''(\mathbf{uvw}) = s(\mathbf{uv})s'(\mathbf{w})$.

or totality [2]:

TS-all: $\mathfrak{M} \models_X All(\mathbf{v})$ iff $X(\mathbf{v}) = M^{|\mathbf{v}|}$.

More in general, all these atoms (and many more besides) can be seen as special cases of the following definition ([17]):

Definition 3 (Dependency Notion). Let $k \in \mathbb{N}$. A k-ary dependency notion **D** is a class, closed under isomorphisms, of models over the signature $\{R\}$, where R is a k-ary relation symbol. For all models $\mathfrak{M}$, all teams X, and all tuples **v** of variables in the domain of X,

$$\mathfrak{M} \models_X \mathbf{Dv} \text{ if and only if } (M, X(\mathbf{v})) \in \mathbf{D}.$$

Given a family $\mathscr{D}$ of dependency notions, we will write **FO**($\mathscr{D}$) for the logic obtained by adding all $\mathbf{D} \in \mathscr{D}$ to the language of first-order logic. We will indicate with $=(\cdot)$ the family of all constancy dependencies $=(\mathbf{v})$ of all arities, with $=(\cdot, \cdot)$ the family of all functional dependency atoms $=(\mathbf{v}, \mathbf{w})$ of all arities, and with *All* the family of all totality atoms *All*($\mathbf{w}$) of all arities; and when necessary, we will indicate the arities as a subscript – for instance, $=(\cdot)_1$ represents the unary constancy atoms $=(v)$ where v is a single variable, and $=(\cdot, \cdot)_{2,2}$ represents the functional dependency atoms of the form $=(v_1 v_2, w_1 w_2)$.

The following notion of *definability* is of central importance for the study of team semantics:

Definition 4 (Definability). Let **D** be a k-ary dependency notion and let $\mathscr{D}$ be a class of dependency notions. Then we say that **D** is *definable* through $\mathscr{D}$ if there exists a formula $\theta(\mathbf{v}) \in \text{FO}(\mathscr{D})$ over the empty vocabulary, where $\mathbf{v} = v_1 \ldots v_k$ is a tuple of k distinct variables, such that

$$\mathfrak{M} \models_X \mathbf{Dv} \text{ if and only if } \mathfrak{M} \models_X \theta(\mathbf{v})$$

for all models $\mathfrak{M}$ and teams X whose domain contains **v**.

It is easy to see that $\mathbf{FO}(=(\cdot)) = \mathbf{FO}(=(\cdot)_1)$: indeed, for any k-tuple $\mathbf{v} = v_1 \ldots v_k$ of variables it is trivial to check that $=(\mathbf{v}) \equiv \bigwedge_{i=1}^{k} =(v_i)$, and hence $=(\cdot)_k$ is definable through $=(\cdot)_1$. On the other hand, in [3] it was shown that

Theorem 1. *For all $k \in \mathbb{N}$, $FO(=(\cdot, \cdot)_{k,1}) \subsetneq FO(=(\cdot, \cdot)_{k+1,1})$,*[1]

in [11] it was shown that a similar result holds for independence atoms, and in [13] it was shown that the same may be said in the case of inclusion atoms too.

What about totality dependencies? We will address this question in Section 4.

All dependencies that we mentioned so far are *first-order* in the following sense:

Definition 5 (First-Order Dependency Notion). A k-ary dependency notion **D** is first-order if and only if there exists a first-order formula $\mathbf{D}^*$ on the signature $\{R\}$ (for R k-ary) such that

$$\mathbf{D} = \{(M, R) : (M, R) \models \mathbf{D}^*\}.$$

---

[1]To be more precise, this results holds if we are allowing models over all signatures. The case in which only models over the empty signature are considered is yet open.

It is easy to see that if $\mathbf{D}$ is first-order, then $\mathfrak{M} \models_X \mathbf{Dv} \Leftrightarrow (M, X(\mathbf{v})) \models \mathbf{D}^*$; but owing to the higher-order nature of team semantics (and in particular, to the second-order quantification implicit in its rules for disjunctions and existential quantifiers) it does not follow from this that these dependencies do not increase the expressive power of first-order logic. For instance, the $\mathbf{FO}(=(\cdot, \cdot)_{1,1})$-sentence

$$\exists x \forall y \exists z (=(z, y) \wedge z \neq x)$$

is true in a model $\mathfrak{M}$ if and only if it is infinite, even though $=(\cdot, \cdot)_{1,1}$ is first-order and corresponds to the sentence $\forall x y y' (Rxy \wedge Rxy' \rightarrow y = y')$.

Therefore, the question arises of whether there exist interesting dependency notions for which this is not the case. More formally, one may ask if there exist nontrivial dependencies which are *strongly first-order* in the following sense:

Definition 6 (Strongly First-Order Dependencies). A k-ary dependency $\mathbf{D}$ is *strongly first-order* if every sentence ϕ of $\mathbf{FO}(\mathbf{D})$ is equivalent (on the level of sentences) to some sentence θ_ϕ of $\mathbf{FO}$, in the sense that

$$\mathfrak{M} \models \phi \text{ if and only if } \mathfrak{M} \models \theta_\phi. \tag{1}$$

Similarly, a family of dependencies $\mathscr{D}$ is strongly first-order if every sentence of $\mathbf{FO}(\mathscr{D})$ is equivalent (in the above sense) to some sentence of $\mathbf{FO}$.

In [8], a positive answer was found for the above question.

Definition 7. A dependency notion $\mathbf{D}$ is *upwards-closed* if $(M, R) \in \mathbf{D}, R \subseteq S \Rightarrow (M, S) \in \mathbf{D}$.

Theorem 2 ([8]). *Let $\mathscr{D}$ be a family of upwards-closed first-order dependencies. Then $\{=(\cdot)\} \cup \mathscr{D}$ is strongly first-order.*

As a consequence, it was shown that – for instance – all the following dependencies are strongly first-order for all arities of $\mathbf{v}$ and $\mathbf{w}$, as is any set containing them (and the constancy atoms $=(\cdot)$):

TS-nonempty: $\mathfrak{M} \models_X NE$ iff $X \neq \emptyset$;

TS-ncon: $\mathfrak{M} \models_X \neq(\mathbf{v})$ iff there exist $s, s' \in X$ such that $s(\mathbf{v}) \neq s'(\mathbf{v})$;

TS-ndep: $\mathfrak{M} \models_X \neq(\mathbf{v}, \mathbf{w})$ iff there exist $s, s' \in X$ with $s(\mathbf{v}) = s'(\mathbf{v})$ but $s(\mathbf{w}) \neq s'(\mathbf{w})$;

TS-geq: For all $n \in \mathbb{N}$, $\mathfrak{M} \models_X |\mathbf{v}| \geq n$ iff $|X(\mathbf{v})| \geq n$;

TS-all: $\mathfrak{M} \models_X All(\mathbf{v})$ iff $X(\mathbf{v}) = M^{|\mathbf{v}|}$;

TS-⊄: $\mathfrak{M} \models_X \mathbf{v} \not\subseteq \mathbf{w}$ iff there exists some $s \in X$ such that for all $s' \in X$, $s(\mathbf{v}) \neq s'(\mathbf{w})$;

TS-⫫: $\mathfrak{M} \models_X \mathbf{v} \not\perp_{\mathbf{u}} \mathbf{w}$ iff there exist $s, s' \in X$ with $s(\mathbf{u}) = s'(\mathbf{u})$ but such that for all $s'' \in X$, $s''(\mathbf{uvw}) \neq s(\mathbf{uv})s'(\mathbf{w})$.

The last two dependencies are not upwards-closed, but as shown in [8] they are definable in terms of constancy atoms and first-order, upwards-closed dependencies.

Definition 8 ($\top$, $\bot$). Let v be any variable. Then we write $\top$ for $\forall v(v = v)$ and $\bot$ for $\exists v(v \neq v)$.

Proposition 2. *For all models $\mathfrak{M}$ and teams X, $\mathfrak{M} \models_X \top$; and furthermore, $\mathfrak{M} \models_X \bot$ if and only if $X = \emptyset$.*

Proof. Obvious.

One thing worth pointing out here is the following: even if a dependency notion **D** is strongly first-order, and thus every sentence ϕ containing it is equivalent (in the sense of Equation (1)) on the level of sentences to some first-order sentence θ_ϕ, it is *not* in general true that this sentence may be freely substituted with θ_ϕ inside other sentences. For instance, the nonemptiness atom *NE* is true in all nonempty teams, and thus it is true in $\{\emptyset\}$, and thus it is equivalent to $\top$ on the level of sentences; but a moment's thought shows that, for instance, $(NE \wedge \bot) \vee \top$ is not equivalent to $(\top \wedge \bot) \vee \top$.

Definition 9 (Dual Negation). Let ϕ be a first-order formula in negation normal form. Then we write $\neg\phi$ as a shorthand for the formula thus obtained

- If ϕ is a positive literal *Rt* or $t_1 = t_2$, $\neg\phi$ is its negation (that is, $\neg Rt$ or $t_1 \neq t_2$);
- If ϕ is a negative literal $\neg Rt$ or $t_1 \neq t_2$, $\neg\phi$ is the corresponding positive literal (that is, *Rt* or $t_1 = t_2$);
- $\neg(\phi \vee \psi) = (\neg\phi) \wedge (\neg\psi)$;
- $\neg(\phi \wedge \psi) = (\neg\phi) \vee (\neg\psi)$;
- $\neg(\exists v\phi) = \forall v(\neg\psi)$;
- $\neg(\forall v\phi) = \exists v(\neg\psi)$;

It is not difficult to see, by structural induction on ϕ, that

Proposition 3. *For all first-order formulas ϕ, all models $\mathfrak{M}$ and all teams X, $\mathfrak{M} \models_X \neg\phi$ if and only if for all $s \in X$ we have that $\mathfrak{M} \models_s \neg\phi$ according to Tarski semantics.*

Definition 10 ($\phi \upharpoonright \theta$). Let $\mathscr{D}$ be any class of dependencies, let $\phi \in \mathbf{FO}(\mathscr{D})$ and let $\theta \in \mathbf{FO}$. Then we write $\phi \upharpoonright \theta$ as a shorthand for

$$(\neg\theta) \vee (\theta \wedge \phi)$$

Definition 11 ($X \upharpoonright \theta$). Let X be a team over some model $\mathfrak{M}$ and let θ be a first-order formula on the signature of $\mathfrak{M}$ and with free variables in the domain of X. Then we write $X \upharpoonright \theta$ for the team

$$\{s \in X : \mathfrak{M} \models_{\{s\}} \theta\} = \{s \in X : \mathfrak{M} \models_s \theta \text{ according to Tarski semantics}\}.$$

Proposition 4 ([8]). *Let $\mathscr{D}$ be any class of dependencies, let $\phi \in \mathbf{FO}(\mathscr{D})$ and let $\theta \in \mathbf{FO}$. Then for all suitable models $\mathfrak{M}$ and teams X,*

$$\mathfrak{M} \models_X \phi \upharpoonright \theta \text{ if and only if } \mathfrak{M} \models_{X \upharpoonright \theta} \phi.$$

Definition 12 (Flattening). Let $\mathscr{D}$ be any class of dependencies and let $\phi \in$ $FO(NE)$. Then we define its *flattening* ϕ^f as the first-order formula obtained by substituting all atoms $\mathbf{Dv}$ in it with $\top$.

Lemma 1. *For all classes of dependencies $\mathscr{D}$, models $\mathfrak{M}$, teams X, and formulas $\phi \in FO(\mathscr{D})$, if $\mathfrak{M} \models_X \phi$, then $\mathfrak{M} \models_X \phi^f$.*

Proof. Trivial.

Lemma 2 ([8]). *Let $\mathscr{D}$ be a class of* upwards-closed *(but not necessarily first-order) dependencies. Then for all models $\mathfrak{M}$, teams X and Y such that $X \subseteq Y$, and $\phi \in FO(\mathscr{D})$, if $\mathfrak{M} \models_X \phi$ and $\mathfrak{M} \models_Y \phi^f$, then $\mathfrak{M} \models_Y \phi$.*

Lemma 3 ([8]). *Let $\mathscr{D}$ be a class of dependencies (not necessarily first-order or upwards-closed) and let $\phi(\mathbf{v})$ be a $FO(=(\cdot), \mathscr{D})$ formula. Then $\phi(\mathbf{v})$ is equivalent to some formula of the form $\exists \mathbf{w}(=(\mathbf{w}) \wedge \psi(\mathbf{w}, \mathbf{v}))$, where $\psi \in FO(\mathscr{D})$ contains the exactly the same instances of $\mathbf{D}$-atoms (for all $\mathbf{D} \in \mathscr{D}$) that ϕ does, and in the same number.*

The following simple result - which allows us to add, essentially for free, the classical disjunction $\sqcup$ to our language – will also be of some use in the rest of this work:

Proposition 5. *Let $\mathscr{D}$ be any class of dependencies and let $\phi \in FO(\mathscr{D}, \sqcup)$. Then ϕ is equivalent to some formula of the form $\bigsqcup_{i=1}^n \psi_i$, where all ψ_i are in $FO(\mathscr{D})$.*

Proof. It suffices to show that the $\sqcup$ connective commutes with all other connectives:

- $(\psi \sqcup \theta) \vee \chi \equiv (\psi \vee \chi) \sqcup (\theta \vee \chi)$: Suppose that $\mathfrak{M} \models_X (\psi \sqcup \theta) \vee \chi$. Then $X = Y \cup Z$ for two Y, Z such that $\mathfrak{M} \models_Y \psi \sqcup \theta$ and $\mathfrak{M} \models_Z \chi$. By the satisfaction conditions for $\sqcup$, we have that $\mathfrak{M} \models_Y \psi$ or $\mathfrak{M} \models_Y \theta$. In the first case we have that $\mathfrak{M} \models_X \psi \vee \chi$ and in the second case we have that $\mathfrak{M} \models_X \theta \vee \chi$, so in either case $\mathfrak{M} \models_X (\psi \vee \chi) \sqcup (\theta \vee \chi)$.

  Conversely, suppose that $\mathfrak{M} \models_X (\psi \vee \chi) \sqcup (\theta \vee \chi)$. Then $\mathfrak{M} \models_X (\psi \vee \chi)$ or $\mathfrak{M} \models_X (\theta \vee \chi)$. In the first case, we have that $X = Y \cup Z$ for two Y and Z such that $\mathfrak{M} \models_Y \psi$ and $\mathfrak{M} \models_Z \chi$; but then $\mathfrak{M} \models_Y \psi \sqcup \theta$ too, and thus $\mathfrak{M} \models_X (\psi \sqcup \theta) \vee \chi$. The case in which $\mathfrak{M} \models_X (\theta \vee \chi)$ is dealt with analogously.
- $(\psi \sqcup \theta) \wedge \chi \equiv (\psi \wedge \chi) \sqcup (\theta \wedge \chi)$: $\mathfrak{M} \models_X (\psi \sqcup \theta) \wedge \chi$ iff $(\mathfrak{M} \models_X \psi$ or $\mathfrak{M} \models_X \theta)$ and $\mathfrak{M} \models_X \chi$ iff $(\mathfrak{M} \models_X \psi$ and $\mathfrak{M} \models_X \chi)$ or $(\mathfrak{M} \models_X \theta$ and $\mathfrak{M} \models_X \chi)$ iff $\mathfrak{M} \models_X (\psi \wedge \chi) \sqcup (\theta \wedge \chi)$.
- $\exists v(\psi \sqcup \theta) \equiv (\exists v \psi) \sqcup (\exists v \theta)$: Suppose that $\mathfrak{M} \models_X \exists v(\psi \sqcup \theta)$. Then there exists a choice function F such that $\mathfrak{M} \models_{X[F/v]} \psi$ or $\mathfrak{M} \models_{X[F/v]} \theta$. In the first case we have that $\mathfrak{M} \models_X \exists v \psi$, and in the second case we have that $\mathfrak{M} \models_X \exists v \theta$; so in either case $\mathfrak{M} \models_X (\exists v \psi) \sqcup (\exists v \theta)$.

  Conversely, suppose that $\mathfrak{M} \models_X (\exists v \psi) \sqcup (\exists v \theta)$. If $\mathfrak{M} \models_X (\exists v \psi)$, then there is a F such that $\mathfrak{M} \models_{X[F/v]} \psi$, and therefore $\mathfrak{M} \models_{X[F/v]} \psi \sqcup \theta$, and therefore $\mathfrak{M} \models_X \exists v(\psi \sqcup \theta)$; and similarly, if $\mathfrak{M} \models_X (\exists v \theta)$ there is a F such that $\mathfrak{M} \models_{X[F/v]} \theta$, and therefore $\mathfrak{M} \models_{X[F/v]} \psi \sqcup \theta$, and therefore $\mathfrak{M} \models_X \exists v(\psi \sqcup \theta)$.

- $\forall v(\psi \sqcup \theta) \equiv (\forall v\psi) \sqcup (\forall v\theta)$: $\mathfrak{M} \models_X \forall v(\psi \sqcup \theta)$ iff $\mathfrak{M} \models_{X[M/v]} (\psi \sqcup \theta)$ iff $(\mathfrak{M} \models_{X[M/v]} \psi$ or $\mathfrak{M} \models_{X[M/v]} \theta)$ iff $(\mathfrak{M} \models_X \forall v\psi$ or $\mathfrak{M} \models_X \forall v\theta)$ iff $\mathfrak{M} \models_X (\forall v\psi) \sqcup (\forall v\theta)$.

Lemma 4. *For all models $\mathfrak{M}$ and sentences $\phi_1, \phi_2 \in$ FO,*

$$\mathfrak{M} \models \phi_1 \sqcup \phi_2 \Leftrightarrow \mathfrak{M} \models \phi_1 \vee \phi_2.$$

Proof. Suppose that $\mathfrak{M} \models \phi_1 \sqcup \phi_2$. Then, by definition, $\mathfrak{M} \models_{\{\emptyset\}} \phi_i$ for some $i \in \{1,2\}$. Suppose, without loss of generality, that $\mathfrak{M} \models_{\{\emptyset\}} \phi_1$; then since ϕ_2 is first-order we have that $\mathfrak{M} \models_\emptyset \phi_2$, and hence $\mathfrak{M} \models_{\{\emptyset\}} \phi_1 \vee \phi_2$. The case for $\mathfrak{M} \models_{\{\emptyset\}} \phi_2$ is analogous. Conversely, suppose that $\mathfrak{M} \models_{\{\emptyset\}} \phi_2 \vee \phi_2$: then $\{\emptyset\} = Y \cup Z$ for two $Y, Z \subseteq \{\emptyset\}$ such that $Y \cup Z = \{\emptyset\}$, $\mathfrak{M} \models_Y \phi_1$ and $\mathfrak{M} \models_Z \phi_2$. Then $Y = \{\emptyset\}$ or $Z = \{\emptyset\}$, and hence $\mathfrak{M} \models \phi_1$ or $\mathfrak{M} \models \phi_2$ and finally $\mathfrak{M} \models \phi_1 \sqcup \phi_2$, as required.

Corollary 1. *Let $\mathscr{D}$ be a strongly first-order class of dependencies. Then every sentence of $FO(\mathscr{D}, \sqcup)$ is equivalent (on the level of sentences) to some sentence of FO.*

Proof. Let $\phi \in$ **FO**$(\mathscr{D}, \sqcup)$. As per the above results, we may assume that ϕ is of the form $\bigsqcup_i \psi_i$, where all ψ_i are **FO**$(\mathscr{D})$-sentences, and hence equivalent (on the level of sentences) to first-order sentences ψ_i'. Now let $\phi' = \bigvee_i \psi_i'$ be our sentence.

3 On the Contradictory Negation

It is known from [19] that *team logic* **FO**$(= (\cdot, \cdot), \sim)$, that is, the logic obtained by adding the contradictory negation *and* functional dependency conditions (of all arities) to the language of first-order logic, is as expressive as second-order logic over sentences; and, furthermore, in [15] it was shown that all second-order properties of teams correspond to the satisfaction conditions of team logic sentences.

But what if we add the contradictory negation to weaker extensions of first-order logic? Or, for that matter, what if we consider **FO**$(\sim)$, that is, the logic obtained by adding *only* the contradictory negation to the language of first-order logic?

In this section, we will prove that

1. Both **FO**$(\sim, =(\cdot))$ and **FO**$(\sim, \neq(\cdot))$ are equivalent to full team logic;
2. **FO**$(\sim) =$ **FO**$(NE, \sqcup) =$ **FO**$(NE, \sqcup, \sim)$;
3. Every sentence of **FO**$(NE, \sqcup)$ is equivalent to some first-order sentence.

Thus, the contradictory negation alone does not suffice to bring the expressive power of our logic beyond that of first-order logic, but as soon as we add even simple strongly first-order dependencies such as constancy or non-constancy we obtain the full expressive power of second-order logic.

Lemma 5. $FO(\sim, =(\cdot)) = FO(\sim, \neq(\cdot))$.

Proof. It suffices to observe that, for any tuple $\mathbf{v}$ of variables, $\neq(\mathbf{v})$ is logically equivalent to $\sim=(\mathbf{v})$ and $=(\mathbf{v})$ is logically equivalent to $\sim\neq(\mathbf{v})$.

Lemma 6. *For any two tuples $\mathbf{v}$, $\mathbf{w}$ of variables, the functional dependence atom* $=(\mathbf{v}, \mathbf{w})$ *is definable in* $FO(\sim, =(\cdot))$.

Proof. Consider the formula

$$\sim (\exists \mathbf{p}\mathbf{q}_1\mathbf{q}_2(=(\mathbf{p})\wedge =(\mathbf{q}_1)\wedge =(\mathbf{q}_2) \wedge \mathbf{q}_1 \neq \mathbf{q}_2 \wedge \sim (\mathbf{v}\mathbf{w} \neq \mathbf{p}\mathbf{q}_1)\wedge \sim (\mathbf{v}\mathbf{w} \neq \mathbf{p}\mathbf{q}_2)) \tag{2}$$

where, for any two tuples of variables $\mathbf{t}$ and $\mathbf{t}'$ with the same length n, $\mathbf{t} \neq \mathbf{t}'$ is a shorthand for $\bigvee_{i=1}^{n} t_i \neq t_i'$ and therefore $\mathfrak{M} \models_Y \sim (\mathbf{t} \neq \mathbf{t}')$ if and only if there exists some $s \in Y$ with $s(\mathbf{t}) = s(\mathbf{t}')$.

It is easy to check that Equation (2) is logically equivalent to $=(\mathbf{v}, \mathbf{w})$, as required.

Corollary 2. $FO(\sim, =(\cdot)) = FO(\sim, \neq(\cdot)) = FO(\sim, =(\cdot, \cdot)) = $ *Team Logic.*

So far so good. But what can we say about $FO(\sim)$? In what follows, we will prove that this logic is *not* more expressive than first-order logic over sentences; indeed, it is equivalent to $FO(NE, \sqcup)$.

Lemma 7. *Let $\psi \in FO(NE)$ and let $\theta \in FO$. Then $\sim (\psi \upharpoonright \theta)$ is logically equivalent to $((\sim \psi) \upharpoonright \theta)$.*

Proof. Suppose that $\mathfrak{M} \models_X \sim (\psi \upharpoonright \theta)$. Then for $Y = \{s \in X : \mathfrak{M} \models_s \theta\}$ we have that $\mathfrak{M} \not\models_Y \psi$. But then $\mathfrak{M} \models_Y \sim \psi$, and thus $\mathfrak{M} \models_X ((\sim \psi) \upharpoonright \theta)$.

Conversely, suppose that $\mathfrak{M} \models_X ((\sim \psi) \upharpoonright \theta)$. Then for Y as above we have that $\mathfrak{M} \not\models_Y \psi$; and therefore, $\mathfrak{M} \not\models_X (\psi \upharpoonright \theta)$, and in conclusion $\mathfrak{M} \models_X \sim (\psi \upharpoonright \theta)$.

Lemma 8. *Let $\phi \in FO(NE)$. Then $\sim \phi$ is equivalent to some formula in* $FO(NE, \sqcup)$.

Proof. We proceed by structural induction on ϕ.

1. Suppose that ϕ is a first-order formula (not necessarily a literal). Then $\sim \phi$ is logically equivalent to $NE \upharpoonright (\neg\phi)$. Indeed, suppose that $\mathfrak{M} \models_X \sim \phi$: then, since ϕ is first-order, there exists a $s \in X$ such that $\mathfrak{M} \not\models_s \phi$ according to Tarski semantics. But then $s \in X \upharpoonright (\neg\phi)$, and thus $\mathfrak{M} \models_X NE \upharpoonright (\neg\phi)$.

   Conversely, suppose that $\mathfrak{M} \models_X NE \upharpoonright (\neg\phi)$. Then the set $X \upharpoonright (\neg\phi)$ is not empty, and therefore there exists some $s \in X$ which satisfies $\neg\phi$ according to Tarski semantics, and finally $\mathfrak{M} \not\models_X \phi$.
2. $\sim NE$ is easily seen to be equivalent to $\bot$, which is true only in the empty team.
3. Suppose that ϕ is of the form $(\psi \vee \theta)$. Then $\sim \phi$ is logically equivalent to

$$((\sim \psi) \upharpoonright \psi^f) \sqcup ((\sim \theta) \upharpoonright \theta^f) \sqcup \sim (\psi^f \vee \theta^f). \tag{3}$$

Indeed, suppose that $\mathfrak{M} \models_X \sim (\psi \vee \theta)$. Then it is not the case that $X = Y \cup Z$ for two Y, Z such that $\mathfrak{M} \models_Y \psi$ and $\mathfrak{M} \models_Z \theta$. In particular, take $Y = X \upharpoonright \psi^f$ and $Z = X \upharpoonright \theta^f$: then $Y \cup Z \neq X$, and hence $\mathfrak{M} \models_X \sim (\psi^f \vee \theta^f)$, or $\mathfrak{M} \not\models_Y \psi$, and hence $\mathfrak{M} \models_X ((\sim \psi) \upharpoonright \psi^f)$, or $\mathfrak{M} \not\models_Z \theta$, and hence $\mathfrak{M} \models_X ((\sim \theta) \upharpoonright \theta^f)$.

Conversely, suppose that $\mathfrak{M} \models_X (\psi \vee \theta)$. Then $X = Y \cup Z$ for two Y, Z such that $\mathfrak{M} \models_Y \psi$ and $\mathfrak{M} \models_Y \theta$. Now take $Y' = X \upharpoonright \psi^f$ and $Z' = X \upharpoonright \theta^f$: by Proposition 1 we have that $\mathfrak{M} \models_Y \psi^f$ and $\mathfrak{M} \models_Z \theta^f$, by Lemma 1 we have that $Y \subseteq Y'$ and $Z \subseteq Z'$, and thus $X = Y' \cup Z'$, and by Lemma 2 we have that $\mathfrak{M} \models_{Y'} \psi$ and $\mathfrak{M} \models_{Z'} \theta$. Therefore $\mathfrak{M} \not\models_X (\sim \psi) \upharpoonright \psi^f$, $\mathfrak{M} \not\models_X (\sim \theta) \upharpoonright \theta^f$, and $\mathfrak{M} \models_X (\psi^f \vee \theta^f)$, so in conclusion X does not satisfy Equation (3).

4. Suppose that ϕ is of the form $(\psi \wedge \theta)$. Then $\sim \phi$ is logically equivalent to $(\sim \psi) \sqcup (\sim \theta)$.

5. Suppose that ϕ is of the form $(\exists v \psi)$. Then $\sim \phi$ is logically equivalent to

$$\sim (\exists v \psi^f) \sqcup \forall v ((\sim \psi) \upharpoonright \psi^f) \tag{4}$$

Indeed, suppose that $\mathfrak{M} \models_X \sim (\exists v \psi)$ and $\mathfrak{M} \models_X \exists v \psi^f$, and consider the choice function F such that $F(s) = \{m : \mathfrak{M} \models_{s[m/v]} \psi^f\}$. $F(s)$ is nonempty for all $s \in X$, since $\mathfrak{M} \models_X \exists v \psi^f$; and therefore, by hypothesis, $\mathfrak{M} \not\models_{X[F/v]} \psi$. But by construction, we have that $X[F/v] = X[M/v] \upharpoonright \psi^f$, and thus $\mathfrak{M} \not\models_{X[M/v]} \psi \upharpoonright \psi^f$, and finally $\mathfrak{M} \models_X \forall v ((\sim \psi) \upharpoonright \psi^f)$.

Conversely, suppose that there exists a choice function $F : X \to \mathscr{P}(M) \backslash \{\emptyset\}$ such that $\mathfrak{M} \models_{X[F/v]} \psi$. Then in particular $\mathfrak{M} \models_{X[F/v]} \psi^f$, and hence $\mathfrak{M} \models_X \exists v \psi^f$ and $\mathfrak{M} \not\models_X \sim (\exists v \psi^f)$; and furthermore, we have that $X[F/v] \subseteq X[M/v] \upharpoonright \psi^f$, and therefore $\mathfrak{M} \models_{X[M/v]} \psi \upharpoonright \psi^f$ and $\mathfrak{M} \not\models_{X[M/v]} (\sim \psi) \upharpoonright \psi^f$. So in conclusion the team X does not satisfy Equation (4).

6. Suppose that ϕ is of the form $(\forall v \psi)$. Then $\sim \phi$ is logically equivalent to $\forall v \sim \psi$: indeed, $\mathfrak{M} \models_X \sim \phi$ iff $\mathfrak{M} \not\models_X \forall v \psi$ iff $\mathfrak{M} \not\models_{X[M/v]} \psi$ iff $\mathfrak{M} \models_{X[M/v]} \sim \psi$ iff $\mathfrak{M} \models_X \forall v \sim \psi$.

We are now equipped to prove the main result of this section:

Theorem 3. *Let $\phi \in FO(NE, \sqcup)$. Then $\sim \phi$ is equivalent to some formula in $FO(NE, \sqcup)$.*

Proof. By Proposition 5, we may assume that ϕ is of the form $\bigsqcup_{i=1}^n \psi_i$, where each ψ_i is in $FO(NE)$. Thus, $\sim \phi$ is logically equivalent to $\bigwedge_{i=1}^n (\sim \psi_i)$; and by the above lemma, if ψ_i is in $FO(NE)$, then $\sim \psi_i$ is in $FO(NE, \sqcup)$, as required. $\qquad\square$

The two following corollaries then follow at once:

Corollary 3. $FO(NE, \sqcup, \sim) = FO(NE, \sqcup)$.

Corollary 4. $FO(\sim) \subseteq FO(NE, \sqcup)$.

We still need to show the other direction of the equivalence between $FO(NE, \sqcup)$ and $FO(\sim)$:

Proposition 6. $FO(NE, \sqcup) \subseteq FO(\sim)$.

Proof. It suffices to show that the nonemptiness atom and the classical disjunction are definable in $FO(\sim)$. As for the former, observe that $\mathfrak{M} \models_X \sim \perp$ if and only if X is nonempty; and for the latter, observe that $\phi \sqcup \psi$ is logically equivalent to $\sim ((\sim \phi) \wedge (\sim \psi))$. $\qquad\square$

Putting everything together, we have that

Theorem 4. $FO(\sim) = FO(NE, \sqcup)$.

Finally, we need to prove that every sentence of $FO(NE, \sqcup)$ is equivalent to some first-order sentence. But this is immediate:

Theorem 5. *Let $\phi \in FO(NE, \sqcup)$ be a sentence. Then ϕ is logically equivalent to some $\phi' \in FO$.*

Proof. By Proposition 5 we may assume that ϕ is of the form $\bigsqcup_{i=1}^{n} \psi_i$, where each ψ_i is a sentence in $FO(NE)$. But then by Theorem 2, each ψ_i is equivalent to some first-order sentence ψ'_i, and thus ϕ is equivalent to the first-order sentence $\bigvee_{i=1}^{n} \psi'_i$.

Corollary 5. *The constancy and inconstancy atoms are not definable in $FO(NE, \sqcup)$.*

Proof. If they were, then we would have that $FO(=(\cdot, \cdot), \sim) \subseteq FO(NE, \sqcup)$; but this is not possible, because $FO(NE, \sqcup)$ is strongly first-order and $FO(=(\cdot, \cdot), \sim)$ is as strong as second-order logic.

4 Arity Hierarchies for Totality Atoms

In this section we will investigate the properties of the k-ary *totality atoms* All_k, and we establish a strict *arity hierarchy* for them.

Let us begin by generalizing a notion from [8]:

Definition 13 (γ-boundedness). Let $\gamma : \mathbb{N} \to \mathbb{N}$ be a function. Then a dependency notion **D** is said to be γ-bounded if for all finite models $\mathfrak{M}$ and teams X, if $\mathfrak{M} \models_X \mathbf{D}$, then there exists a subteam $Y \subseteq X$, $|Y| \leq \gamma(|M|)$, such that $\mathfrak{M} \models_Y \mathbf{D}$.

Proposition 7. *All k-ary dependencies $\mathbf{D}$ are $|M|^k$-bounded.*

Proof. Suppose that $\mathfrak{M} \models_X \mathbf{Dv}$. Then $(M, X(\mathbf{v})) \in \mathbf{D}$; and since $X(\mathbf{v}) \subseteq M^k$, it is clear that $|X(\mathbf{v})| \leq |M|^k$. Now for any $\mathbf{m} \in X(\mathbf{v})$, let $s_{\mathbf{m}} \in X$ be such that $s_{\mathbf{m}}(\mathbf{v}) = \mathbf{m}$, and let $Y = \{s_{\mathbf{m}} : \mathbf{m} \in X(\mathbf{v})\}$. Then $|Y| \leq |M|^k$ and $Y(\mathbf{v}) = X(\mathbf{v})$, and thus $\mathfrak{M} \models_Y \mathbf{Dv}$.

Theorem 6. *Let $\mathscr{D} = \{\mathbf{D}_i : i \in I\}$ be a class of upwards-closed dependencies, for every $\mathbf{D}_i \in \mathscr{D}$ let $\gamma_i : \mathbb{N} \to \mathbb{N}$ be such that $\mathbf{D}_i$ is γ_i-bounded and let $\phi \in FO(\mathscr{D})$ be such that every $\mathbf{D}_i$ occurs k_i times in it. Now define the function $v_\phi : \mathbb{N} \to \mathbb{N}$ as $v_\phi(n) = \Sigma_{i \in I} k_i \gamma_i(n)$. Then ϕ is v_ϕ-bounded, in the sense that*

$$\mathfrak{M} \models_X \phi \Rightarrow \exists Y \subseteq X, |Y| \leq v_\phi(|M|), \mathfrak{M} \models_Y \phi$$

for all finite models $\mathfrak{M}$ and all teams X.

Proof. The proof is by induction, and mirrors the analogous proof from [8].

1. If ϕ is a first-order literal, then it is 0-bounded (since the empty team satisfies it), as required.

2. If ϕ is an atom $\mathbf{Dx}$, then the statement follows at once from the definitions of boundedness.

3. Let ϕ be a disjunction $\psi_1 \vee \psi_2$ then $v_\phi = v_{\psi_1} + v_{\psi_2}$. Suppose now that $\mathfrak{M} \models_X \psi_1 \vee \psi_2$: then $X = X_1 \cup X_2$ for two X_1 and X_2 such that $\mathfrak{M} \models_{X_1} \psi_1$ and $\mathfrak{M} \models_{X_2} \psi_2$. This implies that there exist $Y_1 \subseteq X_1$, $Y_2 \subseteq X_2$ such that $\mathfrak{M} \models_{Y_1} \psi_1$ and $\mathfrak{M} \models_{Y_2} \psi_2$, $|Y_1| \leq v_{\psi_1}(|M|)$ and $|Y_2| \leq v_{\psi_2}(|M|)$. But then $Y = Y_1 \cup Y_2$ satisfies $\psi_1 \vee \psi_2$ and has at most $v_{\psi_1}(|M|) + v_{\psi_2}(|M|)$ elements.

4. If ϕ is a conjunction $\psi_1 \wedge \psi_2$, then again, $v_\phi = v_{\psi_1} + v_{\psi_2}$. Suppose that $\mathfrak{M} \models_X \psi_1 \wedge \psi_2$: then $\mathfrak{M} \models_X \psi_1$ and $\mathfrak{M} \models_X \psi_2$, and therefore by Lemma 1 $\mathfrak{M} \models_X \psi_1^f$ and $\mathfrak{M} \models_X \psi_2^f$; and, by induction hypothesis, there exist $Y_1, Y_2 \subseteq X$ with $|Y_1| \leq v_{\psi_1}(|M|)$, $|Y_2| \leq v_{\psi_2}(|M|)$, $\mathfrak{M} \models_{Y_1} \psi_1$ and $\mathfrak{M} \models_{Y_2} \psi_2$. Now let $Y = Y_1 \cup Y_2$: since $Y \subseteq X$, by Proposition 1 $\mathfrak{M} \models_Y \psi_1^f$ and $\mathfrak{M} \models_Y \psi_2^f$. But $Y_1, Y_2 \subseteq Y$, and therefore by Lemma 2 $\mathfrak{M} \models_Y \psi_1$ and $\mathfrak{M} \models_Y \psi_2$, and in conclusion $\mathfrak{M} \models_Y \psi_1 \wedge \psi_2$.

5. If ϕ is of the form $\exists v \psi$, then $v_\phi = v_\psi$. Suppose that $\mathfrak{M} \models_X \exists v \psi$: then for some F we have that $\mathfrak{M} \models_{X[F/v]} \psi$, and therefore by induction hypothesis there exists a $Z \subseteq X[F/v]$ with $|Z| \leq v_\psi(|M|)$ such that $\mathfrak{M} \models_Z \psi$. For any $h \in Z$, let $\mathfrak{f}(h)$ be a $s \in X$ such that $h \in s[F/v] = \{s[m/v] : m \in F(s)\}$,[2] and let $Y = \{\mathfrak{f}(h) : h \in Z\}$. Now $Z \subseteq Y[F/v] \subseteq X[F/v]$. Since $\mathfrak{M} \models_{X[F/v]} \psi^f$ and $Y[F/v] \subseteq X[F/v]$, we have that $\mathfrak{M} \models_{Y[F/v]} \psi^f$; and since $\mathfrak{M} \models_Z \psi$, this implies that $\mathfrak{M} \models_{Y[F/v]} \psi$ and that $\mathfrak{M} \models_Y \exists v \psi$. Furthermore $|Y| \leq |Z| \leq v_\phi(|M|)$, as required.

6. If ϕ is of the form $\forall v \psi$, then again, $v_\phi = v_\psi$. Suppose that $\mathfrak{M} \models_{X[M/v]} \psi$: again, by induction hypothesis there is a $Z \subseteq X[M/v]$ with $|Z| \leq v_\psi(|M|)$ and such that $\mathfrak{M} \models_Z \psi$. For any $h \in Y$, let $\mathfrak{g}(h)$ pick some $s \in X$ which agrees with h on all variables except possibly v, and let $Y = \{\mathfrak{g}(h) : h \in Z\}$. Similarly to the previous case, $Z \subseteq Y[M/v] \subseteq X[M/v]$: therefore, since $\mathfrak{M} \models_{X[M/v]} \psi^f$ we have that $\mathfrak{M} \models_{Y[M/v]} \psi^f$, and since $\mathfrak{M} \models_Z \psi$ we have that $\mathfrak{M} \models_{Y[M/v]} \psi$. So in conclusion $\mathfrak{M} \models_Y \forall v \psi$, as required, and $|Y| \leq |Z| \leq v_\phi(M)$.

Using some care, we can extend this result to the case of $\mathbf{FO}(=(\cdot), \mathscr{D}, \sqcup)$:

Theorem 7. *Let $\mathscr{D} = \{\mathbf{D}_i : i \in I\}$ be a class of upwards-closed dependencies, for every $\mathbf{D}_i \in \mathscr{D}$ let $\gamma_i : \mathbb{N} \to \mathbb{N}$ be such that $\mathbf{D}_i$ is γ_i-bounded, let $\phi \in \mathbf{FO}(=(\cdot), \mathscr{D}, \sqcup)$ be such that every $\mathbf{D}_i$ occurs k_i times, and let $v_\phi(n) = \Sigma_{i \in I} k_i \gamma_i(n)$. Then ϕ is v_ϕ-bounded, in the sense that*

$$\mathfrak{M} \models_X \phi \Rightarrow \exists Y \subseteq X, |Y| \leq v_\phi(|M|), \mathfrak{M} \models_Y \phi.$$

Proof. By Proposition 5, we can assume that ϕ is of the form $\bigsqcup_{i=1}^n \psi_i$, where all ψ_i are in $\mathbf{FO}(=(\cdot), \mathscr{D})$. Furthermore, by Lemma 3 we can assume that every ψ_i is of the

---

[2] Since $Z \subseteq X[F/v]$, such a s always exists. Of course, there may be multiple ones; in that case, we just pick arbitrarily one.

form $\exists \mathbf{w}_i (=(\mathbf{w}_i) \wedge \theta_i)$, for $\theta_i \in \mathbf{FO}(\mathscr{D})$ and all tuples of variables $\mathbf{w}_i$ are new. Now suppose that $\mathfrak{M} \models_X \phi$: then there exists an $i \in 1 \ldots n$ and a tuple of elements $\mathbf{m} \in M$ such that $\mathfrak{M} \models_{X[\mathbf{m}/\mathbf{w}_i]} \theta_i$. But then there exists a $Y \subseteq X[\mathbf{m}/\mathbf{w}_i]$, $|Y| \leq \nu_{\theta_i}(|M|)$, such that $\mathfrak{M} \models_Y \phi$. Now let Z be the restriction of Y to the domain of X: clearly $Z \subseteq X$ and $|Z| \leq |Y| \leq \nu_{\theta_i}(|M|) \leq \nu_\phi(|M|)$, and furthermore $\mathfrak{M} \models_Z \exists \mathbf{w}_i (=(\mathbf{w}_i) \wedge \theta_i)$ and so in conclusion $\mathfrak{M} \models_Z \phi$.

Theorem 8. *Let $k' > k$, and let $\mathscr{D}$ be a class of k-ary upwards-closed (not necessarily first-order) dependencies. Then $All_{k'}$ is not definable in $\mathbf{FO}(=(\cdot), \mathscr{D}, \sqcup)$.*

Proof. Suppose that $\phi(\mathbf{v}) \in \mathbf{FO}(=(\cdot), \mathscr{D}, \sqcup)$ defines $All_{k'}$. Then, since all dependencies in $\mathscr{D}$ are $|M|^k$-bounded, we have at once that ϕ is $q|M|^k$-bounded for some $q \in \mathbb{N}$. Now let $n \in \mathbb{N}$ be such that $n^{k'} > qn^k$, let M be a model in the empty signature with n elements, let $\mathbf{v}$ be a tuple of k' variables, and let $X = \{\emptyset\}[M/\mathbf{v}]$. Then $M \models_X All_{k'}\mathbf{v}$, and therefore $M \models_X \phi(\mathbf{v})$. But then there must be a $Y \subseteq X$, $|Y| \leq qn^k$, such that $M \models_Y \phi(\mathbf{v})$; and this is not possible, because for such a Y we would have that $M \not\models_Y All_{k'}\mathbf{v}$.

In particular, it follows at once from this that All_{k+1} is not definable in $\mathbf{FO}(=(\cdot), All_k, \sqcup)$. On the other hand if $k' < k$ the operator $All_{k'}\mathbf{v}$ is easily seen to be definable as $\forall \mathbf{w}(All_k \mathbf{v}\mathbf{w})$; therefore

Corollary 6. *For all $k \in \mathbb{N}$, $\mathbf{FO}(=(\cdot), All_k, \sqcup) \subsetneq \mathbf{FO}(=(\cdot), All_{k+1}, \sqcup)$ (and all these logics are equivalent to first-order logic over sentences).*

5   0-ary Dependencies: Escaping the Empty Team

As a limit case of the notion of dependency, we have that

Definition 14. A 0-ary dependency $\mathbf{D}$ is a set of models over the empty signature. For all models $\mathfrak{M}$ and teams X, $\mathfrak{M} \models_X \mathbf{D}$ if and only if $M \in \mathbf{D}$.

If a 0-ary dependency is first-order, we have that $\mathfrak{M} \models_X \mathbf{D}$ if and only if $M \models \mathbf{D}^*$, where $\mathbf{D}^*$ is a sentence over the empty signature; therefore, it is natural to generalize them all to an operator $[\cdot]$ of the form

TS-$[\cdot]$:   For all first-order sentences ϕ in the signature of $\mathfrak{M}$, $\mathfrak{M} \models_X [\phi]$ if and only if $\mathfrak{M} \models \phi$ according to the usual Tarski semantics.

Whenever X is nonempty it follows at once from Proposition 1 that $\mathfrak{M} \models_X [\phi]$ if and only if $\mathfrak{M} \models_X \phi$; but since $\mathfrak{M} \models_\emptyset \phi$ for all first-order sentences ϕ, in first-order logic with team semantics we have no way of verifying whether a given first-order sentence is true of our model when we are considering satisfiability with respect to the empty team. Therefore, we will add this $[\cdot]$ operator to our language. It is easy to see that adding it to a strongly first-order extension of first-order logic does not break the property of being strongly first-order:

Proposition 8. *Let $\mathscr{D}$ be any family of dependencies, and let $\phi \in FO(\mathscr{D}, [\cdot])$. Then ϕ is logically equivalent to some sentence of the form $\bigwedge_i [\theta_i] \wedge \psi$, where $\psi \in FO(\mathscr{D})$.*

Proof. The proof is by induction on ϕ, and it is entirely straightforward. We report only the case of disjunction:

- For all first-order sentences θ_i, θ_j' and all $FO(\mathscr{D})$ formulas ψ_1, ψ_2 we have that $(\bigwedge_i [\theta_i] \wedge \psi_1) \vee (\bigwedge_j [\theta_j'] \wedge \psi_2)$ is logically equivalent to $\bigwedge_i [\theta_i] \wedge \bigwedge_j [\theta_j'] \wedge (\psi_1 \vee \psi_2)$. Indeed, suppose that $X = Y \cup Z$ for two Y, Z such that $\mathfrak{M} \models_Y \bigwedge_i [\theta_i] \wedge \psi_1$ and $\mathfrak{M} \models_Z \bigwedge_j [\theta_j'] \wedge \psi_2$. Then $\mathfrak{M} \models \bigwedge_i \theta_i \wedge \bigwedge_j \theta_j'$, and therefore $\mathfrak{M} \models_X \bigwedge_i [\theta_i] \wedge \bigwedge_j [\theta_j']$; and since $\mathfrak{M} \models_Y \psi$ and $\mathfrak{M} \models_Z \theta$, we also have that $\mathfrak{M} \models_X \psi \vee \theta$, and so in conclusion $\mathfrak{M} \models_X \bigwedge_i [\theta_i] \wedge \bigwedge_j [\theta_j'] \wedge (\psi_1 \vee \psi_2)$.

  The other direction is similar: if $\mathfrak{M} \models \bigwedge_i \theta_i \wedge \bigwedge_j \theta_j'$ and $\mathfrak{M} \models_X \psi_1 \vee \psi_2$, then $X = Y \cup Z$ for two Y and Z such that $\mathfrak{M} \models_Y \psi_1$ and $\mathfrak{M} \models_Z \psi_2$. But then $\mathfrak{M} \models_Y \bigwedge_i [\theta_i] \wedge \psi_1$ and $\mathfrak{M} \models_Z \bigwedge_j [\theta_j] \wedge \psi_2$, and so in conclusion $\mathfrak{M} \models_X (\bigwedge_i [\theta_i] \wedge \psi_1) \vee (\bigwedge_j [\theta_j'] \wedge \psi_2)$.

Therefore we have the following result:

Proposition 9. *Let $\mathscr{D}$ be a strongly first-order class of dependencies and let $\phi \in FO(\mathscr{D}, [\cdot])$ be a sentence. Then ϕ is logically equivalent to some first-order sentence ϕ', in the sense that $\mathfrak{M} \models_{\{\emptyset\}} \phi$ if and only if $\mathfrak{M} \models \phi'$.*

Proof. We may assume that ϕ is on the form $\bigwedge_i [\theta_i] \wedge \psi$, where ψ is a $FO(\mathscr{D})$-sentence. Now since $\mathscr{D}$ is strongly first-order, ψ is equivalent to some first-order ψ'; and since $\{\emptyset\}$ is nonempty, we can take $\phi' = \bigwedge_i \theta_i \wedge \psi$.

6 Unary Dependencies

We will now consider the case of *unary* dependencies, that is, of dependence atoms of arity one. As we will see, *all* first-order unary dependencies are strongly first-order and definable in $\mathbf{FO}(=(\cdot), [\cdot], All_1, \sqcup)$.

In order to prove this we will make use of the following standard result:

Lemma 9. *Let ϕ be a first-order sentence over the vocabulary $\{P\}$, where P is unary. Then ϕ is logically equivalent to a Boolean combination of sentences of the form $\exists^{=k} x P x$ and $\exists^{=k} x \neg P x$.*

Therefore, in order to show that all unary dependencies are in $\mathbf{FO}(=(\cdot), [\cdot], All_1, \sqcup)$ it suffices to show that the following four dependencies are in it:

TS-eq-pos: For all $k \in \mathbb{N}$, $\mathfrak{M} \models_X |v| = k$ iff $|X(v)| = k$;

TS-neq-pos: For all $k \in \mathbb{N}$, $\mathfrak{M} \models_X |v| \neq k$ iff $|X(v)| \neq k$;

TS-eq-neg: For all $k \in \mathbb{N}$, $\mathfrak{M} \models_X |M - v| = k$ iff $|M \backslash X(v)| = k$;

TS-neq-neg: For all $k \in \mathbb{N}$, $\mathfrak{M} \models_X |M - v| \neq k$ iff $|M \backslash X(v)| \neq k$.

Let us prove that this is the case.

Lemma 10. *The nonemptiness atom NE is definable in $\boldsymbol{FO}(All_1)$ as $\forall q All_1 q$.*

Proof. Suppose that $\mathfrak{M} \models_X NE$, that is, $X \neq \emptyset$, and let $s \in X$. Then for all $m \in M$, $s[m/q] \in X[M/1]$, and thus $X[M/q](q) = M$, and thus $\mathfrak{M} \models_X \forall q All_1 q$ as required.

However, if $X = \emptyset$ we have that $X[M/q] = \emptyset$ too, and thus $X[M/q](q) = \emptyset \neq M$, and finally $\mathfrak{M} \not\models_X \forall q All_1 q$.

Definition 15. For all $k \in \mathbb{N}$ and all variables v, we define the following formulas:

$$\phi_{\leq k}(v) = \exists p_1 \dots p_k (\bigwedge_i =(p_i) \wedge \bigvee_{i=1}^{k} v = p_i);$$

$$\phi_{\geq k}(v) = \exists p_1 \dots p_k (\bigwedge_i =(p_i) \wedge \bigwedge_{i \neq j} p_i \neq p_j \wedge \bigwedge_i (NE \upharpoonright v = p_i));$$

$$\psi_{\leq k}(v) = [\exists^{\leq k} x(x = x)] \sqcup \exists p_1 \dots p_k (\bigwedge_i =(p_i) \wedge \exists q(All_1(q) \wedge (\bigvee_i q = p_i \vee q = v)));$$

$$\psi_{\geq k}(v) = (\bot \wedge [\exists^{\geq k} x(x = x)]) \sqcup (NE \wedge \exists p_1 \dots p_k (\bigwedge_i =(p_i) \wedge \bigwedge_{i \neq j} p_i \neq p_j \wedge \bigwedge_{i=1}^{k} v \neq p_i))$$

Proposition 10. *For all $k \in \mathbb{N}$, all variables v, all models $\mathfrak{M}$ and all nonempty teams X whose domain contains v,*

- *$\mathfrak{M} \models_X \phi_{\leq k}(v)$ if and only if $|X(v)| \leq k$;*
- *$\mathfrak{M} \models_X \phi_{\geq k}(v)$ if and only if $|X(v)| \geq k$;*
- *$\mathfrak{M} \models_X \psi_{\leq k}(v)$ if and only if $|M \backslash X(v)| \leq k$;*
- *$\mathfrak{M} \models_X \psi_{\geq k}(v)$ if and only if $|M \backslash X(v)| \geq k$.*

Proof. • Suppose that $\mathfrak{M} \models_X \phi_{\leq k}(v)$ and X is nonempty: then there exist elements $m_1 \dots m_k$ such that for $Y = X[m_1 \dots m_k / p_1 \dots p_k]$, $\mathfrak{M} \models_Y \bigvee_{i=1}^{k} v = p_i$. But then $X(v) \subseteq \{m_1 \dots m_k\}$, and thus $|X(v)| \leq k$. If instead X is empty then trivially $|X(v)| = 0 \leq k$.

Conversely, suppose that $X(v) = \{m_1, \dots m_{k'}\}$ for $k' \leq k$, and let a be an arbitrary element of our model. Then for $Y = X[m_1 \dots m_{k'} a \dots a / p_1 \dots p_k]$ we have that $\mathfrak{M} \models_Y \bigvee_{i=1}^{k} v = p_i$. Thus $\mathfrak{M} \models_X \phi_{\leq k}(v)$, as required.

• Suppose that $\mathfrak{M} \models_X \phi_{\geq k}(v)$. Then there exist distinct elements $m_1 \dots m_k$ such that for $Y = X[m_1 \dots m_k / p_1 \dots p_k]$ and for all $i \in 1 \dots k$, $\mathfrak{M} \models_Y NE \upharpoonright v = p_i$. Thus for all such i there exists a $s \in Y$ with $s(v) = s(p_i) = m_i$, and thus $|X(v)| = |Y(v)| \geq k$.

Conversely, suppose that $\{m_1 \dots m_k\} \subseteq X(v)$, where all m_i are distinct. Now take $Y = X[m_1 \dots m_k / p_1 \dots p_k]$: clearly $\mathfrak{M} \models_Y \bigwedge_i =(p_i) \wedge \bigwedge_{i \neq j} p_i \neq p_j$, and it remains to show that for all i $\mathfrak{M} \models_Y NE \upharpoonright v = p_i$. But $Y \upharpoonright (v = p_i) = \{s \in Y : s(v) = s(p_i) = m_i\}$ is nonempty by hypothesis, and this concludes the proof.

- Suppose that $\mathfrak{M} \models_X \psi_{\leq k}(v)$. If $\mathfrak{M} \models_X [\exists^{\leq k} x(x = x)]$ we have that $|M| \leq k$, from which it follows at once that $|M \backslash X(v)| \leq |M| \leq k$. Otherwise, we can find elements $m_1 \ldots m_k$ such that, for $Y = X[m_1 \ldots m_k / p_1 \ldots p_k]$, there exists a choice function F for which $\mathfrak{M} \models_{Y[F/q]} All_1(q) \wedge (\bigvee_i q = p_i \vee q = v)$. Then $M \backslash X(v)$ must be contained in $\{m_1 \ldots m_k\}$, since q takes all possible values and $s(q) \notin \{m_1 \ldots m_k\} \Rightarrow s(q) = s(v)$.

  Conversely, suppose that $M \backslash X(v) \subseteq \{m_1 \ldots m_k\}$. If $X \neq \emptyset$, let Y be $X[m_1 \ldots m_k / p_1 \ldots p_k]$, and for all $s \in Y$ let $F(s) = \{m_1 \ldots m_k\} \cup \{s(v)\}$. Then $Y[F/q] \models All_1 q$: indeed, if $m \in \{m_1 \ldots m_k\}$, then $m \in F(s)$ for all $s \in Y$, and otherwise $m = s(v)$ for some $s \in Y$ (and hence $m \in F(s)$ for this choice of s). Furthermore, for all $h \in Y[F/q]$, if $h(q) \notin \{m_1 \ldots m_k\}$, then we have that $h(q) = h(v)$, as required. If instead $X = \emptyset$, then $|M| = |M \backslash X(v)| \leq k$, and hence $\mathfrak{M} \models_X \exists^{\leq k} x(x = x)$.

- Suppose that $\mathfrak{M} \models_X \psi_{\geq k}(v)$ and $X \neq \emptyset$. Then there exist distinct elements $m_1 \ldots m_k$ such that for $Y = X[m_1 \ldots m_k / p_1 \ldots p_k]$, $\mathfrak{M} \models_Y \bigwedge_{i=1}^k v \neq p_i$. Therefore $\{m_1 \ldots m_k\} \in M \backslash X$, and thus $|M \backslash X| \geq k$. If instead $X = \emptyset$, then $\mathfrak{M} \models \bot \wedge [\exists^{\geq k} x(x = x)]$ and hence $|M| = |M \backslash X(v)| \geq k$ as required.

  Conversely, suppose that $|M \backslash X(v)| \geq k$. If X is nonempty we can choose elements $m_1 \ldots m_k \in M \backslash X(v)$ and verify that $\mathfrak{M} \models_{X[m_1 \ldots m_k / p_1 \ldots p_k]} \bigwedge_{i \neq j} p_i \neq p_j \wedge \bigwedge_i v \neq p_i$; and if X is empty, then it follows at once that $|M| \geq k$ and hence that $\mathfrak{M} \models_X \bot \wedge [\exists^{\geq k} x(x = x)]$, as required.

Corollary 7. *For all $k \in \mathbb{N}$, the atoms $|v| = k$, $|v| \neq k$, $|M - v| = k$ and $|M - v| \neq k$ are all definable in $\boldsymbol{FO}(=(\cdot), All_1, \sqcup)$.*

Proof. Observe that

- $\mathfrak{M} \models_X |v| = k$ iff $\mathfrak{M} \models_X \phi_{\leq k}(v) \wedge \phi_{\geq k}(v)$;
- $\mathfrak{M} \models_X |v| \neq k$ iff $\mathfrak{M} \models_X \phi_{\leq k-1}(v) \sqcup \phi_{\geq k+1}(v)$;
- $\mathfrak{M} \models_X |M - v| = k$ iff $\mathfrak{M} \models_X \psi_{\leq k}(v) \wedge \psi_{\geq k}(v)$;
- $\mathfrak{M} \models_X |M - v| \neq k$ iff $\mathfrak{M} \models_X \psi_{\leq k-1}(v) \sqcup \psi_{\geq k+1}(v)$

where we let $\phi_{\leq -1} = \psi_{\leq -1} = \bot$.

Putting everything together, we have that

Theorem 9. *Every unary first-order dependency is definable in $\boldsymbol{FO}(=(\cdot), All_1, \sqcup)$.*

Proof. Let $\mathbf{D}$ be a unary first-order dependency and let v be a first-order variable. By definition, $\mathfrak{M} \models_X \mathbf{D} v$ if and only if $(M, X(v)) \models \mathbf{D}^*(P)$, where $\mathbf{D}^*(P)$ is a first-order formula in the vocabulary $\{P\}$ (P unary). But then $\mathbf{D}^*(P)$ is equivalent to a Boolean combination of sentences of the form $\exists^{=k} x P x$ and $\exists^{=k} x \neg P k$; and thus, we may assume that $\mathbf{D}^*(P)$ is of the form $\bigvee_i \bigwedge_j \theta_{ij}$, where each θ_{ij} is $\exists^{=k} x P x$, $\exists^{=k} x \neg P x$, or a negation of a formula of this kind. But then $\mathbf{D} v$ is logically equivalent to

$$\bigsqcup_i \bigwedge_j \theta'_{ij},$$

where

- If θ_{ij} is $\exists^{=k} x P x$, θ'_{ij} is $|v| = k$;
- If θ_{ij} is $\neg\exists^{=k} x P x$, θ'_{ij} is $|v| \neq k$;
- If θ_{ij} is $\exists^{=k} x \neg P x$, θ'_{ij} is $|M - v| = k$;
- If θ_{ij} is $\neg\exists^{=k} x \neg P x$, θ'_{ij} is $|M - v| \neq k$.

Finally, we need to show that every sentence of $\mathbf{FO}(=(\cdot), [\cdot], All_1, \sqcup)$ is equivalent to some first-order sentence. But this is straightforward:

Theorem 10. *Let $\phi \in \mathbf{FO}(= (\cdot), All_1, \sqcup, [\cdot])$ be a sentence. Then ϕ is logically equivalent (on the level of sentences) to some first-order sentence.*

Proof. By Proposition 5, ϕ is equivalent to some sentence of the form $\sqcup_i \psi_i$, for $\psi_i \in \mathbf{FO}(=(\cdot), All_1, [\cdot])$. Observe further that all expressions $[\theta]$ which occur in our formulas are such that θ is a first-order sentence over the empty vocabulary; and therefore, these expressions are trivially upwards-closed first-order dependencies, since for any fixed model they either hold in all teams or in none of them.[3] Then by Theorem 2 and Proposition 9 every such sentence is equivalent (on the level of sentences) to some first-order sentence ψ'_i and thus ϕ is equivalent to $\bigvee_i \psi'_i$.

Putting everything together, we have that

Corollary 8. *Let $\mathbf{D}$ be a unary first-order dependency. Then it is strongly first-order and definable in $\mathbf{FO}(=(\cdot), [\cdot], All_1, \sqcup)$.*

We conclude this section by mentioning an open problem.

Question: Let $k > 1$. Are there any strongly first-order k-ary dependencies which are not definable in $\mathbf{FO}(=(\cdot), [\cdot], All_k, \sqcup)$?

7   Conclusion

Much of the team semantics research has so far focused on formalisms which are greatly more expressive than first-order logic. However, the study of weaker extensions of first-order logic, which do not rise above it insofar as the definability of classes of models is concerned, promises to be also of significant value: not only this investigation offers an opportunity of examining the nature of the boundary between first- and second-order logic, but it also provides us with (comparatively) computationally "safe" classes of dependencies and operators to use in applications.

This work builds on the results of [8] and can only be an initial attempt of making sense of the wealth of these "weak" extensions of first-order logic with team semantics. Much of course remains to be done; but a few distinctive characteristics of this line of investigation may be gleaned already.

---

[3]On the other hand, if θ were a first-order sentence over the nonempty vocabulary, then it would not be a dependency.

- The totality atoms All_k seem to have a role of particular relevance in the theory of strongly first-order dependencies. It remains to be seen whether this role will be preserved by the further developments of the theory; but in any case, the fact that these atoms are the "maximally unbounded" (in the sense of Definition 13) ones for their arities is certainly suggestive, as is the existence of a strict definability hierarchy based on their arities and the fact that all monadic first-order dependencies are definable in terms of the All_1 atom.
- The logic $\mathbf{FO}(\sim) = \mathbf{FO}(NE, \sqcup)$, as the simplest extension of first-order logic with team semantics which is closed under contradictory negation, is also an item of particular interest. As we saw, it suffices to add to it comparatively harmless dependencies such as constancy atoms to obtain the full expressive power of second-order logic; thus, despite its simplicity, this logics appears to be a natural "stopping point" in the family of dependency-based extensions of first-order logic, deserving of a more in-depth study of its properties.
- When working with classes of strongly first-order dependencies, different choices of connectives and operators emerge to the foreground. In particular, the role of the classical disjunction $\phi \sqcup \psi$ in the study of dependence logic and its extensions has been relatively marginal so far; but nonetheless, this connective proved itself of fundamental importance for many of the results of this work. More in general, it appears now that a fully satisfactory account of dependencies and definability cannot be developed if not by integrating it with a general theory of *operators* and *uniform definability* in team semantics. The work of [4, 5, 17] on generalized quantifiers in team semantics seems to be the most natural starting point for such an enterprise; in particular, it would be worthwhile to be able to characterize general families of dependencies *and operators* which do not increase the expressive power of first-order logic (wrt sentences).

Acknowledgements This research was supported by the Deutsche Forschungsgemeinschaft (project number DI 561/6-1). The author thanks an anonymous reviewer for a number of useful corrections and suggestions.

References

1. Abramsky, S., Väänänen, J.: From IF to BI. Synthese **167**, 207–230 (2009). 10.1007/s11229-008-9415-6
2. Abramsky, S., Väänänen, J.: Dependence logic, social choice and quantum physics (2013, in preparation)
3. Durand, A., Kontinen, J.: Hierarchies in dependence logic. CoRR **abs/1105.3324** (2011)
4. Engström, F.: Generalized quantifiers in dependence logic. J. Log. Lang. Inf. **21**(3), 299–324 (2012). doi:10.1007/s10849-012-9162-4
5. Engström, F., Kontinen, J.: Characterizing quantifier extensions of dependence logic. J. Symb. Log. **78**(01), 307–316 (2013)
6. Galliani, P.: Inclusion and exclusion dependencies in team semantics: on some logics of imperfect information. Ann. Pure Appl. Log. **163**(1), 68–84 (2012). doi:10.1016/j.apal.2011.08.005

7. Galliani, P.: The dynamics of imperfect information. Ph.D. thesis, University of Amsterdam (2012). http://dare.uva.nl/record/425951
8. Galliani, P.: Upwards closed dependencies in team semantics. In: Puppis, G., Villa, T. (eds.) Proceedings Fourth International Symposium on Games, Automata, Logics and Formal Verification. EPTCS, vol. 119, pp. 93–106 (2013). doi:http://dx.doi.org/10.4204/EPTCS.119
9. Galliani, P., Hella, L.: Inclusion logic and fixed point logic. In: Rocca, S.R.D. (ed.) Computer Science Logic 2013 (CSL 2013). Leibniz International Proceedings in Informatics (LIPIcs), vol. 23, pp. 281–295. Schloss Dagstuhl–Leibniz-Zentrum fuer Informatik, Dagstuhl, Germany (2013). doi:http://dx.doi.org/10.4230/LIPIcs.CSL.2013.281. http://drops.dagstuhl.de/opus/volltexte/2013/4203
10. Galliani, P.: The doxastic interpretation of team semantics. In: Logic Without Borders: Essays on Set Theory, Model Theory, Philosophical Logic and Philosophy of Mathematics, vol. 5, p. 167. de Gruyter, New York (2015)
11. Galliani, P., Hannula, M., Kontinen, J.: Hierarchies in independence logic. In: Rocca, S.R.D. (ed.) Computer Science Logic 2013 (CSL 2013). Leibniz International Proceedings in Informatics (LIPIcs), vol. 23, pp. 263–280. Schloss Dagstuhl–Leibniz-Zentrum fuer Informatik, Dagstuhl, Germany (2013). doi:http://dx.doi.org/10.4230/LIPIcs.CSL.2013.263. http://drops.dagstuhl.de/opus/volltexte/2013/4202
12. Grädel, E., Väänänen, J.: Dependence and independence. Stud. Logica **101**(2), 399–410 (2013). doi:10.1007/s11225-013-9479-2
13. Hannula, M.: Hierarchies in inclusion logic with lax semantics. In: Logic and Its Applications, pp. 100–118. Springer, Berlin (2015)
14. Hodges, W.: Compositional semantics for a language of imperfect information. J. Interest Group Pure Appl. Log. **5**(4), 539–563 (1997). doi:10.1093/jigpal/5.4.539
15. Kontinen, J., Nurmi, V.: Team logic and second-order logic. In: Ono, H., Kanazawa, M., de Queiroz, R. (eds.) Logic, Language, Information and Computation. Lecture Notes in Computer Science, vol. 5514, pp. 230–241. Springer, Berlin/Heidelberg (2009). doi:10.1007/978-3-642-02261-6_19
16. Kontinen, J., Link, S., Väänänen, J.: Independence in database relations. In: Logic, Language, Information, and Computation, pp. 179–193. Springer, Berlin (2013)
17. Kuusisto, A.: A double team semantics for generalized quantifiers. J. Logic Lang. Inf. **24**(2), 149–191 (2015)
18. Väänänen, J.: Dependence Logic. Cambridge University Press, Cambridge (2007). doi:10.1017/CBO9780511611193
19. Väänänen, J.: Team logic. In: van Benthem, J., Gabbay, D., Löwe, B. (eds.) Interactive Logic. Selected Papers from the 7th Augustus de Morgan Workshop, pp. 281–302. Amsterdam University Press, Amsterdam (2007)

Games for Inclusion Logic and Fixed-Point Logic

Erich Grädel

Abstract One of the most intriguing results on logics of dependence and independence is the tight connection between inclusion logic and the least fixed-point logic LFP. Here we re-examine this connection from a game-theoretic point of view. We study the model-checking games for inclusion logic and for posGFP, the fragment of LFP that uses only (non-negated) greatest fixed points. We show that the evaluation problems for both logics can be represented by a special kind of trap condition in safety games. We then study interpretation arguments for games. In combination with our study of traps for inclusion logic and posGFP, game interpretations will give us a model-theoretic construction of translations between the two logics.

1 Introduction

Modern logics of dependence and independence come with a semantics that, unlike Tarski semantics, is not based on single assignments (mapping variables to elements of a structure) but on sets of such assignments. Sets of assignments with a common domain of variables are called teams. Team semantics was originally introduced by Hodges [10, 11] as a compositional, model-theoretic semantics for the independence-friendly logic IF. In 2007, Väänänen [14] proposed a new approach to logics of dependence and independence. Rather than stating dependencies or independencies as annotations of quantifiers, he suggested to express dependencies as atomic formulae, of the form $=(x_1, \ldots, x_m, y)$, saying that the variable y is functionally dependent on (i.e. completely determined by) the variables $x_1, \ldots, x_m$. Dependence logic is first-order logic together with such dependency atoms. Notice that such dependency statements do not even make sense on a single assignment, but only on larger collection of data, given either by sets of assignments, i.e., teams, or by a table or relation. Besides the functional dependency atoms proposed by Väänänen, there are many other atomic dependence properties that give rise to interesting logics based on team semantics. In [8] we have discussed the notion

E. Grädel (✉)
Math. Foundations of Computer Science, RWTH Aachen University, D-52056 Aachen, Germany
e-mail: graedel@logic.rwth-aachen.de

© Springer International Publishing Switzerland 2016
S. Abramsky et al. (eds.), *Dependence Logic*, DOI 10.1007/978-3-319-31803-5_5

of independence (which is a much more delicate but also more powerful notion than dependence) and introduced independence logics, and Galliani [5] and Engström [4] have studied several logics with team properties based on notions originating in database dependency theory.

Most of the logics of dependence and independence studied so far, including more traditional formalisms such as first-order logic with Henkin quantifiers and IF-logic, have an expressive power that is, at least for sentences, equivalent to the one of existential second-order logic [2, 5, 12–14], and it is rather easy to formalize NP-complete properties of, say, finite graphs in these logics. However, one of the most surprising results on logics with team semantics is the tight connection, established by Galliani and Hella [6], between inclusion logic and the least fixed-point logic LFP. Inclusion logic extends first-order logic (with team semantics) by atomic inclusion dependencies of the form $(\bar{x} \subseteq \bar{y})$, which are true in a team X if every value for $\bar{x}$ in X also occurs as a value for $\bar{y}$ in X. Inclusion logic has been introduced and studied in [5].

The least fixed-point logic LFP, on the other side, is a logic with classical semantics in the sense of Tarski, which extends first-order logic by least and greatest fixed points of definable relational operators. The logic LFP is of fundamental importance in finite model theory and descriptive complexity, for the study of inductive definability, and for the study of logic and games. Fragments of LFP, such as Datalog or the modal μ-calculus are very important in many areas of computer science, including databases, knowledge representation, and verification. See, for instance, [9] for background on least fixed-point logic.

Galliani and Hella showed, by a direct translation via structural induction, that sentences of inclusion logic have the same expressive power as sentences from the fragment of LFP that uses only (non-negated) greatest fixed points, denoted posGFP. It is known that, on finite structures, the full logic LFP collapses to its posGFP-fragment. Hence every property of finite structures that is LFP-definable is also definable in inclusion logic, and vice versa. It follows by the Immerman-Vardi-Theorem that, on *ordered* finite structures, inclusion logic captures polynomial time. For formulae with free variables, the connection between posGFP and inclusion logic is more complicated, due to the different semantics of the two logics. We will discuss this issue in detail in Section 7.

In this article, we re-examine the connection between LFP and inclusion logic from a game-theoretic point of view. We study the model-checking games for inclusion logic and for the posGFP-fragment of least fixed-point logic and use interpretation arguments for games to translate between the two logics.

It is well known that the appropriate games for greatest fixed-point formulae are *safety games*, i.e., games with potentially infinite plays, where Player 0 has just the objective to keep the play inside a safe region or, equivalently, to avoid a given set of losing positions. In our case the positions to avoid are the literals that evaluate to false.

Model-checking games for logics with team semantics are *a priori* quite different. A uniform construction of such games has been presented in [7] in terms of *second-order reachability games* played on trees or forests. Whereas in classical reachability (or safety) games, the winning condition is specified by a set of positions that should be reached (or avoided), a second-order reachability condition is given by a *collection of sets of terminal positions*. Furthermore, a second-order reachability condition does not apply to single plays, but to strategies, and considers the set of all plays that are compatible with the strategy. To be winning, a strategy has to ensure that the set of all terminal positions that are reachable by a play following the strategy forms a winning set. We have shown in [7] that for any logic with team semantics, satisfying some natural basic conditions, the associated model-checking problem can be captured by appropriate second-order reachability games. In particular, this is the case for inclusion logic.

Although second-order reachability games are quite different from safety games, and in general algorithmically more complicated, we shall prove that games played on forests, with a second-order reachability condition of a special form given by a universal-existential statement, can be translated into equivalent safety games, on a transformed game graph that is no longer acyclic. This applies in particular to games for inclusion logic and thus provides safety games for this logic, as an alternative to the second-order reachability games obtained by the generic construction. Further we introduce *I-traps*, a special notion of traps for initial positions in safety games, and prove that this notion faithfully captures evaluation problems with respect to *teams*, for formulae of inclusion logic and also for posGFP-formulae of a special form. On the other side, such traps are definable in both logics in a quite simple way.

We shall then study game interpretations. It is a general observation that the model-checking games for a given formula (from almost any reasonable logic) are uniformly interpretable inside the structure on which the formula is evaluated. In combination with our study of traps for inclusion logic and posGFP, interpretations will give us translations between the two logics. The argument roughly is the following. Given any formula ψ in, say, inclusion logic, we consider the interpretation $J(\psi)$ which, for any structure $\mathfrak{A}$, interprets the safety game for $\mathfrak{A}$ and ψ, denoted $\mathcal{G}_{\mathrm{safe}}(\mathfrak{A}, \psi)$, inside $\mathfrak{A}$. This game has the properties that the teams X that satisfy ψ in $\mathfrak{A}$ coincide with the I-traps in $\mathcal{G}_{\mathrm{safe}}(\mathfrak{A}, \psi)$. On the other side, I-traps in safety games are definable by a formula itrap of the target logic, in this case posGFP. The interpretation $J(\psi)$ maps this formula to another formula $\mathrm{itrap}^{J(\psi)}$, which is also in posGFP, and which essentially expresses in $\mathfrak{A}$ what itrap expresses in the game. From $\mathrm{itrap}^{J(\psi)}$ we then easily get a posGFP-formula that is equivalent to ψ. An analogous translation works in the other direction. We thus obtain a high-level model-theoretic technique for obtaining translations between the two logics, without the need to go through cumbersome structural inductions on the syntax of the formulae.

2 Safety games and traps

There are many models of path-forming games played on graphs. Here we work with a model of turn-based games with two players, called Player 0 and Player 1, that makes explicit not only the sets of positions associated to the two players, but also the initial and terminal positions.

A game graph is a structure $\mathscr{G} = (V, V_0, V_1, T, I, E)$, where $V = V_0 \cup V_1 \cup T$ is the set of positions, partitioned into the sets V_0, V_1 of the two players and the set T of terminal positions, where I is the set of initial positions, and where $E \subseteq V \times V$ is the set of moves. We denote the set of immediate successors of a position v by $vE := \{w : (v, w) \in E\}$ and require that $vE = \emptyset$ if, and only if, $v \in T$. A play from an initial position v_0 is a finite or infinite path $v_0 v_1 v_2 \ldots$ through $\mathscr{G}$ where the successor $v_{i+1} \in v_i E$ is chosen by Player 0 if $v_i \in V_0$ and by Player 1 if $v_i \in V_1$. A play ends when it reaches a terminal node $v_m \in T$. A subgraph of a graph (V, E) is a pair (W, F) with $W \subseteq V$ and $F \subseteq E \cap (W \times W)$.

Definition 1. A *(nondeterministic) strategy* of Player σ in such a game $\mathscr{G}$ is a subgraph $\mathscr{S} = (W, F) \subseteq (V, E)$ satisfying the following conditions:

(1) If $v \in W \cap V_\sigma$, then vF is non-empty.
(2) If $v \in W \cap V_{1-\sigma}$, then $vF = vE$.

Here W is the region of $\mathscr{G}$ on which the strategy is defined, and F is the set of moves that are admitted by the strategy. A strategy $\mathscr{S} = (W, F)$ for Player σ is *deterministic* if $|vF| = 1$ for all $v \in W \cap V_\sigma$. A strategy $\mathscr{S}$ induces the set of those plays from the initial positions in $I \cap W$ whose moves are consistent with F. We call $\mathscr{S}$ well-founded if it does not admit any infinite plays; this is always the case on finite acyclic game graphs, but need not be the case otherwise. We are interested in *winning strategies* according to different winning conditions. Here, we shall mainly consider classical (first-order) safety and reachability conditions, and the second-order reachability conditions introduced in [7].

A *safety condition* for Player 0 is given by a set $L \subseteq V$ of 'losing' positions that Player 0 has to avoid, or dually, by its complement $S = V \setminus L$, the region of safe positions inside of which Player 0 has to keep the play. For convenience in game constructions we do not require that losing positions are terminal (but we could do so since deleting all outgoing edges from losing positions does not change anything relevant in the game). A play in a safety game is won by Player 0 if she can guarantee that the play never reaches a position $v \in L$. If Player 0 can, moreover, ensure, that the play reaches, after a finite number of steps a terminal position $v \in T \setminus L$, then she also wins the associated reachability game. The difference between reachability and safety conditions is relevant only in cases where infinite plays are possible. In first-order games, a winning strategy for a player is a strategy that guarantees that all plays consistent with it are won by that player. For a safety game this amounts to the following:

Definition 2. For a safety game $\mathscr{G}$, with safety condition $S \subseteq V$ and a set $X \subseteq I$ of initial positions, a strategy $\mathscr{S} = (W, F)$ for Player 0 is winning from X if $X \subseteq W \subseteq S$.

Traps. Two notions of fundamental importance for the algorithmic analysis of graph games are *attractors* and *traps*. Intuitively the attractor (for Player 0) of a set $Y \subseteq V$ is the set of all positions from which Player σ has a strategy to ensure that the play reaches Y in a finite number of steps. The dual notion of a *trap* (for Player 1) encompasses those sets $Z \subseteq V$ for which Player 0 has a strategy to guarantee that every play starting at a position in Z remains inside Z.

Notice that in a game $\mathscr{G}$ with $I = V$ and a safety condition $S \subseteq V$ for Player 0, the winning region for Player 0 (i.e. the set of those positions from which she has a winning strategy) is precisely the maximal trap $Z \subseteq S$. For our analysis of games for inclusion logic and fixed-point logic, a specific variant of a trap will be relevant, which we call an I-trap.

Definition 3. For a game $\mathscr{G}$ with a set I of initial positions and a safety condition S for Player 0, an I-*trap* is a set $X \subseteq I$ of initial positions such that Player 0 has a strategy to ensure that every play starting in X remains inside S and avoids $I \setminus X$.

To put it differently, X is an I-trap in $(\mathscr{G}, S)$ if, and only if, Player 0 has a winning strategy from all positions in X for the safety game on $\mathscr{G}$ with losing positions $(V \setminus S) \cup (I \setminus X)$. Notice that an I-trap X is not a trap in $\mathscr{G}$, but it can be viewed as a kind of trap restricted to the set I, in the sense that Player 0 ensures that starting from X the only positions in I that are ever met in the play are those in X.

Clearly, the empty set is a trivial I-trap. Further, a union of I-traps is again an I-trap, so there is a uniquely defined maximal I-trap for every safety game $(\mathscr{G}, S)$.

It is well known that winning regions and winning strategies for reachability and safety games are computable in linear time in the size of the game graph (see, e.g., [1, Chapt. 4]. Further one can, without loss of generality, restrict attention to deterministic strategies.

3 Second-order reachability games

While reachability and safety games are sufficient for many important applications, and in particular for evaluation games of first-order logic and the posGFP and posLFP-fragments of the least fixed-point logic LFP, they are, in general, not adequate for more complicated logics, such as full LFP and logics with team semantics. Model-checking games for the latter can be defined in terms of second-order reachability games.

Definition 4. A *second-order reachability condition* is a collection $\text{Win} \subseteq \mathscr{P}(T)$ defining for each set $U \subseteq T$ of terminal positions whether it is a winning set for Player 0. A *consistent winning strategy* from $X \subseteq I$ for Player 0 for a second-order reachability game $\mathscr{G} = (V, V_0, V_1, T, I, E)$ with winning condition Win is a strategy $\mathscr{S} = (W, F)$ such that

(1) W is the set of nodes that are reachable from X via edges in F.
(2) $W \cap T \in \text{Win}$.

Remark. The condition that W contains only nodes that are reachable from X by edges in the strategy is not needed for winning conditions that are downwards closed, and in particular for classical safety games. However, if Win is not down-wards closed, then this condition is necessary to avoid the inclusion of unreachable nodes which could change a losing set of terminal nodes to a winning one.

As shown in [7], the problem whether a second-order reachability game, given by a finite game graph $\mathcal{G}$ with an oracle for Win, admits a consistent winning strategy for Player 0, is NP-complete. However, there are special cases of second-order reachability conditions for which the associated reachability games are efficiently solvable.

Universal-existential reachability conditions and the translation to safety games

Definition 5. We call a second-order reachability condition Win $\subseteq$ $\mathscr{P}(T)$ *universal-existential* if there exists a relation $R \subseteq T \times T$ such that

$$\text{Win} = \{U \subseteq T : (\forall x \in U)(\exists y \in U)(x, y) \in R\}.$$

We shall see below that, for instance, the model-checking games for inclusion logic are second-order reachability games with universal-existential winning condi-tions. Further, the game graphs of such model-checking games are forests. Let now $\mathcal{G} = (V, V_0, V_1, T, I, E)$ be a second-order reachability game, played on a forest, where I is the set of roots of the forest, with a universal-existential winning condition Win $= \{U \subseteq T : (\forall x \in U)(\exists y \in U)(x, y) \in R\}$.

We want to associate with $(\mathcal{G}, \text{Win})$ a safety game $\mathcal{G}_{\text{safe}}$ such that winning strategies for $(\mathcal{G}, \text{Win})$ from $X \subseteq I$ correspond to winning strategies for $\mathcal{G}_{\text{safe}}$ that ensure that X is an I-trap. The idea is to add to $\mathcal{G}$ moves of Player 0 for pairs $(s, t) \in R$ and moves of Player 1 taking the play back from t to ancestors in the forest. The nodes that Player 0 has to avoid in the safety game $\mathcal{G}_{\text{safe}}$ are the nodes $s \in T$ without outgoing R-edges and the roots in $I \setminus X$.

To make this precise, we duplicate the nodes of $\mathcal{G}$, i.e., we add to $\mathcal{G}$ the set of vertices $V^* := \{v^* : v \in V\}$. In $\mathcal{G}_{\text{safe}}$, we consider nodes $s \in T$ with $sR \neq \emptyset$ as positions of Player 0 and nodes $v^* \in V^*$ as positions of Player 1. We add moves so that Player 0 can move from s to t^* for any $(s, t) \in R$, and Player 1 can move from $v^* \in V^*$ either to v, or to the unique node u^* such that $(u, v) \in E$. We obtain the game graph

$$\mathcal{G}_{\text{safe}} = (V \cup V^*, V_0 \cup \{s \in T : sR \neq \emptyset\}, V_1 \cup V^*, \tilde{T}, I, \tilde{E})$$

with $\tilde{T} := \{s \in T : sR = \emptyset\}$ and

$$\tilde{E} := E \cup \{(s, t^*) : (s, t) \in R\} \cup \{(v^*, u^*) : (u, v) \in E\} \cup \{(v^*, v), v \in V\}.$$

We obtain a safety game $\mathcal{G}_{\text{safe}}$ where Player 0 has to avoid the positions in $\tilde{T}$.

Proposition 6. *Player 0 has a consistent winning strategy for the second-order reachability game $(\mathcal{G}, \text{Win})$ from $X \subseteq I$ if, and only if, X is an I-trap in $\mathcal{G}_{safe}$.*

Proof. Let $\mathcal{S} = (W, F)$ be a consistent winning strategy for Player 0 from X for the game $(\mathcal{G}, \text{Win})$. For $U := W \cap T$ it follows that $U \in \text{Win}$ and hence that $(\forall x \in U)(\exists y \in U)(x, y) \in R$.

We transform $\mathcal{S}$ into a strategy $\tilde{\mathcal{S}} = (\tilde{W}, \tilde{F})$ for $\mathcal{G}_{safe}$ with

$$\tilde{W} := W \cup \{v^* : v \in W\}$$

$$\tilde{F} := F \cup \{(s, t^*) : s, t \in U, (s, t) \in R\} \cup \{(v^*, v) : v \in W\} \cup \{(v^*, u^*) : (u, v) \in F\}.$$

We claim that $\tilde{\mathcal{S}}$ is a winning strategy for $\mathcal{G}_{safe}$ which moreover shows that X is an I-trap. We have to prove that $\tilde{\mathcal{S}}$ is indeed a strategy according to Definition 1 and additionally avoids the positions in $\tilde{T} \cup (I \setminus X)$. Clearly, $\tilde{F} \subseteq \tilde{E} \cap \tilde{W} \times \tilde{W}$ and $X \subseteq W \subseteq \tilde{W}$.

It remains to verify the following conditions:

(1) If $w \in \tilde{W}$ is a node of Player 0, then $w\tilde{F}$ is non-empty.
    For $w \in W \cap V_0$ this is clear since $wF \neq \emptyset$. Otherwise $w = s$ is a node in $W \cap T = U$ so there exists a node $t \in U$ with $(s, t) \in R$ and hence an edge $(s, t^*) \in \tilde{F}$.
(2) If $w \in \tilde{W}$ is a node of Player 1, then $w\tilde{F} = w\tilde{E}$.
    For $w \in W \cap V_1$ we have that $w\tilde{F} = wF = wE = w\tilde{E}$. Otherwise $w = v^*$ for some $v \in W$. Then $w\tilde{E}$ consists of v and, unless v is a root, of the unique node u^* such that $(u, v) \in E$. Since $v \in W$ is reachable from its root by F-edges, also $u \in W$ and $(u, v) \in F$. It follows that edges (v^*, v) and (v^*, u^*) also belong to $\tilde{F}$. Hence in all cases $w\tilde{F} = w\tilde{E}$.
(3) $\tilde{W} \cap (\tilde{T} \cup (I \setminus X)) = \emptyset$.
    Assume that there is a node $w \in \tilde{W} \cap \tilde{T}$. Then $w \in W \cap T = U$ which implies that there exists a node $t \in U$ with $(w, t) \in R$. But then $w \notin \tilde{T}$, contradicting our assumption. Finally, if $w \in W$, then $w \notin I \setminus X$ because all nodes in W are reachable from X by F-edges.

For the converse, consider any winning strategy $\tilde{\mathcal{S}} = (\tilde{W}, \tilde{F})$ from X for Player 0 in the game $\mathcal{G}_{safe}$ which avoids $\tilde{T}$ and $I \setminus X$. Let $\mathcal{S} = (W, F)$ be the strategy for $\mathcal{G}$, with $W = \tilde{W} \cap V$ and $F = (W \times W) \cap E$.

Clearly $\tilde{W}$ cannot contain any position v^* such that v belongs to a tree whose root is in $I \setminus X$ because from such a position, Player 1 can move upwards to that root and win. Hence W only contains nodes that are reachable from a root in X and therefore $\mathcal{S}$ is a strategy from X. To see that $\mathcal{S}$ is winning for the second-order reachability game $\mathcal{G}$, consider the set $U := W \cap T$ of terminal nodes in W. Since $\tilde{\mathcal{S}}$ is winning for the safety game, no node in U is terminal in $\mathcal{G}_{safe}$. Hence $(\forall x \in U)(\exists y \in U)(x, y) \in R$ which means that $U \in \text{Win}$.   □

4 Logics and Their Games

4.1 First-order logic

We assume familiarity with first-order logic (FO) and briefly recall the construction of model-checking games for FO. We consider formulae with relational vocabulary $\tau = \{R_1, \ldots, R_m\}$ and assume that they are presented in negation normal form, i.e., built from literals (atomic formulae and their negations) by means of the propositional connectives $\vee$ and $\wedge$ and quantifiers $\exists$ and $\forall$.

For any such formula $\psi(\bar{x})$, let $\mathcal{T}(\psi)$ be its syntax tree whose nodes are the *occurrences* of the subformulae of ψ, with edges leading from any formula to its immediate subformulae, i.e., from $\varphi \vee \vartheta$ and $\varphi \wedge \vartheta$ to both φ and ϑ and from $\exists y \varphi$ and $\forall y \varphi$ to φ. The leaves of the tree are the nodes associated to literals.

For a formula $\psi(\bar{x})$ and a τ-structure $\mathfrak{A} = (A, R_1, \ldots, R_m)$, the model-checking game $\mathcal{G}(\mathfrak{A}, \psi)$ is obtained by taking an appropriate product of $\mathcal{T}(\psi)$ with the set of assignments mapping variables to elements of $\mathfrak{A}$. More precisely, the positions of the game are the pairs (φ, s) consisting of a node $\varphi \in \mathcal{T}(\psi)$ and an assignment $s :$ free$(\varphi) \to A$. Verifier (Player 0) moves from positions associated with disjunctions and with formulae starting with an existential quantifier. From a position $(\varphi \vee \vartheta, s)$, she moves to either (φ, s') or (ϑ, s'') where s', s'' are the restrictions of s to the free variables of φ and ϑ, respectively. From a position $(\exists y \varphi, s)$, Verifier can move to any position $(\varphi, s[y \mapsto a])$, where a is an arbitrary element of A. Dually, Falsifier (Player 1) makes corresponding moves for conjunctions and universal quantifications. If φ is a literal, then the positions (φ, s) are terminal. The terminal positions are partitioned into the target sets T_0, T_1 of the two players, with $T_0 = \{(\varphi, s) \in T : \mathfrak{A} \models_s \varphi\}$ and $T_1 = \{(\varphi, s) \in T : \mathfrak{A} \models_s \neg\varphi\}$. Notice that since model-checking games for first-order logic are played on forests, it does not matter whether we consider them as reachability games, where Player σ has the objective to reach T_σ, or as safety games, where Player σ seeks to avoid $T_{1-\sigma}$.

4.2 Least fixed-point logic

Least fixed-point logic, denoted LFP, extends first-order logic by least and greatest fixed points of definable relational operators. We will briefly recall some basic definitions here. For a more extensive introduction to LFP, we refer to [9].

Every formula $\psi(R, \bar{x})$, where R is a relation symbol of arity k and $\bar{x}$ is a tuple of k variables, defines, for any structure $\mathfrak{A}$ of matching vocabulary, an update operator $F_\psi^{\mathfrak{A}} : \mathcal{P}(A^k) \to \mathcal{P}(A^k)$ on the class of k-ary relations over the universe A of $\mathfrak{A}$, namely $F_\psi^{\mathfrak{A}} : R \mapsto \{\bar{a} : (\mathfrak{A}, R) \models \psi(R, \bar{a})\}$. If all occurrences of R in ψ are positive, then this operator is monotone in the sense that $R \subseteq R'$ implies $F_\psi^{\mathfrak{A}}(R) \subseteq F_\psi^{\mathfrak{A}}(R')$. It is well known that every monotone operator F has a least fixed-point **lfp**(F) and a greatest fixed-point **gfp**(F), with

$$\mathbf{lfp}(F) = \bigcap \{X : F(X) = X\} = \bigcap \{X : F(X) \subseteq X\}$$

$$\mathbf{gfp}(F) = \bigcup \{X : F(X) = X\} = \bigcup \{X : F(X) \supseteq X\},$$

which can also be constructed by transfinite induction.

LFP is defined by adding to the syntax of first order logic the following *fixed-point formation rule:* If $\psi(R, \bar{x})$ is a formula of vocabulary $\tau \cup \{R\}$, in which the relational variable R occurs only positively, and if $\bar{x}$ is a tuple of variables such that the length of $\bar{x}$ matches the arity of R, then $[\mathbf{lfp}\, R\bar{x} \,.\, \psi](\bar{x})$ and $[\mathbf{gfp}\, R\bar{x} \,.\, \psi](\bar{x})$ are also formulae (of vocabulary τ).

The semantics of least fixed-point formulae in a structure $\mathfrak{A}$, providing interpretations for all free variables in the formula, is the following: $\mathfrak{A} \models [\mathbf{lfp}\, R\bar{x} \,.\, \psi](\bar{a})$ if $\bar{a}$ belongs to the least fixed point of the update operator defined by ψ on $\mathfrak{A}$. Similarly for greatest fixed points.

Note that in formulae $[\mathbf{lfp}\, R\bar{x} \,.\, \psi](\bar{x})$ one may allow ψ to have other free variables besides $\bar{x}$; these are called parameters of the fixed-point formula. However, at the expense of increasing the arity of the fixed-point predicates and the number of variables one can always eliminate parameters. For the construction of model-checking games it is convenient to assume that formulae are parameter-free.

The duality between least and greatest fixed point implies that for any ψ,

$$[\mathbf{gfp}\, R\bar{x} \,.\, \psi](\bar{x}) \equiv \neg[\mathbf{lfp}\, R\bar{x} \,.\, \neg\psi[R/\neg R]](\bar{x}).$$

Using this duality together with de Morgan's laws, every LFP-formula can be brought into *negation normal form*, where negation applies to atoms only.

Example 1 (Definability in safety games). Winning regions of reachability and safety games are definable by LFP-formulae of rather simple form. On game graphs $\mathscr{G}$ where the objective of Player 0 is to keep the play inside a given safe region $S \subseteq V$, and Player 1 wants to reach the set $L = V \setminus S$, the winning regions of the two players are uniformly definable by

$$\mathrm{win}(x) := [\mathbf{gfp}\, Wx \,.\, Sx \wedge (V_0 x \to \exists y(Exy \wedge Wy)) \wedge (V_1 x \to \forall y(Exy \to Wy))](x)$$

$$\mathrm{lose}(x) := [\mathbf{lfp}\, Wx \,.\, Lx \vee (V_1 x \wedge \exists y(Exy \wedge Wy)) \vee (V_0 x \wedge \forall y(Exy \to Wy))](x)$$

A simple modification of this construction gives a definition of I-traps in safety games. Let

$$\mathrm{itrap}(X, x) := [\mathbf{gfp}\, Yx \,.\, Sx \wedge (Ix \to Xx) \wedge (V_0 x \to \exists y(Exy \wedge Yy)) \wedge$$

$$(V_1 x \to \forall y(Exy \to Yy))](x).$$

Then, for every safety game $\mathscr{G}$ and every set $X \subseteq I$ of initial positions we have that

$$(\mathscr{G}, X) \models \forall x(Xx \to \mathrm{itrap}(X, x)) \quad \Leftrightarrow \quad X \text{ is an } I\text{-trap in } \mathscr{G}.$$

The model-checking games for general LFP-formulae are *parity games*. These are games of possibly infinite duration, where each position is assigned a natural number, called its priority, and an infinite play is won by Player 0 if the least priority seen infinitely often in the play is even.

Let ψ be an LFP-formula, which is assumed to be parameter-free, in negation normal form and in which distinct occurrences of fixed-point operators use distinct fixed-point variables. To construct the parity game $\mathscr{G}(\mathfrak{A}, \psi)$, one extends the construction of the first-order model-checking game as follows: For every subformula of ψ of form $\vartheta := [\mathbf{fp}\, R\overline{x} \,.\, \varphi(R, \overline{x})](\overline{x})$ (where $\mathbf{fp}$ is either $\mathbf{lfp}$ or $\mathbf{gfp}$) we add moves from positions (ϑ, s) to (φ, s), and from positions $(R\overline{y}, s)$ to (φ, t) for the assignment t with $t(\overline{x}) = s(\overline{y})$. Since these moves are unique it makes no difference to which of the two players we assign the positions (ϑ, s) and $(R\overline{y}, t)$. Priorities are assigned in such a way that positions $(R\overline{y}, s)$, associated with fixed-point variables, get an even priority if R is a $\mathbf{gfp}$-variable, and an odd priority if R is an $\mathbf{lfp}$-variable. Further R gets a smaller (i.e. more significant) priority than R' if R' depends on R, i.e., if R occurs free in the formula defining R'. All other positions, associated with formulae that are not fixed-point atoms, get maximal (the least significant) priority. Thus the number of priorities needed in a parity game for a fixed-point formula ψ coincides with the alternation depth of least and greatest fixed points in ψ. For details, the proof of correctness, and for algorithmic and model-theoretic results based on parity games, see [9, Chapter 3.3].

4.3   The fragment of positive greatest fixed points

We denote by posGFP the fragment of LFP of formulae in negation normal form such that all its fixed-point operators are greatest fixed points. Since all fixed points are of the same kind, the priority assignment is trivial (all positions get priority 0), and the model-checking game $\mathscr{G}(\mathfrak{A}, \psi)$ for a τ-structure $\mathfrak{A}$ and a formula $\psi \in$ posGFP is a safety game. The positions that Player 0 has to avoid are those of the form (α, s) where α is a τ-atom or a negated τ-atom such that $\mathfrak{A} \models_s \neg\alpha$. From positions associated with a $\mathbf{gfp}$-variable R, the play is taken back to the fixed-point formula that defines R, and infinite plays correspond to successful infinite regeneration sequences of greatest fixed points.

Proposition 7. *For every structure $\mathfrak{A}$, every formula $\psi(\overline{x})$ of posGFP and every assignment $s : \text{free}(\psi) \to A$ we have that $\mathfrak{A} \models_s \psi(\overline{x})$ if, and only if, Player 0 has a winning strategy for the safety game $\mathscr{G}(\mathfrak{A}, \psi)$ from the initial position (ψ, s).*

For the relationship between inclusion logic and fixed-point logic, we shall consider sentences in posGFP of vocabulary $\tau \cup \{X\}$ of the form

$$\vartheta := \forall \overline{x}(X\overline{x} \to \varphi(\overline{x}))$$

such that X occurs only positively in φ. The model-checking game $\mathscr{G}((\mathfrak{A}, X), \vartheta)$ is a safety game with initial position $(\vartheta, \emptyset)$ and safety condition

$$S(X) = \{(\eta, s) : \text{if } \eta \text{ is a } \tau\text{-literal, then } \mathfrak{A} \models_s \eta \text{ and}$$

$$\text{if } \eta = X\bar{y} \text{ then } s(\bar{y}) \in X \text{ and}$$

$$\text{if } \eta = \neg X\bar{x} \text{ then } s(\bar{x}) \notin X\}.$$

We modify these model-checking games by eliminating every explicit reference to the relation X and associate the model-checking problem of whether $(\mathfrak{A}, X) \models \vartheta$ with a trap condition for a modified game $\mathscr{G}^*(\mathfrak{A}, \varphi)$. To do this, we identify every position of form $(X\bar{y}, t)$ with the position $(\varphi(\bar{x}), s)$ such that $s(\bar{x}) = t(\bar{y})$; this means that every edge in the game graph to a position $(X\bar{y}, t)$ is replaced by an edge to $(\varphi(\bar{x}), s)$, and the node $(X\bar{y}, t)$ is deleted. The set I of initial positions now consists of all pairs of form $(\varphi(\bar{x}), s)$ and the safety condition is simplified to

$$S^* := \{(\eta, s) : \text{ if } \eta \text{ is a } \tau\text{-literal, then } \mathfrak{A} \models_s \eta\}.$$

Given any interpretation for the relation X, let $X^* \subseteq I$ be the set of positions (φ, s) where $s(\bar{x}) \in X$.

Proposition 8. *The resulting game $\mathscr{G}^*(\mathfrak{A}, \varphi)$ has the property that*

$$(\mathfrak{A}, X) \models \forall \bar{x}(X\bar{x} \to \varphi(\bar{x})) \iff X^* \text{ is an } I\text{-trap in } \mathscr{G}^*(\mathfrak{A}, \varphi).$$

4.4   Logics with team semantics

Let $\mathfrak{A}$ be a structure of vocabulary τ with universe A. An *assignment* (into $\mathfrak{A}$) is a map $s : \mathscr{V} \to A$ whose domain $\mathscr{V}$ is a set of variables. Given such an assignment s, a variable y, and an element $a \in A$ we write $s[y \mapsto a]$ for the assignment with domain $\mathscr{V} \cup \{y\}$ that updates s by mapping y to a. A *team* is a set of assignments with the same domain. For a team X, a variable y, and a function $F : X \to \mathscr{P}(A)$, we write $X[y \mapsto F]$ for the set of all assignments $s[y \mapsto a]$ with $s \in X$ and $a \in F(s)$. Further we write $X[y \mapsto A]$ for the set of all assignments $s[y \mapsto a]$ with $s \in X$ and $a \in A$.

Team semantics, for a logic L, defines whether a formula $\psi \in L$ is satisfied by a team X in a structure $\mathfrak{A}$, written $\mathfrak{A} \models_X \psi$. We always assume formulae to be in negation normal form and require that the domain of X contains all free variables of ψ. Further we shall always make sure that the *locality principle* holds, saying that the meaning of a formula can only depend on the variables actually occurring in it. More precisely, if $Y = X \upharpoonright \text{free}(\psi)$ is the restriction of the team X to the free variables of ψ, then $\mathfrak{A} \models_X \psi$ if, and only if, $\mathfrak{A} \models_Y \psi$. A special case is the empty team which satisfies all formulae: $\mathfrak{A} \models_\emptyset \psi$ for all $\mathfrak{A}$ and all ψ. The locality principle implies that a sentence ψ (i.e. a formula without free variables) is true for a non-empty team X if, and only if, it is true for the team $\{\emptyset\}$ consisting only of the empty assignment. Thus, as it should be and as in logics with Tarski semantics, the truth of a sentence just depends on the structure in which it is evaluated. For sentences we then write $\mathfrak{A} \models \psi$ to denote that $\mathfrak{A} \models_{\{\emptyset\}} \psi$. This allows us to directly compare the expressive power of sentences between logics with team semantics and logics with Tarski semantics. For open formulae, the situation is different and will be discussed later.

For the operators of first-order logic (FO) the semantic rules are the following:

(1) If ψ is an atom $x = y$ or $Rx_1 \ldots x_m$ or the negation of such an atom, then $\mathfrak{A} \models_X \psi$ if, and only if, $\mathfrak{A} \models_s \psi$ (in the sense of Tarski semantics) for all $s \in X$.
(2) $\mathfrak{A} \models_X (\varphi \wedge \vartheta)$ if, and only if, $\mathfrak{A} \models_X \varphi$ and $\mathfrak{A} \models_X \vartheta$.
(3) $\mathfrak{A} \models_X (\varphi \vee \vartheta)$ if, and only if, there exist teams Y, Z with $X = Y \cup Z$ such that $\mathfrak{A} \models_Y \varphi$ and $\mathfrak{A} \models_Z \vartheta$.
(4) $\mathfrak{A} \models_X \forall y\varphi$ if, and only if, $\mathfrak{A} \models_{X[y \mapsto A]} \varphi$.
(5) $\mathfrak{A} \models_X \exists y\varphi$ if, and only if, there is a map $F : X \to (\mathscr{P}(A) \setminus \{\emptyset\})$ such that $\mathfrak{A} \models_{X[y \mapsto F]} \varphi$.

Remark. Clause (5) giving semantics to existential quantifiers might seem surprising at first sight since it permits the choice of an arbitrary non-empty set of witnesses for an existentially quantified variable rather than a single witness (for each $s \in X$). What we use here has been called *lax semantics* in [5], as opposed to the more common *strict semantics*. For disjunctions (clause (3)) there is also a strict variant, requiring that the team X is split into *disjoint* subteams Y and Z. For first-order logic, and also for dependence logic, the difference is immaterial since the two semantics are equivalent. However, this is not the case for other logics of dependence and independence, in particular for independence logic and inclusion logic. In these cases, only the lax semantics is appropriate since it preserves the locality principle whereas the strict semantics violates this principle. In game-theoretic terms the difference between strict and lax semantics corresponds to the difference between deterministic and nondeterministic strategies, and it turns out that model-checking games for inclusion logic and independence logic do not admit deterministic winning strategies.

For first-order logic itself, team semantics does not provide anything new since a first-order formula is true for a team X if, and only if, it is true in the sense of Tarski semantics, for all individual assignments $s \in X$:

$$\mathfrak{A} \models_X \psi \ \Leftrightarrow \ \mathfrak{A} \models_{\{s\}} \psi \ \Leftrightarrow \ \mathfrak{A} \models_s \psi.$$

This changes radically, when atomic statements on teams which express properties of dependence or independence are added to the logic. The most common examples of such properties are the following:

Dependence: A dependence atom has the form $=(x_1 \ldots, x_m, y)$. It is true in a team X if all assignments s, s' in X that agree on the variables $x_1, \ldots, x_m$

also have the same value for y. Dependence logic is first-order logic with dependence atoms. An important property of dependence logic is downwards closure: If a formula is satisfied by a team X, then it is also satisfied by all subteams $Y \subseteq X$. Formulae of dependence logic are equivalent to sentences of existential second-order logic, with an additional predicate for the team that may occur only negatively; see [12, 14] for further results.

Independence: Independence atoms come in several variants. Intuitively two variables x and y are independent, denoted $x \perp y$, if acquiring more knowledge about one does not provide any additional knowledge about the other, which means that values for (x, y) appear in all conceivable combinations: if values (a, b) and (a', b') occur for (x, y), then so do (a, b') and (a', b). To make this sufficiently general, we proposed in [8] the general conditional independence atom $\bar{y} \perp_{\bar{x}} \bar{z}$, for arbitrary tuples $\bar{x}, \bar{y}, \bar{z}$ of variables, which is true in team X if, and only if, for all assignments $s, s' \in X$ such that $s(\bar{x}) = s'(\bar{x})$ there is an assignment $s'' \in X$ with $s''(\bar{x}) = s(\bar{x})$, $s''(\bar{y}) = s(\bar{y})$ and $s''(\bar{z}) = s'(\bar{z})$. Independence logic is strictly more powerful than dependence logic and it is not downwards closed for teams. Galliani [5] has shown that independence logic is equivalent with existential second-order logic. Furthermore the conditional independence atoms can be eliminated in favour of pure independence atoms $\bar{x} \perp \bar{y}$.

Exclusion and inclusion: An exclusion atom $(\bar{x} \mid \bar{y})$ expresses that the values of $\bar{x}$ in the given team are disjoint from the values of $\bar{y}$. Inclusion atoms $(\bar{x} \subseteq \bar{y})$ state that all values for $\bar{x}$ in the given team occur also as values for $\bar{y}$ in X. It has been proved by Galliani [5] that first-order logic with both inclusion and exclusion atoms is equivalent with independence logic.

There are many other variants of atomic dependence or independence properties. In [7] we have shown that there is a uniform construction of model-checking games for logics with team semantics, based on the notion of a second-order reachability game. For every formula $\psi(\bar{x})$ (which we always assume to be in negation normal form) and every structure $\mathfrak{A}$ we define the game $\mathscr{G}(\mathfrak{A}, \psi)$ as follows. The game graph is defined in precisely the same way as in the case of first-order logic. In particular, $\mathscr{G}(\mathfrak{A}, \psi)$ is a forest, consisting of trees with roots (ψ, s) for all assignments $s : \text{free}(\psi) \to A$. In the case that ψ is a sentence, we only have the empty assignment to consider, and the game graph is a tree. Given a team X of assignments $s : \text{free}(\psi) \to A$, the relevant set of initial positions is $I(X) := \{(\psi, s) : s \in X\}$.

Although the game graphs for logics with team semantics are defined as for first-order logic, the winning conditions are very different. Indeed, model-checking games for logics with team semantics are special cases of second-order reachability games.

To describe the second-order winning condition, we observe that any set W of nodes in a model-checking game for ψ associates to a formula $\varphi \in \mathcal{T}(\psi)$ a team

$$\text{Team}(W, \varphi) := \{s : \text{free}(\varphi) \to A : (\varphi, s) \in W\}.$$

We now say that a set U of terminal positions is a winning set if, for every literal α,

$$\mathfrak{A} \models_{\text{Team}(U,\alpha)} \alpha.$$

Notice that for literals α for which no pair (α, s) appears in U, this is trivially satisfied because for the logics that we consider here, the empty team satisfies all formulae.

Described more abstractly, the model-checking game for ψ on $\mathfrak{A}$ consists of the game graph $\mathcal{G}(\mathfrak{A}, \psi) = (V, V_0, V_1, T, I, E)$ and the second-order reachability condition Win consisting of all sets $U \subseteq T$ such that

$$\mathfrak{A} \models_{\text{Team}(U,\alpha)} \alpha, \text{ for all literals } \alpha.$$

Thus, a consistent winning strategy $S = (W, F)$ of Player 0 for $\mathcal{G}(\mathfrak{A}, \psi)$, from the set $I(X) \subseteq I$ of those initial positions that are associated with a team X, has the property that, for every literal φ, the team $\text{Team}(S, \varphi) := \text{Team}(W, \varphi) = \{s : (\varphi, s) \in W\}$ satisfies φ. As proved in [7] this then extends beyond the literals to all formulae in $\mathcal{T}(\psi)$ and in particular to the formula ψ itself. Let L be a logic with team semantics.

Theorem 9. *For every structure $\mathfrak{A}$, every formula $\psi(\bar{x}) \in L$, and every team X with domain free(ψ) we have that $\mathfrak{A} \models_X \psi$ if, and only if, Player 0 has a consistent winning strategy $S = (W, F)$ for $\mathcal{G}(\mathfrak{A}, \psi)$ from $I(X)$, with $\text{Team}(S, \psi) = X$.*

Proof. We proceed by induction on ψ. First, let ψ be a literal. The game $\mathcal{G}(\mathfrak{A}, \psi)$ is just the set of isolated nodes (ψ, s) for all possible assignments s. If $\mathfrak{A} \models_X \psi$ then let $W_\psi = \{(\psi, s) : s \in X\}$ and $F_\psi = \emptyset$. Clearly $S_\psi = (W_\psi, F_\psi)$ is a consistent winning strategy in $\mathcal{G}(\mathfrak{A}, \psi)$ with $\text{Team}(S_\psi, \psi) = X$. If $\mathfrak{A} \not\models_X \psi$, then for any consistent winning strategy S with $\mathfrak{A} \models_{\text{Team}(S,\psi)} \psi$ it must be the case that $\text{Team}(S, \psi) \neq X$.

Next suppose that $\psi = \eta \vee \vartheta$. If $\mathfrak{A} \models_X \eta \vee \vartheta$, then there exist teams Y, Z with $X = Y \cup Z$ such that $\mathfrak{A} \models_Y \eta$ and $\mathfrak{A} \models_Z \vartheta$. By induction hypothesis there are consistent winning strategies $S_\eta = (W_\eta, F_\eta)$ in $\mathcal{G}(\mathfrak{A}, \eta)$ and $S_\vartheta = (W_\vartheta, F_\vartheta)$ in $\mathcal{G}(\mathfrak{A}, \vartheta)$ with $\text{Team}(S_\eta, \eta) = Y$ and $\text{Team}(S_\vartheta, \vartheta) = Z$. We obtain a consistent winning strategy $S_\psi = (W_\psi, F_\psi)$ in $\mathcal{G}(\mathfrak{A}, \psi)$ by setting $W_\psi := W_\eta \cup W_\vartheta \cup \{(\psi, s) : s \in X\}$ and $F_\psi := F_\eta \cup F_\vartheta \cup \{((\psi, s), (\eta, s')) : s \in Y, s' = s \restriction_{\text{free}(\eta)} \} \cup \{((\psi, s), (\vartheta, s')) : s \in Z, s' = s \restriction_{\text{free}(\vartheta)} \}$. Obviously $\text{Team}(S_\psi, \psi) = X$ and since $X = Y \cup Z$ the strategy S_ψ admits, from every point $(\psi, s) \in W_\psi$ at least one edge to either (η, s') or (ϑ, s'). Conversely, every consistent winning strategy $S_\psi = (W_\psi, F_\psi)$ for Player 0 with $\text{Team}(S_\psi, \psi) = X$ induces a decomposition $X = Y \cup Z$ where Y contains those $s \in X$ such that F_ψ admits a move from (ψ, s) to $(\eta, s \restriction_{\text{free}(\eta)})$ and analogously for Z and ϑ. By induction hypothesis it follows that $\mathfrak{A} \models_Y \eta$ and $\mathfrak{A} \models_Z \vartheta$ and therefore $\mathfrak{A} \models_X \psi$.

The arguments for $\psi = \eta \wedge \vartheta$ are analogous (and in fact even simpler).

Let us now consider formulae $\psi = \exists y \varphi$. If $\mathfrak{A} \models_X \psi$, then there is a function $F : X \to (\mathscr{P}(A) \backslash \{\emptyset\})$ such that $\mathfrak{A} \models_{X[y \mapsto F]} \varphi$. By induction hypothesis, Player 0 has a consistent winning strategy $S_\varphi = (W_\varphi, F_\varphi)$ with $\mathrm{Team}(S_\varphi, \varphi) = X[y \mapsto F]$. We obtain a consistent winning strategy $S_\psi = (W_\psi, F_\psi)$ by setting $W_\psi := W_\varphi \cup \{(\psi, s) : s \in X\}$ and $F_\psi = F_\varphi \cup \{((\psi, s), (\varphi, s[y \mapsto a])) : s \in X, a \in F(s)\}$. Obviously, $\mathrm{Team}(S_\psi, \psi) = X$. Conversely, a consistent winning strategy $S_\psi = (W_\psi, F_\psi)$ with $\mathrm{Team}(S_\psi, \psi) = X$ requires that from every node (ψ, s) with $s \in X$ the set $(\psi, s)F_\psi$ of admissible successor nodes is non-empty. Let $F(s) := \{a \in A : (\varphi, s[y \mapsto a]) \in (\psi, s)F_\psi\}$. By induction hypothesis $\mathfrak{A} \models_{X[y \mapsto F]} \varphi$ and hence $\mathfrak{A} \models_X \psi$.

Again the arguments for formulae $\forall y \varphi$ are analogous. $\qquad\square$

4.5 Inclusion logic

We now turn to inclusion logic, which is a specific case of a logic with team semantics. We recall the definition.

Definition 10. A team X satisfies an *inclusion atom* $\bar{x} \subseteq \bar{y}$ if for all $s \in X$ there is an $s' \in X$ with $s(\bar{x}) = s'(\bar{y})$. Inclusion logic is the extension of first-order logic with team semantics by inclusion atoms.

By structural induction, it is easy to verify that formulae $\varphi(\bar{x})$ of inclusion logic are closed under union of teams. For every structure $\mathfrak{A}$ and any collection $\{X_i : i \in I\}$ of teams such that $\mathfrak{A} \models_{X_i} \varphi$ for all $i \in I$ we also have that $\mathfrak{A} \models_X \varphi$ for $X = \bigcup\{X_i : i \in I\}$. Thus, there exists, for every structure $\mathfrak{A}$ a unique maximal team $X_{\max}$ with $\mathfrak{A} \models_{X_{\max}} \varphi$.

We next exhibit important examples for the power of inclusion logic, showing that winning regions, traps, and I-traps of safety games are definable in this logic. Further the trap-formula also reveals the technique of copying values from one variable to another one which is often necessary for dealing with disjunctions in the intended way. To simplify notation we identify a relation $Y \subseteq V^k$ with the team of all those assignments $s : \{x_1, \ldots, x_k\} \to V$ such that $s(\bar{x}) := (s(x_1), \ldots, s(x_k)) \in Y$.

For safety games $\mathscr{G} = (V, V_0, V_1, T, I, E)$ with safety condition $S \subseteq V$, we construct the formulae

$$\mathrm{trap}(x) := Sx \wedge \exists z (z \subseteq x \wedge (V_0 x \to \exists y (Exy \wedge y \subseteq z)) \wedge (V_1 x \to \forall y (Exy \to y \subseteq z))),$$

$$\mathrm{itrap}(x) := \exists y (x \subseteq y \wedge \mathrm{trap}(y) \wedge (Iy \to y \subseteq x)).$$

Here (and elsewhere) implications $(\alpha \to \varphi)$, for first-order literals α, are just meant as a different notation for $(\neg \alpha \vee \varphi)$.

Proposition 11. *For every game graph $\mathscr{G}$, every safety condition $S \subseteq V$ and every set $X \subseteq V$, we have that $(\mathscr{G}, S) \models_X \mathrm{trap}(x)$ if, and only if, X is a trap for Player 1, i.e., Player 0 has a winning strategy that keeps every play from X inside X. Further $X \subseteq I$ is an I-trap in $(\mathscr{G}, S)$ if, and only if, $(\mathscr{G}, S) \models_X \mathrm{itrap}(x)$.*

Proof. A set $X \subseteq V$ is a trap for Player 1 in $\mathscr{G}$ if, and only if, $X \subseteq S$ and there exists a set of edges $F \subseteq (X \times X) \cap E$ such that, for all $v \in V_0 \cap X$ there exists a node $w \in X$ with $(v, w) \in F$, for all $v \in V_1 \cap X$ and all edges $(v, w) \in E$ it holds that also $w \in X$. We claim that given such an X and F we can show that $(\mathscr{G}, S) \models_X \text{trap}(x)$. Let XX be the team of assignments $s : (x, z) \to (v, v')$ such that $v, v' \in X$. It suffices to prove that $\mathscr{G} \models_{XX} (V_0 x \to \exists y (Exy \wedge y \subseteq z))$ and $\mathscr{G} \models_{XX} V_1 x \to \forall y (Exy \to y \subseteq z))$. For the first claim we split the team XX into the subteams $\overline{V_0}X = \{s \in XX : s(x) \notin V_0\}$ and $V_0 X = \{s \in XX : s(x) \in V_0\}$. Trivially, $\overline{V_0}X$ satisfies $\neg V_0 x$. To prove that $V_0 X$ satisfies $\exists y (Exy \wedge y \subseteq z)$, the team $V_0 X$ is expanded to $V_0 XY = \{s : (x, z, y) \mapsto (v, v', w) : v \in V_0, v' \in X, (v, w) \in F\}$ and we claim that $\mathscr{G} \models_{V_0 XY} Exy \wedge y \subseteq z$. Since $F \subseteq E$, the atom Exy is clearly satisfied, and for the inclusion atom we find, for each $s : (x, z, y) \mapsto (v, v', w)$ the assignment $s' : (x, z, y) \mapsto (v, w, w)$ so that $s(y) = s'(z)$ and s' is in $V_0 XY$ as well (because $w \in X$). The reasoning for $\mathscr{G} \models_{XX} V_1 x \to \forall y (Exy \to y \subseteq z))$ is analogous. We get a team $V_1 XX$ which has to be universally expanded, by values for y, to a team $V_1 XY = \{s : (x, z, y) \mapsto (v, v', w) : v \in V_1, v' \in X, w \in V\}$. This team is then split into, on the one hand, the subteam of those assignments with $(v, w) \notin E$, to satisfy the literal $\neg Exy$, and, on the other hand, the remaining set of assignments. But in the remaining team all assignments $s : (x, z, y) \mapsto (v, v', w)$ satisfy $(v, w) \in F$ and hence $w \in X$. Thus we can again map s to $s' : (x, z, y) \mapsto (v, w, w)$ to make sure that $y \subseteq z$ is satisfied. Notice that without copying all values for x in X also as values for z this reasoning would not work.

For the converse, assume that $(\mathscr{G}, S) \models_X \text{trap}(x)$. Clearly $X \subseteq S$. If X were not a trap, then there would be a node $v \in X$ such that either $v \in V_0$ and no edge from v leads to a node in X, or $v \in V_1$ and at least one edge from v leaves X. In both cases the formula is false for X, hence we would have a contradiction.

Finally $(\mathscr{G}, S) \models_X \text{itrap}(x)$ if, and only if, there is a trap Y in $(\mathscr{G}, S)$ such that $X \subseteq Y$ and $Y \cap I \subseteq X$. This is the case if, and only if X is an I-trap in $(\mathscr{G}, S)$. $\square$

Let us now look at games for inclusion logic. In the second-order reachability games for formulae of inclusion logic, the terminal positions are associated either with first-order literals or with inclusion atoms of form $\overline{x} \subseteq \overline{y}$.

Proposition 12. *The winning conditions of second-order reachability model-checking games for inclusion logic are universal-existential.*

Proof. The winning condition in a game $\mathscr{G}(\mathfrak{A}, \psi)$ consists of those sets $U \subseteq T$ such that, for every literal φ,

$$\mathfrak{A} \models_{\text{Team}(U, \varphi)} \varphi.$$

We have to find a relation $R \subseteq T \times T$ such that a set $U \subseteq T$ is winning if, and only if, $(\forall x \in U)(\exists y \in U)R(x, y)$. Positions in T are either of the form (φ, s) where φ is a first-order literal, or of the form $(\overline{x} \subseteq \overline{y}, s)$. We define R to contain all loops $((\varphi, s)(\varphi, s))$ for first-order literals φ such that $\mathfrak{A} \models_s \varphi$, and all edges $((\overline{x} \subseteq \overline{y}, s), (\overline{x} \subseteq \overline{y}, t))$ such that $t(\overline{y}) = s(\overline{x})$. Then clearly, for all $U \subseteq T$ it holds that

$$\mathfrak{A} \models_{\text{Team}(U,\varphi)} \varphi \text{ for all literals } \varphi \quad \Leftrightarrow \quad (\forall x \in U)(\exists y \in U)R(x,y). \qquad \square$$

By the construction in Sect. 3 we thus obtain safety games for inclusion logic and associate the teams satisfying the formula with the I-traps in the game. For further reference, let us describe this in a more detailed way. Given a structure $\mathfrak{A}$ and a formula $\psi(\bar{x})$ of inclusion logic we obtain a safety game $\mathcal{G}_{\text{safe}}(\mathfrak{A}, \psi)$, with the set I of initial positions consisting of all pairs (ψ, s) with assignments $s : \text{free}(\psi) \to A$, and the safety winning condition S excluding the pairs (φ, s) where φ is a first-order literal with $\mathfrak{A} \models_s \neg\varphi$. The game consists of

- the forest $\mathcal{G} = \mathcal{G}(\mathfrak{A}, \psi)$ defined as for first-order logic, with positions of form (φ, s),
- a copy $\mathcal{G}^*$ of this forest, with positions $(\varphi, s)^*$, in which Player 1 either moves upwards the forest, or from $(\varphi, s)^*$ to the corresponding position (φ, s) in $\mathcal{G}$, and
- moves of Player 0 from positions in $\mathcal{G}$ associated to inclusion atoms, into $\mathcal{G}^*$. Such moves go from $(\bar{x} \subseteq \bar{y}, s)$ to positions $(\bar{x} \subseteq \bar{y}, t)^*$ subject to the condition that $t(\bar{y}) = s(\bar{x})$.

The objective of Player 0 in this game is to keep the play inside S and to avoid the initial positions (ψ, s) with $s \notin X$. We formulate this in terms of I-traps.

Proposition 13. *For every structure $\mathfrak{A}$, every formula $\psi(\bar{x})$ of inclusion logic, and every team X with domain $\text{free}(\psi)$, we have that $\mathfrak{A} \models_X \psi$ if, and only if $X^* := \{(\psi, s) \in I : s \in X\}$ is an I-trap in $\mathcal{G}_{\text{safe}}(\mathfrak{A}, \psi)$.*

5 Interpretations

The notion of an interpretation is fundamental in mathematical logic. Interpretations are used to define a copy of a structure inside another one, and thus permit us to transfer definability, decidability, and complexity results between theories. Here we shall use interpretations of model-checking games as a method to embed one logic inside another and to provide normal forms and complete problems for logics. A bit of care is necessary for the application of interpretations in the context of team semantics.

The interpretations that we consider are classical first-order interpretations (with Tarski semantics); in fact we are interested in interpretations that are given by very simple formulae, namely quantifier-free ones. However, we shall apply these simple interpretations as translations among formulae of more powerful logics, such as greatest fixed-point logic and inclusion logic.

For every first-order formula $\varphi(x_1, \ldots x_k)$ and every structure $\mathfrak{A}$, we write $\varphi^{\mathfrak{A}}$ for the relation defined by $\varphi(\bar{x})$ on $\mathfrak{A}$, i.e., $\varphi^{\mathfrak{A}} := \{\bar{a} \in A^k : \mathfrak{A} \models \varphi(\bar{a})\}$.

Definition 14. Let L be a fragment of first-order logic, let σ, τ be vocabularies, where $\tau = \{R_1, \ldots, R_m\}$ is relational, and let r_i be the arity of R_i. A *(k-dimensional) $L[\sigma, \tau]$-interpretation* is given by a sequence I of formulae in $L(\sigma)$ consisting of

- $\delta(x_1, \ldots, x_k)$, called the domain formula,
- $\varepsilon(x_1, \ldots, x_k, y_1, \ldots, y_k)$, called the equality formula, and,
- for every relation symbol $R \in \tau$ (of arity r), a formula $\psi_R(\bar{x}_1, \ldots, \bar{x}_r)$ (of arity kr).

An $L[\sigma, \tau]$-interpretation induces two mappings: one between structures and the other one between formulae. For a τ-structure $\mathfrak{B}$ and a σ-structure $\mathfrak{A}$, we say that I *interprets* $\mathfrak{B}$ in $\mathfrak{A}$ (in short, $I(\mathfrak{A}) = \mathfrak{B}$) if there exists a surjective map $h : \delta^{\mathfrak{A}} \to B$, called the *coordinate map*, such that

- for all $\bar{a}, \bar{a}' \in \delta^{\mathfrak{A}}$,

$$\mathfrak{A} \models \varepsilon(\bar{a}, \bar{a}') \iff h(\bar{a}) = h(\bar{a}');$$

- for every relation R of $\mathfrak{B}$ and all $\bar{a}_1, \ldots, \bar{a}_r \in \delta^{\mathfrak{A}}$,

$$\mathfrak{A} \models \psi_R(\bar{a}_1, \ldots, \bar{a}_r) \iff (h(\bar{a}_1), \ldots, h(\bar{a}_r)) \in R,$$

i.e., $h^{-1}(R) = (\delta^{\mathfrak{A}})^r \cap \psi_R^{\mathfrak{A}}$.

Hence $I = \langle \delta, \varepsilon, \psi_{R_1}, \ldots, \psi_{R_m} \rangle$ defines (together with the function $h : \delta^{\mathfrak{A}} \to B$) an interpretation of $\mathfrak{B} = (B, R_1, \ldots, R_m)$ in $\mathfrak{A}$ if, and only if, $\varepsilon(\bar{x}, \bar{y})$ defines a congruence on the structure $(\delta^{\mathfrak{A}}, \psi_{R_1}^{\mathfrak{A}}, \ldots, \psi_{R_m}^{\mathfrak{A}})$ and h is an isomorphism between the quotient structure $(\delta^{\mathfrak{A}}, \psi_{R_1}^{\mathfrak{A}}, \ldots, \psi_{R_m}^{\mathfrak{A}})/\varepsilon^{\mathfrak{A}}$ and $\mathfrak{B}$.

Besides the mapping $\mathfrak{A} \mapsto I(\mathfrak{A})$ from σ-structures to τ-structures, I also defines a mapping from τ-formulae to σ-formulae. With every τ-formula φ it associates a σ-formula φ^I, which is obtained by replacing every variable x by a k-tuple $\bar{x}$ of variables, by replacing every quantifier Qx by a quantifier $Q\bar{x}$ over k-tuples, relativized to $\delta(\bar{x})$, by replacing equalities $u = v$ by $\varepsilon(\bar{u}, \bar{v})$, and by replacing every atom $Ru_1 \ldots u_r$ by the corresponding formula $\psi_R(\bar{u}_1, \ldots, \bar{u}_r)$. In the case of fixed-point formulae or second-order formulae, we may have relation variables Y (of some arity r) which we have to translate into corresponding relation variables Y^* of arity kr.

Most of the common logics (with Tarski semantics), including FO, LFP and its gfp-fragment, second-order logic, etc., are closed under interpretations, that is, for every formula φ and every (first-order)-interpretation I, also φ^I is a formula of the same logic.

The semantics of these transformations is described by the Interpretation Lemma, which we formulate for formulae $\varphi(Y_1, \ldots, Y_m, x_1, \ldots, x_n)$ which may contain free relation variables Y_i and free element variables x_j. A k-dimensional interpretation I with coordinate map $h : \delta^{\mathfrak{A}} \to B$ induces for every relation $Y_i \subseteq B^r$ the relation $Y_i^* := h^{-1}(Y_i) \subseteq A^{kr}$ and for every assignment $s : \{x_1, \ldots, x_n\} \to B$, the set of assignments $h^{-1}s$ consisting of all $t : \{x_{ij} : 1 \leq i \leq n, 1 \leq j \leq k\} \to A$ such that, for all $i \leq n$, $t(x_{i1}, \ldots x_{ik}) \in h^{-1}(s(x_i))$.

Lemma 15 (Interpretation Lemma). *Let I be an $L[\sigma, \tau]$-interpretation with coordinate map h, let $\mathfrak{A}$ be a σ-structure, and let $\varphi = \varphi(Y_1, \ldots, Y_m, x_1, \ldots, x_n)$ be a formula of vocabulary τ with free relation variables $Y_1, \ldots, Y_m$ and free element variables $x_1, \ldots, x_n$. Then for every tuple $Y_1, \ldots Y_m$ of relations over $I(\mathfrak{A})$, every assignment $s : \{x_1, \ldots, x_n\} \to B$ and every assignment $t \in h^{-1}(s)$ we have that*

$$(\mathfrak{A}, Y_1^*, \ldots, Y_m^*) \models_t \varphi^I \;\Leftrightarrow\; (I(\mathfrak{A}), Y_1, \ldots, Y_m) \models_s \varphi.$$

In particular, for every τ-sentence φ, we have that $\mathfrak{A} \models \varphi^I \Leftrightarrow I(\mathfrak{A}) \models \varphi$.

For formulae with team semantics, the translation is a little bit more delicate, since we need to consider the transformations of teams under interpretations and make sure that the atomic properties of teams are compatible with these transformations. In the presence of congruences, this may require to change the atomic formulae.

For inclusion logic, however, such complications do not arise. The interpretation I translates an inclusion statement $\eta := (x_1, \ldots, x_m \subseteq y_1, \ldots, y_m)$ on m-tuples of variables into an inclusion statement $\eta^I := (\bar{x}_1, \ldots \bar{x}_m \subseteq \bar{y}_1, \ldots, \bar{y}_m)$ on mk-tuples. For a team X of assignments s mapping variables x_i into the interpreted structure $I(\mathfrak{A})$, we get the team $h^{-1}(X) = \bigcup\{h^{-1}(s) : s \in X\}$ taking values in $\mathfrak{A}$.

Lemma 16. *An inclusion atom $\eta := (\bar{x} \subseteq \bar{y})$ holds in a team X with values in $I(\mathfrak{A})$ if, and only if, η^I holds in the team $h^{-1}(X)$.*

Proof. For simplicity of notation we just consider inclusion atoms of form $\eta := (x \subseteq y)$. Suppose that the translated inclusion statement $(\bar{x} \subseteq \bar{y})$ holds in $h^{-1}(X)$. Take any $s \in X$. We have to prove that there exists a $s' \in X$ with $s'(y) = s(x)$. For every $t \in h^{-1}(s)$ there exists a $t' \in h^{-1}(X)$ with $t'(\bar{y}) = t(\bar{x})$. For $s' = h(t') \in X$ we have that $s'(y) = h(t'(\bar{y})) = h(t(\bar{x})) = s(x)$.

For the converse, assume that $(x \subseteq y)$ holds in X. For every $t \in h^{-1}(X)$ we can choose some $s = h(t) \in X$. By assumption there exists an $s' \in X$ with $s'(y) = s(x)$. Since $h(t(\bar{x})) = s(x) = s'(y)$ there exists a $t' \in h^{-1}(s') \subseteq h^{-1}(X)$ with $t'(\bar{y}) = t(\bar{x})$. Thus $(\bar{x} \subseteq \bar{y})$ holds in $h^{-1}(X)$.

As above this extends to a translation from arbitrary formulae φ of inclusion logic on $I(\mathfrak{A})$ into formulae φ^I of inclusion logic on $\mathfrak{A}$.

Lemma 17 (Interpretation Lemma for Inclusion Logic). *Let I be a quantifier-free first-order interpretation, mapping a structure $\mathfrak{A}$ to $I(\mathfrak{A})$ with coordinate map h, and let $\varphi(\bar{x})$ be a formula of inclusion logic. Then for any team X with values in $I(\mathfrak{A})$ we have that*

$$\mathfrak{A} \models_{h^{-1}(X)} \varphi^I \;\Leftrightarrow\; I(\mathfrak{A}) \models_X \varphi.$$

6   Interpretability of game graphs

We now prove that for every formula there is a uniform interpretation of the model-checking games for that formula in the structure in which the formula is evaluated. We first explain the construction for first-order formulae, but it immediately carries over to stronger logics such as the least fixed-point logic LFP and its fragments, and to logics with team semantics.

Proposition 18. *For every formula $\psi(\bar{x}) \in \mathrm{FO}(\tau)$ there exists a quantifier-free interpretation I_ψ which, for every τ-structure $\mathfrak{A}$ with at least two elements, interprets the game graph $\mathscr{G}(\mathfrak{A}, \psi)$ in $\mathfrak{A}$.*

Proof. Let $\mathrm{Sf}(\psi)$ be the set of subformulae of $\psi(\bar{x})$ and let $\mathscr{T}(\psi) = (\mathrm{Sf}(\psi), E_\psi)$ be its syntax-tree. Let $x_1 \ldots, x_k$ be the variables occurring in ψ. Recall that an *equality type* in m-variables $u_1, \ldots, u_m$ is a maximally consistent conjunction of equalities $u_i = u_j$ and inequalities $u_i \neq u_j$. Let E_m be the set of equality types in $u_1 \ldots u_m$ (up to equivalence). Choose m sufficiently large so that $|E_m| \geq |\mathrm{Sf}(\psi)|$ and fix for every formula $\varphi \in \mathrm{Sf}(\psi)$ a separate equality type $e_\varphi \in E_m$.

A node (φ, s) of the game graph $\mathscr{G}(\mathfrak{A}, \psi)$ is represented in $\mathfrak{A}$ by the class of all $(m + k)$-tuples $(\bar{c}, \bar{a})$ such that $\bar{c}$ has equality type e_φ and $a_i = s(x_i)$ for all $x_i \in \mathrm{free}(\varphi)$. Thus, the domain and equality formulae of the interpretation I_ψ are

$$\delta(\bar{u}, \bar{x}) := \bigvee_{\varphi \in \mathrm{Sf}(\psi)} e_\varphi(\bar{u}),$$

$$\varepsilon(\bar{u}, \bar{x}; \bar{v}, \bar{y}) := \bigvee_{\varphi \in \mathrm{Sf}(\psi)} \left(e_\varphi(\bar{u}) \wedge e_\varphi(\bar{v}) \wedge \bigwedge_{x_i \in \mathrm{free}(\varphi)} x_i = y_i \right).$$

The relations V_0, V_1, E, T_0, T_1 of the game graph $\mathscr{G}(\mathfrak{A}, \psi)$ are clearly representable by quantifier-free formulae in $\mathfrak{A}$. Actually the formulae for V_0, V_1, E are pure equality formulae. Explicitly,

$$\psi_{V_\sigma}(\bar{u}, \bar{x}) := \bigvee_{\varphi \text{ belongs to Player } \sigma} e_\varphi(\bar{u}),$$

$$\psi_E(\bar{u}, \bar{x}; \bar{v}, \bar{y}) := \bigvee_{(\varphi, \vartheta) \in E_\psi} \left(e_\varphi(\bar{u}) \wedge e_\vartheta(\bar{v}) \wedge \bigwedge_{x_i \in \mathrm{free}(\varphi) \cap \mathrm{free}(\vartheta)} x_i = y_i \right).$$

Finally, the formulae defining the target sets T_0 and T_1 of the two players are

$$\psi_{T_0}(\bar{u}, \bar{x}) := \bigvee_{\varphi \text{ is a literal}} e_\varphi(\bar{u}) \wedge \varphi(\bar{x}),$$

$$\psi_{T_1}(\bar{u}, \bar{x}) := \bigvee_{\varphi \text{ is a literal}} e_\varphi(\bar{u}) \wedge \neg\varphi(\bar{x}).$$

This completes the definition of I_ψ. Clearly, for every structure $\mathfrak{A}$ with more than one element, we have the coordinate map $h : \delta^{\mathfrak{A}} \to \mathscr{G}(\mathfrak{A}, \psi)$ that maps every tuple $(\overline{c}, \overline{a}) \in \delta^{\mathfrak{A}}$ to the unique node (φ, s) such that $\overline{c}$ has equality type e_φ and $s(x_i) = a_i$ for all free variables x_i of φ. $\qquad\square$

Obviously this construction is not limited to model-checking games for first-order logic. Indeed, with only trivial modifications it also works for many other logics, including the following cases:

(1) Least fixed-point logic LFP and the associated parity games.
(2) The posGFP-fragment of LFP and the associated safety games.
(3) Logics with team semantics and the associated second-order reachability games.
(4) The safety games for inclusion logic (and other logics with universal-existential dependencies).

We briefly describe the modifications. For a parity game $\mathscr{G}(\mathfrak{A}, \psi)$ for an LFP-formula ψ one has to take into account the additional edges from fixed-point formulae and fixed-point atoms to the formulae defining the fixed point (as explained in Sect. 4.2), which changes the syntax tree to a syntax graph. In addition, the model-checking game $\mathscr{G}(\mathfrak{A}, \psi)$ has unary relations P_i associated to the priorities. These are defined by formulae $\psi_{P_i}(\overline{u}, \overline{x})$ which are just the disjunctions over all equality types $e_\varphi(\overline{u})$ for the fixed-point atoms $\varphi = R\overline{x}$ which have priority i. For the safety games associated to posGFP-formulae the construction is similar, but instead of priorities we need a formula for the safety condition, of the form

$$\psi_S(\overline{u}, \overline{x}) := \bigvee_{\varphi \text{ is a literal}} e_\varphi(\overline{u}) \to \varphi(\overline{x}),$$

saying that Player 0 has to avoid positions (φ, s) where φ is a literal with $\mathfrak{A} \models_s \neg\varphi$.

In what follows we shall need such game interpretations for the safety games associated with formulae of posGFP and for formulae of inclusion logic. The following proposition is an immediate consequence of the arguments given above and the constructions given in Sect. 3 and Sect. 4.5.

Proposition 19. *For every formula $\psi(\overline{x})$ of inclusion logic, there is a quantifier-free first-order interpretation $I(\psi)$ which, for every structure $\mathfrak{A}$, interprets the safety game $\mathscr{G}_{safe}(\mathfrak{A}, \psi)$ in $\mathfrak{A}$.*

An analogous statement holds for the safety games associated with posGFP, based on the construction of Sect 4.3.

Game interpretations can be very useful to give high-level arguments for transformations of formulae among different logics (as we are going to show in the next section) and for establishing normal forms, without the need to go through a cumbersome structural induction over formulae. Let us illustrate this for the posGFP-fragment of least fixed-point logic.

Consider the formula win(x) that defines the winning region of Player 0 in safety games $\mathscr{G}$. For any formula ψ of posGFP, possibly with deeply nested gfp-operators, the interpretation $I = I(\psi)$ induces a translation of win(x) into a formula win$^I(\bar{u}, \bar{x})$ such that, for every structure $\mathfrak{A}$, every subformula $\varphi(\bar{y})$ of ψ, and every assignment $s : \text{free}(\varphi) \to A$ we have that

$$\mathfrak{A} \models_s \varphi(\bar{x}) \ \Leftrightarrow \ \text{Player 0 wins } \mathscr{G}(\mathfrak{A}, \psi) \text{ from position } (\varphi, s)$$

$$\Leftrightarrow \ \mathscr{G}(\mathfrak{A}, \psi) \models \text{win}((\varphi, s))$$

$$\Leftrightarrow \ \mathfrak{A} \models \text{win}^I(\bar{c}, \bar{a}) \text{ for one, and hence all, tuples } (\bar{c}, \bar{a}) \in h^{-1}((\varphi, s))$$

$$\Leftrightarrow \ \mathfrak{A} \models_s \exists\bar{u}\exists\bar{y}\Big(e_\varphi(\bar{u}) \wedge \bigwedge_{x_i \in \text{free}(\varphi)} y_i = x_i \wedge \text{win}^I(\bar{u}, \bar{y})\Big).$$

Since the interpretation I is quantifier-free, the transformation win$(x) \ \mapsto \ win^I(\bar{u}, \bar{x})$ does not change the structure very much; in particular it does not change the number of **gfp**-operators and the alternation depth of existential and universal quantifiers. Since the winning region of safety games is definable in posGFP by a formula with just one application of a **gfp**-operator to a disjunction (or conjunction) of an existential and a universal formula, the same is true for the translated formula.

Proposition 20. *Every formula in posGFP is equivalent to a formula with a single application of a gfp-operator to a positive Boolean combination of purely existential and purely universal first-order formulae.*

It is known that, on finite structures, every LFP-formula is equivalent to a posGFP-formula. Thus, on finite structures, the normal form given by this proposition is in fact a normal form for LFP, a result first observed by Dahlhaus [3]. On arbitrary structures, however, this is not true since least fixed points are not expressible by greatest ones without negation, and the alternation hierarchy of least and greatest fixed points is strict.

7 Least fixed-point logic versus inclusion logic

We now are ready to explain in what sense inclusion logic and positive greatest fixed-point logic are 'equivalent'. For sentences, this is an unproblematic statement: for every sentence ψ of inclusion logic there is a sentence φ of posGFP, and vice versa, such that, for every structure $\mathfrak{A}$ we have that $\mathfrak{A} \models \psi$ if, and only if, $\mathfrak{A} \models \varphi$. An informal proof of this goes as follows. The model-checking games for a sentence ψ in any of these two logics are safety games, and these are interpretable by an interpretation $I(\psi)$ in the structure in which the sentence ψ is evaluated. Now take a formula of the other of the two logics, saying that Player 0 wins the given safety game (from the root), and transform it, by means of $I(\psi)$ back into a sentence on $\mathfrak{A}$, which is equivalent to ψ. (We shall give a precise argument for a more general statement below.)

For formulae with free variables, the situation is more complicated since formulae of logics with team semantics, such as inclusion logic, define different semantic objects than formulae of logics with Tarski semantics, such as posGFP. A formula $\varphi(\bar{x})$ (of vocabulary τ and arity k) with Tarski semantics defines a *query* Q_φ, a function that associates with every τ-structure $\mathfrak{A}$ the k-ary relation $\varphi^{\mathfrak{A}} := \{\bar{a} : \mathfrak{A} \models \varphi(\bar{a})\}$. Of course, the relation $\varphi^{\mathfrak{A}}$ can be identified with a team, namely the team

$$X_\varphi^{\mathfrak{A}} := \{s : \bar{x} \mapsto \bar{a} : \bar{a} \in \varphi^{\mathfrak{A}}\} = \{s : \mathfrak{A} \models_s \varphi\}.$$

For the general definition of a query, one has to require invariance under isomorphism.

Definition 21. A query (of vocabulary τ and arity k) is a function Q, that associates with every τ-structure a team $Q(\mathfrak{A})$ of assignments $s : \{x_1, \ldots, x_k\} \rightarrow A$ such that, for any isomorphism $h : \mathfrak{A} \rightarrow \mathfrak{B}$ between two τ-structures, we also have that $hQ(\mathfrak{A}) = Q(\mathfrak{B})$, which means that for any assignment $s : \{x_1, \ldots, x_k\} \rightarrow A$ we have that $s \in Q(\mathfrak{A})$ if, and only if, $h \circ s \in Q(\mathfrak{B})$.

On the other side, a formula $\psi(\bar{x})$ (again of vocabulary τ and arity k) with team semantics defines what we call a *team query*. It associates with every structure the set $\llbracket \psi \rrbracket^{\mathfrak{A}}$ of all teams X such that $\mathfrak{A} \models_X \varphi$. For a general definition, we again have to impose isomorphism invariance.

Definition 22. A *team query* (of vocabulary τ and arity k) is a function TQ that associates with every τ structure $\mathfrak{A}$ a set $\mathrm{TQ}(\mathfrak{A})$ of teams with domain $\{x_1, \ldots, x_k\}$ and values in A, such that, for every isomorphism $h : \mathfrak{A} \rightarrow \mathfrak{B}$ between two τ-structures and every team X with values in A, we have that $X \in \mathrm{TQ}(\mathfrak{A})$ if, and only if $hX \in \mathrm{TQ}(\mathfrak{B})$. We say that a team query TQ is L-definable (for a logic L with team semantics) if there exists a formula $\psi \in L$ such that $\llbracket \psi \rrbracket^{\mathfrak{A}} = \mathrm{TQ}(\mathfrak{A})$ for all structures $\mathfrak{A}$.

Definability of team queries can also be considered in logics with Tarski semantics, by means of *sentences* of a vocabulary that is expanded by a relation representing the team.

Definition 23. A team query TQ of vocabulary τ is defined by a sentence ψ of vocabulary $\tau \cup \{X\}$ if, for every τ-structure $\mathfrak{A}$ we have that $\mathrm{TQ}(\mathfrak{A}) = \{X : (\mathfrak{A}, X) \models \psi\}$.

In this sense it has been shown in [5, 12] that

- the team queries definable in dependence logic are precisely those that are definable by an existential second-order sentence in which the predicate for the team occurs only negatively, and
- the team queries definable in independence logic (or inclusion-exclusion logic) are precisely those definable by existential second-order sentences.

Notice that, by the closure under unions of teams, there exists for every formula $\varphi(\bar{x})$ of inclusion logic and every structure $\mathfrak{A}$ a unique maximal team $X_{\max} :=$ $\max\{X : \mathfrak{A} \models_X \varphi\}$. It follows that every team query TQ that is definable in inclusion logic induces a query max TQ associating with $\mathfrak{A}$ the maximal team $X \in \mathrm{TQ}(\mathfrak{A})$. For the same reason it follows that inclusion logic cannot be equivalent with greatest fixed-point logic in the same sense in which independence logic is equivalent with existential second-order logic. Indeed sentences $\psi(X)$ of posGFP need not be closed under union (of relations for X), and there need not be a well-defined maximal relation X satisfying ψ.

The relationship between inclusion logic and positive greatest fixed-point logic can be made precise in several ways:

(1) The queries definable in posGFP are precisely those that occur as the maximum of a team query that is definable in inclusion logic.
(2) The team queries definable in inclusion logic are those definable by sentences in posGFP (of expanded vocabulary) of the form $\forall\bar{x}(X\bar{x} \to \psi(X,\bar{x}))$ where the team predicate X occurs only positively in ψ.
(3) The post-fixed points of posGFP-definable relational operators coincide with the team queries definable in inclusion logic.

Theorem 24. *For every formula $\psi(\bar{x})$ of inclusion logic there is a formula $\varphi(X,\bar{x})$ in* posGFP, *with only positive occurrences of X, such that, for every structure $\mathfrak{A}$ and every team X we have that*

$$\mathfrak{A} \models_X \psi(\bar{x}) \iff (\mathfrak{A},X) \models \forall\bar{x}(X\bar{x} \to \varphi(X,\bar{x})) \iff F_\varphi^{\mathfrak{A}}(X) \supseteq X.$$

In particular, for all assignments s, we have that $\mathfrak{A} \models_s [\mathbf{gfp}\, X\bar{x}.\varphi(X,\bar{x})](\bar{x})$ if, and only if, $s \in \max\{X : \mathfrak{A} \models_X \psi(\bar{x})\}$.

Proof. Let $I(\psi)$ be the quantifier-free first-order interpretation which, for every structure $\mathfrak{A}$ and every team X, interprets the safety game $\mathscr{G}_{\mathrm{safe}}(\mathfrak{A},\psi)$ in $\mathfrak{A}$, and let h be the associated coordinate map. Recall that $\mathfrak{A} \models_X \psi(\bar{x})$ if, and only if, $X^* = \{(\psi,s) : s \in X\}$ is an I-trap in $\mathscr{G}_{\mathrm{safe}}(\mathfrak{A},\psi)$.

Further, let $\mathrm{itrap}(X^*,z)$ be the posGFP-formula, from Example 1, defining I-traps in safety games, in the sense that $(\mathscr{G},X^*) \models \forall z(X^*z \to \mathrm{itrap}(X^*,z))$ if, and only if, X^* is an I-trap in $\mathscr{G}$.

By the Interpretation Lemma we get a formula $\mathrm{itrap}^{I(\psi)}(Y,\bar{z})$, which is also in posGFP such that, for $Y = h^{-1}(X^*)$,

$$(\mathfrak{A},Y) \models \forall\bar{z}(Y\bar{z} \to \mathrm{itrap}^{I(\psi)}(Y,\bar{z})) \iff (\mathscr{G}_{\mathrm{safe}}(\mathfrak{A},\psi),X^*) \models \forall z(X^*z \to \mathrm{itrap}(X^*,z)).$$

For an assignment s with domain $\mathrm{free}(\psi) = \{x_1, \ldots x_k\}$ the tuples $(\bar{c},\bar{a})$ in $\mathfrak{A}$ that interpret the position (ψ,s) are those satisfying $e_\psi(\bar{c}) \wedge \bigwedge_{i=1}^k a_i = s(x_i)$. (Notice that the length of $\bar{a}$ is in general, greater than the length of $\bar{x}$.) Thus we may write tuples $\bar{z}$ representing game positions as $\bar{z} = (\bar{u},\overline{xx}')$ and observe that $Y = h^{-1}(X^*) = \{(\bar{u},\overline{xx}') : (\mathfrak{A},X) \models e_\psi(\bar{u}) \wedge X\bar{x}\}$. Let now $\mathrm{itrap}^*(X,\bar{u},\bar{x},\bar{x}')$ be the formula that is obtained from $\text{itrap}^{J(\psi)}(Y, \overline{uxx}')$ by replacing all atoms $Y\overline{v}yy'$ by the formula $(e_\psi(\overline{v}) \wedge X\overline{y})$. Finally, we put

$$\varphi(X, \overline{x}) := \forall \overline{u}\forall \overline{x}'(e_\psi(\overline{u}) \rightarrow \text{itrap}^*(X, \overline{uxx}')).$$

Notice that all occurrences of X in $\varphi(X, \overline{x})$ are positive. Putting everything together, we have that

$$\mathfrak{A} \models_X \psi(\overline{x}) \Leftrightarrow (\mathfrak{A}, X) \models \forall \overline{x}(X\overline{x} \rightarrow \varphi(X, \overline{x})). \qquad \square$$

Corollary 25. *If* TQ *is a team query that is definable in inclusion logic, then the query* max TQ *is definable in* posGFP. *Moreover,* TQ *is definable by a sentence of* posGFP *(of an expanded vocabulary by a relation for the team).*

For the converse relationship we establish the following result:

Theorem 26. *For every formula* $\varphi(X, \overline{x})$ *in* posGFP, *with only positive occurrences of* X, *there is a formula* $\psi(\overline{x})$ *in inclusion logic such that, for every structure* $\mathfrak{A}$ *and every team* X *we have that*

$$(\mathfrak{A}, X) \models \forall \overline{x}(X\overline{x} \rightarrow \varphi(X, \overline{x})) \Leftrightarrow \mathfrak{A} \models_X \psi(\overline{x}).$$

Proof. Let $\mathscr{G}^*(\mathfrak{A}, \varphi)$ be the game from Proposition 8 such that $(\mathfrak{A}, X) \models \forall \overline{x}(X\overline{x} \rightarrow \varphi(X, \overline{x}))$ if, and only if, $X^* = \{(\varphi, s) : s(\overline{x}) \in X\}$ is an I-trap in $\mathscr{G}^*(\mathfrak{A}, \varphi)$. Further, let $J(\varphi)$ be the quantifier-free interpretation with coordinate map h which, for every structure $\mathfrak{A}$, interprets $\mathscr{G}^*(\mathfrak{A}, \varphi)$ in $\mathfrak{A}$.

Let $\text{itrap}(x)$ be the formula of inclusion logic, constructed in Sect. 4.5, such that for all games $\mathscr{G}$, $[\text{itrap}(x)]^{\mathscr{G}}$ is the set of teams that define an I-trap in $\mathscr{G}$. By the Interpretation Lemma we get a formula $\text{itrap}^{J(\varphi)}(\overline{u}, \overline{z})$, which is also in inclusion logic, such that $[\text{itrap}^{J(\varphi)}(\overline{u}, \overline{z})]^{\mathfrak{A}}$ is the set of all teams $Z = h^{-1}(Y)$ where Y defines an I-trap of $\mathscr{G}^*(\mathfrak{A}, \varphi)$. Notice that such a Z consists of all assignments $t' : (\overline{u}, \overline{z}) \mapsto (\overline{c}, \overline{a})$ such that $h(\overline{c}, \overline{a}) = t(x)$ for some $t \in Y$.

For the specific team that defines X^* we can write $\overline{z} = \overline{y}y'$ and get, that $Z(X) = h^{-1}(X^*)$ is the set of all assignments $(\overline{u}, \overline{x}x') \mapsto (\overline{c}, \overline{a}a')$ such that $e_\psi(\overline{c})$ and $\overline{a} \in X$. We now set

$$\psi(\overline{x}) := \forall \overline{u}\forall \overline{x}'(e_\psi(\overline{u}) \rightarrow \text{itrap}^{J(\varphi)}(\overline{u}, \overline{xx}'))$$

so that $\mathfrak{A} \models_X \psi(\overline{x})$ if, and only if, $\mathfrak{A} \models_{Z(X)} \text{itrap}^{J(\varphi)}(\overline{u}, \overline{xx}')$. Putting everything together, we have

$$(\mathfrak{A}, X) \models \forall X(X\overline{x} \rightarrow \varphi(\overline{x})) \Leftrightarrow X^* \text{ is an } I\text{-trap in } \mathscr{G}^*(\mathfrak{A}, \varphi)$$

$$\Leftrightarrow \mathscr{G}^*(\mathfrak{A}, \varphi) \models_{X^*} \text{itrap}(x)$$

$$\Leftrightarrow \mathfrak{A} \models_{Z(X)} \text{itrap}^{J(\varphi)}(\overline{u}, \overline{xx}')$$

$$\Leftrightarrow \mathfrak{A} \models_X \psi(\overline{x}). \qquad \square$$

Corollary 27. *For every formula $\varphi(\bar{x})$ in posGFP there is a formula $\psi(\bar{x})$ in inclusion logic such that, for every structure $\mathfrak{A}$ and every team X we have that*

$$\mathfrak{A} \models_s \varphi(\bar{x}) \text{ for all } s \in X \iff \mathfrak{A} \models_X \psi(\bar{x}).$$

Corollary 28. *For every query Q that is definable in posGFP, the team query $\mathscr{P}(Q)$, which associates with every structure $\mathfrak{A}$ the power-set of the team $Q(\mathfrak{A})$, is definable in inclusion logic.*

Corollary 29. *The posGFP-definable queries Q are precisely those, for which there exists a team query TQ, that is definable in inclusion logic with $\max TQ = Q$.*

References

1. Apt, K., Grädel, E. (eds.): Lectures in Game Theory for Computer Scientists. Cambridge University Press, Cambridge (2011)
2. Blass, A., Gurevich, Y.: Henkin quantifiers and complete problems. Ann. Pure Appl. Logic **32**, 1–16 (1986)
3. Dahlhaus, E.: Skolem normal forms concerning the least fixed point. In: Börger, E. (ed.) Computation Theory and Logic. Lecture Notes in Computer Science, vol. 270, pp. 101–106. Springer, Berlin (1987)
4. Engström, F.: Generalized quantifiers in dependence logic. J. Logic Lang. Inf. **21**, 299–324 (2012)
5. Galliani, P.: Inclusion and exclusion in team semantics — on some logics of imperfect information. Ann. Pure Appl. Logic **163**, 68–84 (2012)
6. Galliani, P., Hella, L.: Inclusion logic and fixed-point logic. In: Computer Science Logic 2013. Leibniz International Proceedings in Informatics (LIPIcs), vol. 23, pp. 281–295 (2013)
7. Grädel, E.: Model-checking games for logics of incomplete information. Theor. Comput. Sci. **493**, 2–14 (2013)
8. Grädel, E., Väänänen, J.: Dependence and independence. Stud. Logica **101**(2), 399–410 (2013)
9. Grädel, E., et al.: Finite Model Theory and Its Applications. Springer, Berlin (2007)
10. Hodges, W.: A Shorter Model Theory. Cambridge University Press, Cambridge (1997)
11. Hodges, W.: Logics of imperfect information: why sets of assignments? In: van Benthem, J., Löwe, B., Gabbay, D. (eds.) Interactive Logic. Texts in Logic and Games, vol. 1, pp. 117–134. Amsterdam University Press, Amsterdam (2007)
12. Kontinen, J., Väänänen, J.: On definability in dependence logic. J. Log. Lang. Inf. **18**, 317–241 (2009)
13. Mann, A., Sandu, G., Sevenster, M.: Independence-Friendly Logic. A Game-Theoretic Approach. London Mathematical Society Lecture Notes Series, vol. 386. Cambridge University Press, Cambridge (2012)
14. Väänänen, J.: Dependence Logic. Cambridge University Press, Cambridge (2007)

Remarks on Compositionality

Wilfrid Hodges

Abstract The paper makes some historical remarks about compositionality, the role it played in the foundations of Dependence Logic, and more broadly how the idea of compositionality has developed over the last millennium. There are also some brief remarks explaining why, in the author's view, the idea of compositionality is unlikely to be helpful for putting the subtle non-uniform-definability results of Galliani and Yang into a broader context.

Jouko Väänänen very kindly invited me to give a talk at Dagstuhl on compositionality and its relation to Dependence Logic. My own work in this area has long ago been absorbed (magnificently) into the background ideas of Dependence Logic, and the interesting advances are all being made now by fresher minds than mine. So I confine myself to some brief remarks, mostly historical. I thank the referee for helpful comments.

1 Compositionality in the background of dependence logic

The beginning of my own involvement with the logic of dependence was when Johan van Benthem invited me to comment on the chapter 'Game-Theoretical Semantics' by Jaakko Hintikka and Gabriel Sandu [7], in the *Handbook of Logic and Language*, where they give a good exposition of the IF logic that they had introduced in [6]. I knew something about games in semantics, but I had never studied IF logic. Reading the chapter, I was intrigued by their claim that there is no compositional truth definition for IF logic. They remark that there are ways of ensuring a compositional semantics by changing the language or the notion of subformula, but that these ways are unnatural. They say that 'the very idea of quantifier independence in IF logic violates the principle of compositionality' ([7] p. 370).

W. Hodges (✉)
Herons Brook, Sticklepath, Okehampton, Devon EX20 2PY, UK
e-mail: wilfrid.hodges@btinternet.com

© Springer International Publishing Switzerland 2016
S. Abramsky et al. (eds.), *Dependence Logic*, DOI 10.1007/978-3-319-31803-5_6

That seemed to me suspicious, because they gave no definition of 'composition-ality' that was anywhere near precise enough to make this a well-defined question. Also there was no indication that they had explored possible alternatives to Tarski's style of truth definition. I told them so, and suggested a possible way of getting a truth definition that could reasonably be counted as compositional. Hintikka later told me that he had discussed the matter with Väänänen and they had agreed that nothing along the lines of my suggestion would work. (Väänänen has no recollection of this conversation!) I am very much in debt to Hintikka for his remark, because there is nothing like a direct challenge for getting one's neurons working. It also made me pin down as precisely as I could what sense of 'compositionality' is involved.

A peculiar feature of Hintikka and Sandu's IF logic was that their game semantics gave a semantic interpretation only to sentences, not to subformulas with free variables. You can meaningfully say 'There is an x, not depending on y, such that for every y, Axy'. But then if you remove the quantifier, you find yourself trying to give a meaning to 'x doesn't depend on y'. What could it mean to say that 6 is not dependent on 13, for example? Of course with Dependence Logic in place, we wonder why anybody ever got stuck at that problem. But before we had the solution, the problem presented itself in a more abstract form: given a semantics on sentences, how to extend it to parts of sentences? Could that be done compositionally? I remember being very impressed by some fieldwork by linguists, for example, Maria Bittner, which posed exactly this problem for some natural languages that were being investigated.

Actually more was needed than just having a precise sense of 'compositional'. With the definition in place, it was clear that there are trivial solutions, like taking the meaning of any sentence to be the sentence itself. So a key part of the work was to identify a 'best' compositional semantics extending the semantics on sentences, and see what properties could be proved for it. For this I found the computer scientists' notion of full abstraction very helpful, particularly because it gave a precise sense to Frege's remarks about looking for the meaning of a word in the interconnections of the word within the sentence.

Perhaps the single most important thing was the existence theorem; a tidied up version can be found, for example, in [9]. The theorem says that, subject to some very minimal conditions, a fully abstract solution of the extension problem can always be found. It's a familiar fact for mathematicians, as I guess it is for any kind of explorer, that if you know for sure that a certain thing exists, then you are much more willing to put in the effort needed to find it.

Fortunately it worked out [8], and trump semantics (a forerunner of team semantics) turned out to be exactly the required fully abstract semantics for a language closely related to that of IF logic. The main work was done in a seminar on game semantics at Queen Mary College in London. I particularly thank Graham White in that seminar for his refusal to take anything for granted.

It also turned out that slight adjustments to the language could lead to serious complications in the semantics. I believe Hintikka himself enjoyed thinking up adjustments that made the compositional semantics harder to find—though the existence theorem still guarantees that there always is a fully abstract compositional semantics out there somewhere. It came to light later that we had overlooked a complication that results from nesting of quantifiers with the same variable. Xavier Caicedo, Francien Dechesne and Theo M. V. Janssen [2] did a first-rate job of sorting out that glitch, though I think the most sensible reaction with logics like IF logic is probably to ban that kind of nesting.

For a while it worried me that the existence theorem seemed to be isolated in the literature. Leaving aside theorems of Gödel, any idea worth having is formulated independently by at least two people, normally in slightly different forms that point in slightly different directions. Fortunately the isolation was only temporary, and two independent discoveries have come to light. As far as I can see, both are special cases of the result in [9]. But one, given in 1986 by Willem Blok and Don Pigozzi [1], is really very close to the result in [9], and I must apologise to Blok and Pigozzi for not having referenced their work earlier. One could view my result as a generalisation of theirs, got by dropping the assumption that a language is an algebra, and identifying the exact conditions needed in this more general case. My *fregean cover* generalises their *Leibniz operator*, and my relation of *having the same fregean value* generalises their *Leibniz congruence*. The other independent discovery is an unpublished but archived preprint of Shalom Lappin and Wlodek Zadrozny [12], which is the special case where the initial assignment of meanings is already defined on the whole language but need not be compositional.

It would still be very welcome to be told that the whole scheme coincides with, say, something known to category theorists in another context.

2 Compositional translation?

A number of people have asked how the notion of compositional translation fits into the general scheme of compositionality. This question came to the fore when Pietro Galliani [4] and Fan Yang [20] proved some results to the effect that we can translate from one kind of Dependence Logic to another, but not uniformly. These results are agreeably subtle. Jouko Väänänen asked me whether the notion of compositionality could illuminate what is going on with results like these.

The answer seems to be No. I regret not having any new mathematical facts to report here, but sometimes a negative observation that blocks off dead ends can be useful.

It will be helpful to have a quick summary of the definitions in [9], as follows.

Definition 1. (a) By a *constituent structure* we mean an ordered pair of sets $(\mathbb{E}, \mathbb{F})$, where the elements of $\mathbb{E}$ are called the *expressions* and the elements of $\mathbb{F}$ are called the *frames*, such that the four conditions below hold. (Here and below, e, f etc. are expressions; F, $G(\xi)$ etc. are frames.)

    1. $\mathbb{F}$ is a set of nonempty partial functions on $\mathbb{E}$.

2. (Nonempty Composition) If $F(\xi_1, \ldots, \xi_n)$ and $G(\eta_1, \ldots, \eta_m)$ are frames, $1 \leqslant i \leqslant n$ and there is an expression

$$F(e_1, \ldots, e_{i-1}, G(f_1, \ldots, f_m), e_{i+1}, \ldots, e_n),$$

then

$$F(\xi_1, \ldots, \xi_{i-1}, G(\eta_1, \ldots, \eta_m), \xi_{i+1}, \ldots, \xi_n)$$

is a frame.

3. (Nonempty Substitution) If $F(e_1, \ldots, e_n)$ is an expression, $n > 1$ and $1 \leqslant i \leqslant n$, then

$$F(\xi_1, \ldots, \xi_{i-1}, e_i, \xi_{i+1}, \ldots, \xi_n)$$

is a frame.

4. (Identity) There is a frame $1(\xi)$ such that for each expression e, $1(e) = e$.

(b) We say that an expression e is a *constituent* of an expression f if f is $G(e)$ for some frame G; e is a *proper constituent* of f if e is a constituent of f and $e \neq f$.

The definitions of *expression, constituent* etc. apply to a language L when we fix a constituent structure for L.

Definition 2. (a) Let $\simeq$ be an equivalence relation on the set of expressions. We say that $\simeq$ is *compositional* if for every pair of expressions $F(e_1, \ldots, e_n)$ and $F(f_1, \ldots, f_n)$,

if $e_1 \simeq f_1$ and $\ldots$ and $e_n \simeq f_n$ then $F(e_1, \ldots, e_n) \simeq F(f_1, \ldots, f_n)$.

(b) Let ϕ be a function defined on the set of expressions. We say that ϕ is *compositional* if for each expression $F(e_1, \ldots, e_n)$, the value

$$\phi(F(e_1, \ldots, e_n))$$

is determined by F and the values $\phi(e_i)$.

The two notions of 'compositional' in this definition are interdefinable. For example, $\simeq$ is compositional if and only if there is a compositional function ϕ defined on the set of expressions, such that $e \simeq f$ if and only if $\phi(e) = \phi(f)$.

The situation with compositional translation is that we have two languages, say L^1 and L^2, each with its own constituent structure $(\mathbb{E}^1, \mathbb{F}^1)$ and $(\mathbb{E}^2, \mathbb{F}^2)$. A translation from L^1 to L^2 is a map $\tau : \mathbb{E}^1 \longrightarrow \mathbb{E}^2$ which preserves meaning. To formalise 'preserves meaning' we can introduce meaning maps defined on $\mathbb{E}^1$ and $\mathbb{E}^2$ and require that the obvious diagram commutes; but we won't need this.

There is a notion of 'compositional translation' in the literature, for example, in the Rosetta book [18]. In our setting, to say that τ is a compositional translation is to say that for every frame $F^1(\xi_1, \ldots, \xi_n)$ of L^1 there is a corresponding frame $(\tau F^1)(\xi_1, \ldots, \xi_n)$ of L^2 such that for every expression $e = F^1(f_1, \ldots, f_n)$ of L^1,

$$\tau(e) = (\tau F^1)(\tau(f_1), \ldots, \tau(f_n)). \tag{1}$$

At least it seems to me that this is the correct rendering of the notion in [18] into the framework of [9]. (See Peters and Westerståhl [16] p. 427 for a closely similar definition of 'compositional translation' and a good discussion of its properties.)

Now Fan Yang [20] proves that if L^1 and L^2 are, respectively, the languages of **PID** and **PD**, and F^1 is the frame

$$(\xi_1 \vee \xi_2),$$

then there is no frame (τF^1) of **PD** that will serve to make (1) above true. Her notion of 'context' seems to be an exact match for my notion of 'frame'. So she shows that in the Rosetta sense, there is no compositional translation from **PID** to **PD**.

However, the Rosetta notion of compositionality of translations is not the notion of compositionality of Definition 2 above, which agrees with Partee et al. [15] p. 318. For that notion the requirement for the translation τ to be compositional is that for every frame $F^1(\xi_1, \ldots, \xi_n)$ of L^1 and all expressions $e = F^1(e_1, \ldots, e_n)$ and $f = F^1(f_1, \ldots, f_n)$,

$$\tau(e_1) = \tau(f_1), \ldots, \tau(e_n) = \tau(f_n) \Rightarrow \tau(e) = \tau(f).$$

If τ is injective on expressions—and Yang doesn't discuss this, but it might well be—then the equations on the left of $\Rightarrow$ imply that $e_1 = f_1, \ldots, e_n = f_n$ and so $e = f$; so in this case the compositionality condition is completely trivial.

Even if we try to make the question more interesting by choosing a single translation in **PD** for each logical equivalence class of expressions of **PID**, the compositionality condition still has very little bite. The non-uniform translation of $\vee$ into **PD** still works out as compositional in our sense. Spelling it out from Yang's preprint [20], we have

$$\tau(\phi \vee \psi) = \bigwedge \{\Theta_Z^\sharp : Z \subseteq 2^n, (\exists X \subseteq 2^n)(\exists Y \subseteq 2^n)(X \models \phi, Y \models \psi, X \cup Y \subseteq Z)\},$$

where $\Theta_Z^\sharp$ has the definition that she quotes from Taneli Huuskonen. We need only check that this definition of $\tau(\phi \vee \psi)$ is invariant when we replace ϕ by a logically equivalent ϕ' and ψ by a logically equivalent ψ'. But this is clear: the condition for ϕ and ϕ' to be logically equivalent is precisely that they are satisfied by the same teams X.

We are not quite home yet; the formulas ϕ and ψ don't fully determine $\tau(\phi \vee \psi)$ because they don't determine n. But this is easily dealt with: let n be the number of distinct propositional variables occurring in ϕ or in ψ.

These disappointing outcomes illustrate the fact that with compositionality, definitions have a habit of playing out in ways that weren't expected or intended. (See Peters and Westerståhl [16] pp. 421–428 for a fuller discussion. They prove some facts connecting compositionality with compositional translation, but—at least to my eye—don't uncover any common idea underlying the two notions.)

3   The history of compositionality

The remainder of this note looks back at some events in the history of the notion of compositionality. Most of this is completely irrelevant to Dependence Logic, but please indulge me. Compositionality is an important idea in the background of logic, and it deserves to be better understood from every angle.

In my chapter [9] I offered 'Aristotelian compositionality' as a name for the view that there is a correspondence between the parts of a sentence and the parts of the meaning of the sentence, so that each part X of the sentence corresponds to a part Y of the meaning, and Y is the meaning of X. This is a completely different kind of compositionality from the modern one discussed above, which (as mentioned above) is a formalisation of the one offered in Partee et al. [15] p. 318. The modern version has no room for any notion of 'parts of a meaning'.

Although one can draw a loose parallel with notions in Plato and Aristotle, the earliest known statement of full-blooded Aristotelian compositionality was given by the Arabic logician Al-Fārābī in the 10th century, and it was taken up by the Persian-Arabic logician Ibn Sīnā in the 11th. A version of Aristotelian compositionality appears also in Abelard in 12th century France; see Irene Rosier-Catach [19]. (A paper of Martin Lenz [13] has Abelard and compositionality in the title, but the contents—though interesting—are not in fact about compositionality.) In [9] (p. 246) I said: 'Since there is no known line of influence from Al-Fārābī to Abelard, it seems likely that similar formulations existed in the writings of their common source, namely the Aristotelian commentators of the late Roman Empire'. Since then there has been no progress at all in finding a common source. In December 2012 I asked Sir Richard Sorabji about it, mentioning the possibility that the 3rd century AD Palestinian logician Porphyry might be relevant. Richard passed on my question to 'probably the best people anywhere to answer your questions', namely Peter Adamson, Jonathan Barnes, Marwan Rashed and Carlos Steel. From the total lack of any response I deduce that totally nothing is known.

There are, as it happens, two small indications that the notions might actually be home-grown both in the Arabic world and in Abelard. The first indication is that Abelard's formulation is substitutional whereas the Arabic formulations are functional. In other words, Abelard tells us that if we replace a component of a sentence by a new component with the same meaning, the meaning of the whole sentence stays the same; cf. Definition 2(a) above. This style continues in the West, for example, in Leibniz's discussions of substitution *salva veritate* [10] (see Chapter 2 'The principle of substitutivity *salva veritate*'), and we find it also in

Frege. (Frege's debt to the western tradition passing through Leibniz explains why Blok and Pigozzi said Leibniz where I said Frege.) I don't recall formulations in the Arabic writers that use the notion of substitution; rather we read that the meaning of the whole sentence is built up from parts that are the meanings of the sentence parts, which is closer to the style of Definition 2(b) above. Versions of this view can be found in the Arabic linguists too, not just in the logicians.

The second indication is that there is a possible explanation of the Arabic 'Aristotelian compositionality' that rests on semantic assumptions shared between at least some of the Arabic logicians and at least some of the Arabic linguists. So it may have been an original Arabic invention. Manuela Giolfo and I [5] compared the views of the 11th century logician Ibn Sīnā and the 10th century linguist Al-Sīrāfī on some issues that interested both of them, and we found the following. They both had a tendency to think of meanings as being in the first instance the intentions of a person making a statement. So, for example, the dictionary meaning of a word would be a kind of average of the intentions of speakers in the relevant community when they deliberately include the word in a sentence. But then presumably if you say what you intended to say, every word in your sentence was put there with an intention, and that intention is the meaning of the word in context. So there is automatically a structural one-one correspondence between the parts of the sentence and the meanings assigned to the parts by the speaker of the sentence. It's crude but it does yield a form of Aristotelian compositionality.

We turn now to the modern Partee-style compositionality. There is a widespread belief that the name 'compositionality' was first introduced for this concept in Los Angeles by one or other of Rudolf Carnap and Richard Montague. Lots of people have assured me of this, but I have never seen the slightest evidence in support. In fact all the publicly available evidence points to the name having been introduced at MIT in Boston around 1960, in connection with the adaptation of Tarski's truth definition to natural languages. Jerrold Katz and Jerry Fodor put it in print in 1963 [11]. At the Dagstuhl meeting Dag Westerståhl pointed out that the same notion is presented under the name 'compositional mapping' in a paper of Hilary Putnam delivered in Oxford in 1960 but not published till later [17].

In [9] I mentioned what seems to be the first appearance of a Partee-style compositional semantics, in a paper of Tarski in 1930 that contains traces of an early version of the truth definition. Barbara Partee reports (personal communication, to Theo M. V. Janssen and to me) that she gave a talk soon after 1970 in which she explained the notion of compositionality; Tarski was in the audience and he told her after the talk that his truth definition was not compositional. Some of us would have liked to have been there to discuss this with him. The best guess at what he meant seems to be that for him it was significant that the clauses of the truth definition take the form 'Assignment $\bar{a}$ satisfies formula ϕ in structure M' rather than 'the semantic value of ϕ is such-and-such'. For Tarski in the 1930s this was important because it allowed him to state the clauses of the truth definition with fewer existential assumptions from higher-order logic; by the 1970s it would have been just a matter of historical accuracy. But I don't think there is any doubt that Tarski's truth definition, either in the 1930s version or in the later model-theoretic version, really should be described as compositional. Ian Chiswell and I presented the model-theoretic version as Appendix B in our text [3] in a format that makes the compositionality immediate.

Although there is nothing in the record to suggest that Montague originated the notion of compositionality, its role in the collaboration between Montague and Partee at the beginning of the 1970s was probably the main cause of it becoming widely known. (For philosophers Donald Davidson advertised it too.) I won't say any more about this, because there are two current pieces of historical research that could instantly supersede anything I said. One is a project of Barbara Partee [14] to write a historical account of Richard Montague's work on semantics. The other is a projected intellectual biography of Montague by Ivano Caponigro.

References

1. Blok, W., Pigozzi, D.: Protoalgebraic logics. Stud. Logica **45**, 337–369 (1986)
2. Caicedo, X., Dechesne, F., Janssen, T.M.V.: Equivalence and quantifier rules for logic with imperfect information. Log. J. IGPL **17**, 91–129 (2009)
3. Chiswell, I., Hodges, W.: Mathematical Logic. Oxford University Press, Oxford (2007)
4. Galliani, P.: Epistemic operators and uniform definability in dependence logic. Stud. Logica **101**, 367–397 (2013)
5. Giolfo, M.E.B., Hodges, W.: Syntax and meaning in Sīrāfī and Ibn Sīnā. Romano-Arabica **xiii**, 81–97 (2013)
6. Hintikka, J., Sandu, G.: Informational independence as a semantical phenomenon. In: Fenstad, D.E., et al. (eds.) Logic, Methodology and Philosophy of Science VIII, pp. 571–589. Elsevier, Amsterdam (1989)
7. Hintikka, J., Sandu, G.: Game-theoretical semantics. In: Van Benthem, J., Ter Meulen, A. (eds.) Handbook of Logic and Language, pp. 361–410. Elsevier, Amsterdam (1997)
8. Hodges, W.: Compositional semantics for a language of imperfect information. Log. J. IGPL **5**, 539–563 (1997)
9. Hodges, W.: Formalizing the relationship between meaning and syntax. In: Werning, M., Hinzen, W., Machery, E. (eds.) The Oxford Handbook of Compositionality, pp. 245–261. Oxford University Press, Oxford (2012)
10. Ishiguro, H.: Leibniz's Philosophy of Logic and Language, 2nd edn. Cambridge University Press, Cambridge (1990)
11. Katz, J., Fodor, J.: The structure of a semantic theory. Language **39**, 170–210 (1963)
12. Lappin, S., Zadrozny, W.: Compositionality, synonymy, and the systematic representation of meaning (January 9, 2000). https://archive.org/details/arxiv-cs0001006 cited 25 Feb 2014
13. Lenz, M.: Are thoughts and sentences compositional? A controversy between Abelard and a pupil of Alberic on the reconciliation of ancient theses on mind and language. In: Marenbon, J. (ed.) The Many Roots of Medieval Logic: The Aristotelian and the Non-Aristotelian Traditions, pp. 39–58. Brill, Leiden (2007)
14. Partee, B.: Montague's "linguistic" work: motivations, trajectory, attitudes. http://semanticsarchive.net/sub2012/Partee.pdf (2016). Cited 17 April 2016
15. Partee, B., Ter Meulen, A., Wall, R.: Mathematical Methods in Linguistics. Kluwer, Dordrecht (1990)
16. Peters, S., Westerståhl, D.: Quantifiers in Language and Logic. Clarendon Press, Oxford (2006)
17. Putnam, H.: Do true assertions correspond to reality? In: Putnam, H. (ed.) Mind, Language and Reality: Philosophical Papers, vol. 2, pp. 70–84. Cambridge University Press, Cambridge (1975). (original 1960)

18. Rosetta, M.T.: Compositional Translation. Kluwer, Dordrecht (1994)
19. Rosier-Catach, I.: La notion de translation, le principe de compositionalité et l'analyse de la prédication accidentelle chez Abélard. In: Biard, J. (ed.) Langage, Sciences, Philosophie au XIIIe Siècle, pp. 125–164. Vrin, Paris (1999)
20. Yang, F.: Uniform definability of connectives in **PD**. Preprint

Independence in Model Theory

Åsa Hirvonen

Abstract This article is a survey of independence notions in model theory. I give an overview of various approaches to independence, properties of independence as well as the phenomenon, that in a model class, if there is a well-behaved notion of independence, it tends to be unique.

1 Introduction

This paper gives a survey of independence notions in model theory. When can they be defined, what properties do they have and which are the properties aimed for? The main idea with $a \downarrow_A B$ is that "B does not give more information on a than A does". In various contexts this is defined differently, and the properties it satisfies vary. The starting point is Shelah's forking relation defined for first order theories. The idea has been extended to more general frameworks, but generalisation comes at a cost: when moving to more general frameworks the nature of independence shifts from giving a detailed view on a formula level to a coarser notion on complete types.

There is not a standardised set of axioms for independence, but properties studied vary depending on what can be achieved in various contexts as well as on different authors' different viewpoints. It is to be noted that, while *independence logic* [10] studies syntactic axiomatisations of independence, the axioms used in model theory are semantic by nature. The phenomenon studied, however, is the same. Hyttinen and Paolini [14] have shown that the independence notion studied in independence logic can be reduced to the model theoretic notion of dividing independence in atomless Boolean algebras.

The reason model theorists are interested in independence is classification: given a class of structures, is there a set of invariants that will determine the structure up to isomorphism? In vector spaces over a fixed field the linear dimension is such an invariant. For algebraically closed fields of a fixed characteristic the transcendence degree determines the structure. A common feature of these two examples is that

Å. Hirvonen (✉)
Department of Mathematics and Statistics, University of Helsinki,
P.O. Box 68, 00014, Helsinki, Finland
e-mail: asa.hirvonen@helsinki.fi

© Springer International Publishing Switzerland 2016
S. Abramsky et al. (eds.), *Dependence Logic*, DOI 10.1007/978-3-319-31803-5_7

the invariant is a dimension given by a notion of independence: linear independence in the vector spaces and algebraic independence in the fields. In the 1970s Shelah defined the notion of forking independence as a generalisation of these notions. A thorough study of the properties of forking was included in the book [29] which described the methods and results so far achieved in Shelah's quest to separate all first order axiomatisable structures into two classes: the classifiable ones and the 'bad' ones, a work whose conclusion was published in the second edition of the book [34].

The two main ingredients in classification results are an independence notion and a notion of generation. Both parts are studied and developed in the area of model theory known as *stability theory*. In this survey, however, I concentrate on the independence notions. Also, I do not cover all variants and approaches to independence, but try to give an overview of the most important ones.

The second edition of Shelah's book [34] culminated in a proof of Shelah's Main Gap theorem: there is a strict dividing line, narrowed down to a few properties, such that on one side are the classifiable theories, with models built on certain backbone tree structures, on the other side are the non-classifiable theories with the maximum number of models, hard to distinguish from one another. However, the story was not finished here. On the non-classifiable side, there were still structures to study. In some of these forking could be used, for others new independence notions were developed that better suited studying these structures. There was also research in the inverse direction: classifying theories according to how well-behaved independence notions they admit.

An even larger work ground opened up as attention turned to non-elementary classes, i.e. classes not axiomatisable by first order theories. Classification techniques have been developed, e.g., in classes of substructures of a given model and classes of atomic models of a given theory. To develop a general framework for fragments of $\mathcal{L}_{\kappa,\omega}$ without getting entangled with set theoretic assumptions on κ, Shelah defined a very general framework for the study of various non-elementary classes, that of *abstract elementary classes* [33]. These have been the object of an ever increasing interest and are now the main area of interest in non-elementary classification theory.

2   Notation and preliminaries

This section shortly introduces the notation and basic notions used in this paper. A reader familiar with basic model theoretic conventions and definitions may skip this section.

I will use α, β, δ for ordinals, κ, λ, μ for cardinals and k, l, m, n for natural numbers; i, j are indexes that may run over natural numbers, ordinals or any given index set. In general A, B, C will be sets M, N models. I will use the same symbol for a model and its universe. If A and B are sets, I will use the shorthand AB for

$A \cup B$. Tuples of elements are written $\mathbf{a}, \mathbf{b}, \mathbf{c}$. $\mathbf{x}, \mathbf{y}, \mathbf{z}$ are tuples of variables. I will write $\mathbf{a} \in A$ as a shorthand for $\mathbf{a} \in A^{\text{length}(\mathbf{a})}$ and $\mathbf{ab}$ is used as a shorthand for concatenation $\mathbf{a}^\frown \mathbf{b}$.

I will mainly look at classes of models. When there is a theory T, it will be a complete first order theory with infinite models and the class will be the class of models of T. I will also consider more general classes.

When there is a language (logic) present, types will be sets of formulae. (We will see another notion of type in the last section.) Let M be a model, $A \subseteq M$ and $\mathbf{a} \in M$. The *type of* $\mathbf{a}$ *over* A *in* M is

$$\text{tp}(\mathbf{a}/A; M) = \{\varphi(\mathbf{x}, \mathbf{c}) : \mathbf{c} \in A, M \models \varphi(\mathbf{a}, \mathbf{c})\}.$$

An n-type over A in M is a set p of formulae of the form $\varphi(\mathbf{x}, \mathbf{a})$ where $\mathbf{x} = (x_0, \ldots, x_{n-1})$ and $\mathbf{a} \in A \subseteq M$, such that p is consistent with M, i.e. p is finitely satisfiable in M. An n-type p is a *complete type* over A if for any formula $\varphi(\mathbf{x}, \mathbf{a})$ as above, either $\varphi \in p$ or $\neg\varphi \in p$. $S^m(A)$ denotes the set of complete m-types over A, $S(A) = \bigcup_{m<\omega} S^m(A)$. If $p \in S(A)$ and $B \subseteq A$, then I denote by $p \upharpoonright B$ the set

$$\{\varphi(\mathbf{x}, \mathbf{a}) \in p : \mathbf{a} \in B\}.$$

A tuple $\mathbf{a} \in M$ is said to *realise* a type p over $A \subseteq M$ if $p \subseteq \text{tp}(\mathbf{a}/A; M)$. If p is an n-type in M it can be realised in an elementary extension of M.

To avoid having to switch to a larger model every now and then to find realisations of types, model theorists often work inside a large, saturated and strongly homogeneous enough[1] *monster model* $\mathfrak{M}$, i.e. one realising all types over small enough parameter sets, and where small enough partial elementary maps extend to automorphisms. In the monster model $\text{tp}(\mathbf{a}/A) = \text{tp}(\mathbf{b}/A)$ if and only if there is an automorphism f of $\mathfrak{M}$ fixing A pointwise (i.e. $f \in \text{Aut}(\mathfrak{M}/A)$) such that $f(\mathbf{a}) = \mathbf{b}$. Using a monster also has the advantage that all models studied can be elementarily embedded in $\mathfrak{M}$, so to simplify notation, I assume that all models are elementary submodels of $\mathfrak{M}$.

The notation $(A_i)_{i \in I} \equiv_C (B_i)_{i \in I}$ is used to denote that there is $f \in \text{Aut}(\mathfrak{M}/C)$ such that $f(A_i) = B_i$ for all $i \in I$.

A model class $\mathbb{K}$ is *stable in* λ if no model in the class satisfies more than λ types over parameter sets (or models) of cardinality λ. $\mathbb{K}$ is stable if it is stable in at least one cardinality, superstable if it is stable in every $\lambda \geq \lambda_0$ for some λ_0. The strongest stability property is ω-stability, as this implies the class is stable in all (infinite) cardinalities.

A class is *categorical in* λ if it, up to isomorphism, has exactly one model of cardinality λ.

---

[1]A model M is κ-*saturated*, if it realises all types over sets $A \subset M$ with $|A| < \kappa$. It is *strongly* κ-*homogeneous* if given two (possibly infinite) tuples $\mathbf{a}$ and $\mathbf{b}$ of the same type and of length strictly less than κ, there is an automorphism of M mapping $\mathbf{a}$ to $\mathbf{b}$.

A sequence $(\mathbf{a}_i)_{i\in I}$ is *indiscernible over A* if for any $n < \omega$ and any $\mathbf{a}_{i_0}, \ldots, \mathbf{a}_{i_{n-1}}$ and $\mathbf{a}_{j_0}, \ldots, \mathbf{a}_{j_{n-1}}$ with $i_0 < \cdots < i_{n-1}, j_0 < \cdots < j_{n-1}$, we have

$$\mathrm{tp}(\mathbf{a}_{i_0} \ldots \mathbf{a}_{i_{n-1}}/A) = \mathrm{tp}(\mathbf{a}_{j_0} \ldots \mathbf{a}_{j_{n-1}}/A).$$

A set is indiscernible if it is an indiscernible sequence regardless of how it is ordered.

3 Forking independence

Modern classification theory was born in 1965 when Morley [23] proved his famous categoricity transfer theorem (previously known as Łos' conjecture): If a countable theory is categorical in one uncountable cardinality, then it is categorical in all uncountable cardinalities. Morley used a notion of rank in his proof to develop a notion of dimension in the categorical theories. He singled out a class of *totally transcendental* theories (now called ω-stable), in which this rank was always bounded. This was the starting point of stability theory, a branch that since came to be largely dominated by the work of Shelah. He first divided up all first order theories according to their degree of stability (in order of increasing stability: unstable, stable, superstable and ω-stable). In the 1970s Shelah defined the forking relation, generalising the idea of Morley's ranks. The basic theory on stability, forking and their properties was presented in Shelah's book [29].

The original definition of forking is

Definition 1. Let U be an index set and p a (possibly incomplete) type in $\bar{x}_U$. The type *p forks over A* if there are formulae $\varphi_0(\mathbf{x}_0; \mathbf{a}_0), \ldots, \varphi_{n-1}(\mathbf{x}_{n-1}; \mathbf{a}_{n-1})$, with $\mathbf{x}_k \subseteq \mathbf{x}_U$ for $k < n$, such that

1. $p \vdash \bigvee_{k<n} \varphi_k(\mathbf{x}_k; \mathbf{a}_k)$,
2. $\varphi_k(\mathbf{x}_k; \mathbf{a}_k)$ *divides over A* for each $k < n$, i.e. there are $m_k < \omega$ and sequences $\mathbf{a}_{k,l}$, $l < \omega$, such that

   a. $\mathrm{tp}(\mathbf{a}_{k,l}/A) = \mathrm{tp}(\mathbf{a}_k/A)$,
   b. $\{\varphi_k(\mathbf{x}_k; \mathbf{a}_{k,l}) : l < \omega\}$ is m_k-inconsistent, i.e. for every $w \subset \omega$, $|w| = m_k$, the formula $\exists \mathbf{x} \bigwedge_{l\in w} \varphi(\mathbf{x}; \mathbf{a}_{k,l})$ is contradictory.

The connection between forking and rank is that the notion 'p does not fork over B' was designed to generalise 'p and $p \upharpoonright B$ have the same rank'.

Shelah describes in his book that he set out to find a notion realising the following properties:

1. There is a (minimal) cardinality $\kappa(T) \leq |T|^+$ such that for every $p \in S^m(A)$ there is $B \subseteq A$ with $|B| < \kappa(T)$ such that p does not fork over B.
2. For every $B \subseteq A$ with $|B| < \kappa(T)$, $\{p \in S^m(A) : p \text{ does not fork over } B\}$ has cardinality $\leq 2^{|T|}$.
3. If $p \in S^m(A)$ forks over B, then for some finite $C \subset A$, $p \upharpoonright (BC)$ forks over B.

4. If $p \in S^m(A_\delta)$, $(A_i)_{i \leq \delta}$ is increasing, and $p \restriction A_{i+1}$ forks over A_i, then T is unstable in λ whenever $\lambda^{|\delta|} > \lambda$.

5. If p does not fork over B and $\mathrm{dom}(p) \subseteq A$, then there is q s.t. $p \subseteq q \in S^m(A)$ and q does not fork over B.

A first attempt to satisfy the properties above was to study the notion of splitting[2], but it only satisfies (4) for $\lambda < 2^{|\delta|}$. The next attempt was to study strong splitting[3]. However, the problem with this notion is a dependence on the parameters appearing in the type, which might violate (5). To fix this property, the definition was based on the idea that a type p should fork over B if each complete extension of p (over some parameter set) splits strongly over B.

To see forking (or rather nonforking) as an independence notion, one defines the *nonforking independence relation*:

$$\mathbf{a} \downarrow_A B$$

if $\mathrm{tp}(\mathbf{a}/AB)$ does not fork over A. Note, however, that although we here look at tuples of a model, forking is really a property of their types. Also this notion is restrictive in the sense, that it only considers complete types.

Theorem 1 (Shelah [29]). *For any first order theory T the nonforking independence relation satisfies*

- *(invariance) If* $\mathbf{a} \downarrow_A B$ *and* f *is an automorphism, then* $f(\mathbf{a}) \downarrow_{f(A)} f(B)$.
- *(monotonicity) If* $A \subseteq A' \subseteq B' \subseteq B$ *and* $\mathbf{a} \downarrow_A B$, *then* $\mathbf{a} \downarrow_{A'} B'$.
- *(extension) If* $\mathbf{a} \downarrow_A B$, *then for any set* C *there is* $\mathbf{b}$ *such that* $\mathrm{tp}(\mathbf{b}/AB) = \mathrm{tp}(\mathbf{a}/AB)$ *and* $\mathbf{b} \downarrow_A BC$.
- *(finite character)* $A \not\downarrow_B C$ *if and only if for some finite* $\mathbf{a} \in A$ *and* $\mathbf{c} \in C$, $\mathbf{a} \not\downarrow_B \mathbf{c}$
- *(reflexivity) If* $\mathbf{a} \not\subseteq \mathrm{acl}(A)$, *then* $\mathbf{a} \not\downarrow_A \mathbf{a}$.[4]

Theorem 2 (Shelah [29]). *If T is stable, then (in addition to the previous list) we have*

- *(existence) For all* $\mathbf{a}$ *and* A, $\mathbf{a} \downarrow_A A$.
- *(locality) There is* $\kappa(T) \leq |T|^+$ *such that there are no increasing sequence* A_i, $i < \kappa(T)$ *and* $\mathbf{a}$ *such that for all* $i < \kappa(T)$, $\mathbf{a} \not\downarrow_{A_i} A_{i+1}$, *equivalently: for each* $\mathbf{a}$ *and* A *there is* $B \subseteq A$ *of cardinality* $< \kappa(T)$ *such that* $\mathbf{a} \downarrow_B A$.
- *(symmetry)* $\mathbf{a} \downarrow_A \mathbf{b}$ *if and only if* $\mathbf{b} \downarrow_A \mathbf{a}$.
- *(transitivity) If* $A \subseteq B \subseteq C$, $\mathbf{a} \downarrow_B C$ *and* $\mathbf{a} \downarrow_A B$, *then* $\mathbf{a} \downarrow_A C$.
- *(pair) Assume* $A \subseteq B$. *Then* $\mathbf{a} \downarrow_A B$ *and* $\mathbf{b} \downarrow_{A\mathbf{a}} B$ *if and only if* $\mathbf{ab} \downarrow_A B$.

---

[2] A type p *splits over* B if there are $\mathbf{b}, \mathbf{c}$ such that $\mathrm{tp}(\mathbf{b}/B) = \mathrm{tp}(\mathbf{c}/B)$ but $\varphi(\mathbf{x}; \mathbf{b}), \neg\varphi(\mathbf{x}; \mathbf{c}) \in p$ for some formula φ.

[3] A type p *strongly splits over* B if there is a set $\{\mathbf{b}_i : i < \omega\}$ indiscernible over B, such that $\varphi(\mathbf{x}; \mathbf{b}_0), \neg(\mathbf{x}; \mathbf{b}_1) \in p$ for some formula φ.

[4] $\mathrm{acl}(A)$ is the *algebraic closure of A*, i.e. the union of all finite sets definable over A.

- *(boundedness) For any $A \subseteq B$ and $\mathbf{a}$, there are at most $2^{|T|}$ tuples $\mathbf{b}$ satisfying* $\mathrm{tp}(\mathbf{b}/A) = \mathrm{tp}(\mathbf{a}/A)$ *and* $\mathbf{b} \downarrow_A B$.
- *(stationarity over models) If M is a model, $M \subset A$, $\mathrm{tp}(\mathbf{a}/M) = \mathrm{tp}(\mathbf{b}/M)$, $\mathbf{a} \downarrow_M A$ and $\mathbf{b} \downarrow_M A$, then $\mathrm{tp}(\mathbf{a}/A) = \mathrm{tp}(\mathbf{b}/A)$.*

The use of these properties, as often is done in classification proofs, is sometimes referred to as *independence calculus*.

4 Pregeometries

In 1930 van der Waerden [37] extracted a list of axioms for algebraic dependence in fields. Some years later Whitney [38] studied similar properties of linear dependence of vector spaces and named the abstract structures satisfying certain independence axioms *matroids*. What is nowadays called matroids[5] or *pregeometries* are structures with a closure property satisfying van der Waerden's axioms:

Definition 2. A nonempty set X with a closure map $cl : \mathscr{P}(X) \longrightarrow \mathscr{P}(X)$ is a *pregeometry* if it satisfies

1. (reflexivity) $A \subseteq \mathrm{cl}(A)$
2. (monotonicity) if $A \subseteq B$, then $\mathrm{cl}(A) \subseteq \mathrm{cl}(B)$
3. (finite character) if $a \in \mathrm{cl}(A)$, then for some finite $A_0 \subseteq A$, $a \in \mathrm{cl}(A_0)$
4. (symmetry or exchange) if $a \in \mathrm{cl}(Ab) - \mathrm{cl}(A)$, then $b \in \mathrm{cl}(Aa)$
5. (transitivity) if $A \subseteq \mathrm{cl}(B)$ and $B \subseteq \mathrm{cl}(C)$, then $A \subseteq \mathrm{cl}(C)$

The corresponding independence notion here is 'not dependent', i.e. a is independent from A if $a \notin \mathrm{cl}(A)$. To define relativised independence ('independent *over A*') one defines localisations of the closure

$$\mathrm{cl}_A(B) = \mathrm{cl}(B \cup A).$$

In a pregeometry one can always define dimensions, a feature Baldwin took as a starting point when presenting model theoretic independence in his book on stability theory [3]. However, he pointed out that forking independence, although specialising to linear independence and algebraic independence, is not in general a generalisation of the notion of pregeometric independence, as it is stronger in some respects but weaker in others. But forking is more generally applicable, as it allows one to find (families of) dimensions in a varied class of theories.

In strongly minimal theories[6] algebraic closure acl defines a pregeometry and thus gives dimensions. In general the closure operation forking defines (a is in the closure of A over B if $a \not\downarrow_B A$) is not a pregeometry. The types, among whose realisations we do get a pregeometry, are called regular.

---

[5]Sometimes the term *matroid* is reserved for finite pregeometries.
[6]T is *strongly minimal* if all definable subsets of models of T are either finite or co-finite.

5 Ranks and independence

Since Morley [23] defined his rank function, a large variety of ranks have been defined, some of them on formulae, others on types. A good introduction to ranks can be found in [3, Chapter VII], but this of course does not treat the multitude of ranks defined in various contexts more recently. Ranks give a way of measuring the complexity of definable sets. Also, in the first order case, different stability classes can be characterised by the rank functions they admit. Most often the ranks are defined by variations to the following schema:

Schema 1. Let p be a (partial) type.

1. $R(p) \geq 0$ if p is consistent (or realised in a given monster model).
2. for α a limit, $R(p) \geq \alpha$ if $R(p) \geq \beta$ for all $\beta < \alpha$.
3. $R(p) \geq \alpha + 1$ if there is some large enough (the exact demand on largeness varies from one rank to another) collection of types q_i, $i \in I$, such that

   - each q_i is an extension of p, possibly of a certain given form (e.g. extension by a given formula).
   - $R(q_i) \geq \alpha$ for all q_i
   - the collection of q_is is contradictory in some way (e.g. $\bigcup_{i \in I} q_i$ is not realised in $\mathfrak{M}$ or $\{q_i : i \in I\}$ is n-contradictory for some n).

In general, if the rank function is bounded, it defines an independence notion. The most straightforward way to define an independence notion is to say that **a** is independent of B over A ($A \subseteq B$) if $R(\mathrm{tp}(a/B)) = R(\mathrm{tp}(a/A))$. However, usually one wishes to build in various locality and finite character properties in the definition.

6 Uniqueness

A perhaps surprising property of independence is that well-behaved independence notions are unique: in a given class (or monster model), if one can define independence in two different ways, in the cases where both are well behaved, these will coincide.

This can be seen most clearly in theories with a natural independence notion: The nonforking independence developed by Shelah coincides with linear independence in vector spaces and algebraic independence in fields. Independence notions developed for more general classes show the same behaviour: e.g. Berenstein and Buechler [5] have studied expansions of Hilbert spaces and shown that dividing freeness coincides with orthogonality.

Another instance of this phenomenon is the correspondence of two different approaches to independence. Lascar introduced a notion of *heir*, corresponding to certain extensions of definable types. This approach was developed further by Lascar and Poizat [22] and is the chosen approach to free extensions in, e.g., Poizat's book on model theory [26] and Pillay's Introduction to stability theory [25]. But these extensions correspond exactly to nonforking extensions of definable types, and thus the approaches define the same notion in stable theories.

For superstable theories an abstract uniqueness result was first established by Lascar [21]. He generalised the rank notion from Morley's paper [23] and studied various ordinal-valued rank notions. He demanded them to satisfy five natural axioms, corresponding to the axioms of monotonicity, isomorphism invariance, extension, bounded number of free extensions and finite character. Assuming superstability he proved that the relation 'to have the same rank' for complete types and their complete extensions was independent on the particular rank used. He gave an equivalent characterisation for this, using the notion of heir and noted that for a stable theory T, heirs and nonforking extensions are the same.

Harnik and Harrington gathered together the results on forking up to then in [12]. The paper's approach to forking is close to that of Lascar and Poizat and it generalises Lascar's uniqueness result on the relation of having the same ranks. They noted that any rank R defines a relation $\sqsubseteq$ by: $p \sqsubseteq q$ if and only if $p \subset q$ and $R(p) = R(q)$. They then proved that such a relation is unique if it is well behaved and that this implies that the theory has to be stable:

Theorem 3 (Harnik, Harrington [12]). *Assume that a relation $\sqsubseteq$ between complete types of T satisfies:*

1. If $p \sqsubseteq q$, then $p \subset q$ and if f is elementary, then $p \sqsubseteq q$ if and only if $fp \sqsubseteq fq$.
2. If $p \subset q \subset r$, then

- *$p \sqsubseteq q \sqsubseteq r$ implies that $p \sqsubseteq r$,*
- *$p \sqsubseteq r$ implies that $p \sqsubseteq q$,*
- *$p \sqsubseteq r$ implies that $q \sqsubseteq r$.*

3. If $p \in S(A)$ and $A \subset B$, then $p \sqsubseteq q$ for some $q \in S(B)$.
4. There is a cardinal κ such that if $p \in S(A)$, then $p \upharpoonright A_0 \sqsubseteq p$ for some $A_0 \subset A$ with $|A_0| < \kappa$.
5. For any p there is a cardinal λ such that there are at most λ mutually contradictory types q such that $p \sqsubseteq q$.

Then T is stable and this relation is unique.

7 Towards greater generality I: unstable classes

The success of forking inspired attempts to generalise the method outside stable first order theories. This included two directions: one was to loosen the demand of stability of the theory, the other direction was to generalise the techniques of forking to non-elementary classes.

In [28] Shelah proved that a theory is unstable if and only if it has the strict order property[7] or it has the independence property[8]. Avoiding these properties give different directions for generalising forking. One direction pursued was simple theories, which are a subclass of theories without the strict order property. Another class, the so-called NIP theories, are theories that do not have the independence property.

Of course the uniqueness result (Theorem 3) by Harnik and Harrington meant that one would have to give up some of the properties of independence.

7.1 Simple theories

Simple theories were defined by Shelah in [30], in a search for meaningful dividing lines within the class of unstable theories. The original definition of simplicity was that the theory lacks the tree property:

Definition 3. 1. A theory T has the *tree property* if there are a formula $\varphi(\mathbf{x}, \mathbf{y})$, $k < \omega$, a model M of T and sequences $\mathbf{a}_\eta \in M$ ($\eta \in {}^{\omega>}\omega$) such that for any $\eta \in {}^{\omega>}\omega$ the set $\{\varphi(\mathbf{x}; \mathbf{a}_{\eta^\frown\langle l\rangle}) : l < \omega\}$ is k-contradictory (i.e. no subset of cardinality k is satisfied in M) but for every $\eta \in {}^\omega\omega$, the set $\{\varphi(\mathbf{x}; \mathbf{a}_{\eta\restriction n}) : n < \omega\}$ is consistent.
2. A theory without the tree property is *simple*.

So the tree property says that there is a tree where every branch is consistent, but the set of successors of any node is inconsistent. Stable theories are simple, but the simple theories also include unstable ones.

Shelah ([30]) proved that simplicity was equivalent to local character:

Theorem 4 (Shelah [30]). *T is simple if and only if there is $\kappa < |T|^+$ such that for every $p \in S^m(A)$, p does not fork over some $B \subseteq A$, $|B| < \kappa$.*

The success model theorists had with forking in specific simple theories led Kim [19] to study forking in general in this context. He proved that for simple first order theories, forking and dividing coincide

Theorem 5 (Kim). *Let T be simple. Then any partial (or complete) type p forks over a set A if and only if p divides over A.*

In [30] Shelah had proved that forking is reasonably well behaved in simple theories. Kim extended the results proving that symmetry and transitivity hold, thus

---

[7]A theory T has the *strict order property* if there is a formula $\varphi(\mathbf{x}; \mathbf{y})$ such that for every $n < \omega$ there are $\mathbf{a}_0, \ldots, \mathbf{a}_{n-1}$ satisfying: if $k, l < n$, then $\mathfrak{M} \models \exists \mathbf{x}(\neg\varphi(\mathbf{x}; \mathbf{a}_k) \wedge \varphi(\mathbf{x}, \mathbf{a}_l)) \Leftrightarrow k < l$.

[8]A theory T has the *independence property* if there is a formula $\varphi(\mathbf{x}; \mathbf{y})$ and for every $n < \omega$ there are sequences $\mathbf{a}_0, \ldots, \mathbf{a}_{n-1}$ such that: for every $w \subset n$, $\mathfrak{M} \models \exists \mathbf{x}(\bigwedge_{k<n} \varphi(\mathbf{x}; \mathbf{a}_k)^{\text{if } (k \in w)})$, where $\varphi^0 = \varphi$ and $\varphi^1 = \neg\varphi$.

proving that forking in a simple theory satisfies all the usual properties of forking except boundedness of the set of nonforking extensions of a complete type.

Theorem 6 (Kim [19]). *If T is a simple complete first order theory, then forking independence satisfies*

1. *Invariance under automorphisms.*
2. *Symmetry. The type* tp($\mathbf{a}/A\mathbf{b}$) *does not fork over A if and only if* tp($\mathbf{b}/A\mathbf{a}$) *does not fork over A.*
3. *Transitivity. Let $A \subseteq B \subseteq C$. Then* tp($\mathbf{a}/C$) *does not fork over A if and only if* tp($\mathbf{a}/B$) *does not fork over A and* tp($\mathbf{a}/C$) *does not fork over B.*
4. *Existence. Any partial type p over A does not fork over A.*
5. *Extension. If a partial type p over B does not fork over A, then there is a complete type p' over B containing p such that p' does not fork over A.*
6. *Finite Character. The type* tp($\mathbf{a}/B$) *does not fork over A if and only if for each finite tuple $\mathbf{b} \in B$, the type* tp($\mathbf{a}/A\mathbf{b}$) *does not fork over A.*
7. *Local Character. For any complete type p over B, there is a subset A of B such that $|A| \leq |T|$ and p does not fork over A.*
8. *Reflexivity. If $\mathbf{a} \not\subseteq$ acl(A), then* tp($\mathbf{a}/A\mathbf{a}$) *forks over A.*
9. *Monotonicity. Let p, q be partial types and let $q \subseteq p$. If p does not fork over A, then q does not fork over A either.*

Kim and Pillay continued the investigation of forking in simple theories and proved

Theorem 7 (Kim, Pillay [20]).

1. *A theory T is simple if and only if T has a notion of independence $\downarrow$ satisfying*

    a. *invariance under automorphisms*
    b. *local character (with $\kappa \leq |T|^{+}$)*
    c. *finite character*
    d. *existence*
    e. *extension*
    f. *symmetry*
    g. *transitivity*
    h. *Independence Theorem over a model: For M a model, $p \in S(M)$, $A \supset M$, $B \supset M$ and $A \downarrow_M B$, if $p_1 \in S(A)$ and $p_2 \in S(B)$ are nonforking extensions of p, then there is $q \in S(AB)$ extending $p_1 \cup p_2$ such that q is a nonforking extension of p.*

2. *If there is an independence notion as above, then it coincides with nonforking, i.e. $a \downarrow_A B$ ($A \subseteq B$) if and only if* tp(a/B) *does not fork over A.*

So simplicity is equivalent to forking having local character and further to there being any independence notion satisfying the properties listed in Theorem 7(1).

7.2 NIP theories

Forking has also been studied in theories without the independence property, so-called *NIP theories*, also known as *dependent theories*. The independence property was defined by Shelah in [28] and the knowledge of NIP theories was developed by turns by Shelah and Poizat. Adler [1] gives a recent update on the state of research on NIP theories containing an axiomatic approach to independence in NIP theories.

In NIP theories forking independence satisfies the boundedness of free extensions property, but for unstable NIP theories symmetry fails. It still seems to be a very useful notion in the so-called o-minimal theories, see [8].

7.3 Thorn-forking

Thorn-forking was defined by Onshuus in [24]. It gives rise to a well-behaved independence notion in a large class of theories, called 'rosy', which include both the simple and o-minimal theories, but also theories outside these classes. In the stable and o-minimal case, the notion agrees with forking.

Definition 4. 1. A formula $\varphi(\mathbf{x}, \mathbf{a})$ *strongly divides* over A if tp$(\mathbf{a}/A)$ is non-algebraic and $\{\varphi(\mathbf{x}, \mathbf{a}') : \mathbf{a}' \models \mathrm{tp}(\mathbf{a}/A)\}$ is k-inconsistent for some $k \in \mathbb{N}$.
2. A formula $\varphi(\mathbf{x}, \mathbf{a})$ þ-*divides* (thorn divides) over A if there is a tuple $\mathbf{c}$ such that $\varphi(\mathbf{x}, \mathbf{a})$ strongly divides over $A\mathbf{c}$.
3. A formula þ-*forks* (thorn-forks) over A if it implies a finite disjunction of formulae which þ-divide over A.
4. A type p þ-forks over A if there is a formula in p which þ-forks over A.

Adler [2] has given a nice geometric exposition of both forking and þ-forking. He studied the following list of axioms:

1. (invariance) If $A \downarrow_C B$ and $(A', B', C') \equiv (A, B, C)$, then $A' \downarrow_{C'} B'$.
2. (monotonicity) If $A \downarrow_C B$, $A' \subseteq A$ and $B' \subseteq B$, then $A' \downarrow_C B'$.
3. (base monotonicity) Suppose $D \subseteq C \subseteq B$. If $A \downarrow_D B$, then $A \downarrow_C B$.
4. (left transitivity) Suppose $D \subseteq C \subseteq B$. If $B \downarrow_C A$ and $C \downarrow_D A$, then $B \downarrow_D A$.
5. (normality) $A \downarrow_C B$ implies $AC \downarrow_C B$.
6. (extension) If $A \downarrow_C B$ and $\hat{B} \supseteq B$, then there is $A' \equiv_{BC} A$ such that $A' \downarrow_C \hat{B}$.
7. (finite character) If $A_0 \downarrow_C B$ for all finite $A_0 \subseteq A$, then $A \downarrow_C B$.
8. (local character) For every A there is a cardinal $\kappa(A)$ such that for any set B there is a subset $C \subseteq B$ of cardinality $|C| < \kappa(A)$ such that $A \downarrow_C B$.
9. (anti-reflexivity) $\mathbf{a} \downarrow_B \mathbf{a}$ implies $\mathbf{a} \in \mathrm{acl}(B)$.
10. (full existence) For any A, B and C there is $A' \equiv_C A$ such that $A' \downarrow_C B$.
11. (symmetry) $A \downarrow_C B \Leftrightarrow B \downarrow_C A$.

The list of axioms defines the same notion as the first 7 axioms by Kim and Pillay, so adding the independence theorem over models gives the characterisation of forking independence in simple theories.

Adler points out that forking independence $\downarrow^f$ can be seen as the relation one gets from dividing independence $\downarrow^d$ by adding the extension property:

$$A \underset{C}{\overset{f}{\downarrow}} B \Leftrightarrow \left(\text{for all } \hat{B} \supseteq B \text{ there is } A' \equiv_{BC} A \text{ such that } A' \underset{C}{\overset{d}{\downarrow}} \hat{B} \right).$$

Adler continues to show that similarly one can add base monotonicity to the independence notion one gets from algebraic closures

$$A \underset{C}{\overset{a}{\downarrow}} B \Leftrightarrow \mathrm{acl}(AC) \cap \mathrm{acl}(BC) = \mathrm{acl}(C).$$

Now adding base monotonicity yields M-dividing independence $\downarrow^M$:

$$A \underset{C}{\overset{M}{\downarrow}} B \Leftrightarrow \left(\begin{array}{l} \text{for any } C' \text{ such that } C \subseteq C' \subseteq \mathrm{acl}(BC): \\ \mathrm{acl}(AC') \cap \mathrm{acl}(BC') = \mathrm{acl}(C') \end{array} \right).$$

Then adding extensions yields þ-forking:

$$A \underset{C}{\overset{\text{þ}}{\downarrow}} B \Leftrightarrow \left(\text{for all } \hat{B} \supseteq B \text{ there is } A' \equiv_{BC} A \text{ such that } A' \underset{C}{\overset{M}{\downarrow}} \hat{B} \right).$$

Adler proved in [2] that þ-forking is the weakest independence notion in the sense that if there is any independence notion $\downarrow$ satisfying his axioms, then $\downarrow^{\text{þ}}$ also satisfies them and $A \downarrow_C B$ implies $A \underset{C}{\overset{\text{þ}}{\downarrow}} B$. Dually he proves that forking independence is the strongest independence notion. For stable theories the two notions are the same, but García, Onshuus and Usvyatsov [9] point out that if the theory admits a definable order, forking and thorn-forking are very different (but in this case they of course do not satisfy all the independence axioms).

8 Towards greater generality II: non-elementary classes

Another direction to generalise the notion of forking was not to reduce the stability assumptions but to move outside elementary classes.

8.1 Finite diagrams or homogeneous model theory

Homogeneous model theory was initiated in the late 1960s by Shelah's article on finite diagrams [27]. The framework corresponds to working with a strongly μ-homogeneous (but not saturated) monster model $\mathfrak{M}$, for some suitably large μ, studying the class of its elementary substructures. The most significant difference to studying elementary classes is that in a homogeneous class one does not have a compactness theorem.

A class $\mathbb{K}$ in this context is called λ-stable, if $\mathfrak{M}$ realises only λ complete 1-types over any parameter set of cardinality λ. Stability and superstability are defined correspondingly as $\mathfrak{M}$ being λ-stable for some λ or from some λ onwards. Note that a homogeneous monster may be stable although its theory is not.

In the late 1990s Hyttinen and Shelah [15] studied strong splitting in the framework of a stable homogeneous class and defined an independence relation based on it:

Definition 5 (Hyttinen, Shelah). Assume $\mathbb{K}$ is a stable homogeneous class.

1. Denote by $\kappa(\mathbb{K})$ the least cardinal κ such that there is no strongly splitting sequence of length κ.
2. We write $\mathbf{a} \downarrow_A B$ if there is $C \subseteq A$ of power $< \kappa(\mathbb{K})$ such that for all $D \supseteq AB$ there is $\mathbf{b}$ which satisfies $\mathrm{tp}(\mathbf{b}/AB) = \mathrm{tp}(\mathbf{a}/AB)$ such that $\mathrm{tp}(\mathbf{b}/D)$ does not split strongly over C. For an arbitrary set C, $C \downarrow_A B$ means $\mathbf{a} \downarrow_A B$ for all finite $\mathbf{a} \in C$.

The independence notion satisfies many of the properties of nonforking:

Theorem 8 (Hyttinen, Shelah [15]). *In a stable homogeneous class $\downarrow$ satisfies*

- *If $A \subseteq A' \subseteq B' \subseteq B$ and $\mathbf{a} \downarrow_A B$, then $\mathbf{a} \downarrow_{A'} B'$.*
- *If $A \subseteq B$, $\mathbf{a} \downarrow_A A$ and $\mathrm{tp}(\mathbf{a}/A)$ is unbounded[9], then there is $\mathbf{b}$ such that $\mathbf{b} \downarrow_A B$ and $Lstp(b/A) = Lstp(a/A)$[10].*
- *If $A \subseteq B$, $\mathbf{a} \not\downarrow_A B$, $\mathbf{a} \downarrow_A A$, and $\mathrm{tp}(\mathbf{a}/A)$ is unbounded, then there is some finite $B' \subseteq B$ such that $\mathbf{a} \not\downarrow_A B'$.*
- *For all $\mathbf{a}, \mathbf{b}$ and A, $\mathbf{b} \downarrow_A A$ and $\mathbf{a} \downarrow_A \mathbf{b}$ implies $\mathbf{b} \downarrow_A \mathbf{a}$. By finite character this generalises to: if $A \downarrow_B C$ and $C \downarrow_B B$, then $C \downarrow_B A$.*
- *If $\mathbf{b} \downarrow_A D$ and $\mathbf{c} \downarrow_{Ab} D$, then $\mathbf{bc} \downarrow_A D$.*
- *If $\mathbf{a} \downarrow_A \mathbf{c}$, $\mathbf{b} \downarrow_A \mathbf{c}$ and $Lstp(\mathbf{a}/A) = Lstp(\mathbf{b}/A)$, then $\mathrm{tp}(\mathbf{a}/A\mathbf{c}) = \mathrm{tp}(\mathbf{b}/A\mathbf{c})$.*
- *If $C \downarrow_A B$ and $D \downarrow_{AC} B$, then $CD \downarrow_A B$.*
- *If $A \subseteq B$, $\mathbf{a} \downarrow_A B$, $\mathbf{a} \downarrow_B C$, $C \downarrow_B B$ and $B \downarrow_A A$, then $\mathbf{a} \downarrow_A C$.*

Hyttinen and Shelah define *a*-saturation as follows:

Definition 6. 1. $SE^n(A)$ is the set of all equivalence relations E in $\mathfrak{M}^n$ that are A-invariant (i.e. for any automorphism fixing A pointwise $\mathbf{a}E\mathbf{b}$ if and only if $f(\mathbf{a})Ef(\mathbf{b})$) and have a bounded (i.e. $< |\mathfrak{M}|$) number of equivalence classes.
2. $\mathbf{a}$ and $\mathbf{b}$ have the same *Lascar strong type over* A, $Lstp(\mathbf{a}/A) = Lstp(\mathbf{b}/A)$ if $\mathbf{a}E\mathbf{b}$ for every $E \in SE^{\mathrm{length}(\mathbf{a})}$
3. A model M is *strongly $F_\kappa^{\mathfrak{M}}$-saturated* if for all $A \subseteq M$ of cardinality $< \kappa$ and $\mathbf{a} \in \mathfrak{M}$ there is $\mathbf{b} \in M$ such that $Lstp(\mathbf{b}/A) = Lstp(\mathbf{a}/A)$.
4. *a*-saturated means strongly $F_{\kappa(\mathbb{K})}^{\mathfrak{M}}$-saturated.

---

[9]A type is *unbounded* if it has $|\mathfrak{M}|$ many realisations.

[10]Lascar strong types (Lstp) are defined in Definition 6.

Over a-saturated models the independence notion is even better behaved.

Theorem 9 (Hyttinen, Shelah [15]).

1. If M is a-saturated, then for all $\mathbf{a}$, $\mathbf{a} \downarrow_M M$.
2. Complete types over a-saturated models are stationary.

Hyttinen and Lessmann [13] defined a rank for superstable homogeneous classes as follows:

Definition 7 (Hyttinen, Lessmann [13]). For a complete type p in finitely many variables over finitely many parameters, $R(p) \geq \alpha$ is defined by induction on α as follows:

- $R(p) \geq 0$ if p is realised in $\mathfrak{M}$.
- $R(p) \geq \alpha$, if $R(p) \geq \beta$ for each $\beta < \alpha$ when α is a limit ordinal.
- $R(p) \geq \alpha + 1$, if for each λ, there exists a family $\{r_i : i < \lambda\}$ such that

  - r_i is a complete type over finitely many parameters extending p, for each $i < \lambda$,
  - $\bigcup_{i \in S} r_i$ is not realised in $\mathfrak{M}$, for each infinite $S \subseteq \lambda$,
  - $R(r_i) \geq \alpha$, for each $i < \lambda$.

$R(p) = \alpha$ if $R(p) \geq \alpha$ but it is not the case that $R(p) \geq \alpha + 1$. $R(p) = \infty$ if $R(p) \geq \alpha$ for all α. Finally, when q is a complete type over infinitely many parameters,

$$R(q) = \min\{R(q \restriction B) : B \subseteq \mathrm{dom}(q) \text{ finite}\}.$$

With this rank they defined an independence relation, improving the independence properties from [15].

Definition 8. If B and C are sets and $\mathbf{a}$ a finite sequence

$$\mathbf{a} \downarrow_B C$$

is defined to hold if there exists a finite set $B' \subseteq B$ such that for all D containing BC, there exists $\mathbf{a}' \models \mathrm{tp}(\mathbf{a}/BC)$ such that $R(\mathrm{tp}(\mathbf{a}'/B'D)) = R(\mathrm{tp}(\mathbf{a}'/B'))$.

$A \downarrow_B C$ if for all finite $\mathbf{a} \in A$, $\mathbf{a} \downarrow_B C$.

Theorem 10 (Hyttinen, Lessmann [13]). *In a superstable homogeneous class the independence notion $\downarrow$ defined above satisfies*

- *(Restricted local character) If $\mathbf{a} \downarrow_B C$, then $\mathbf{a} \downarrow_{B'} C$ for some finite $B' \subseteq B$.*
- *(Monotonicity) If $B \subseteq B_1 \subseteq C_1 \subseteq C$ and $\mathbf{a} \downarrow_B C$, then $\mathbf{a} \downarrow_{B_1} C_1$.*
- *(Extension) If $\mathbf{a} \downarrow_B C$ and $D \supseteq C$, then $\mathbf{a}' \downarrow_B D$ for some $\mathbf{a}' \models \mathrm{tp}(\mathbf{a}/CB)$.*
- *(Transitivity) If $B \subseteq C \subseteq D$, then $\mathbf{a} \downarrow_B D$ if and only if $\mathbf{a} \downarrow_B C$ and $\mathbf{a} \downarrow_C D$.*
- *(Invariance) $\downarrow$ is invariant under automorphisms.*
- *(Symmetry, partial) If M is a-saturated, then $\mathbf{a} \downarrow_M \mathbf{b}$ if and only if $\mathbf{b} \downarrow_M \mathbf{a}$.*
- *(Symmetry over extension bases) If $\mathbf{a} \downarrow_A A$ and $\mathbf{b} \downarrow_A A$, then $\mathbf{a} \downarrow_A \mathbf{b}$ if and only if $\mathbf{b} \downarrow_A \mathbf{a}$.*

So $\downarrow$ has all properties of nonforking in the first order case, except extension (or existence in this case, since the relation has built-in extensions), i.e., whether $\mathbf{a} \downarrow_A A$ holds for all finite $\mathbf{a}$ and sets A.

Hyttinen and Lessmann also gave an example, contributed to Shelah, showing that the dependence developed is in a sense best possible: there exists a superstable (actually ω-stable) homogeneous monster $\mathfrak{M}$ which does not admit a dependence relation satisfying the properties in Theorem 10 and in addition extension over all sets. This proves that in homogeneous classes ω-stability does not imply simplicity.

They also proved the following characterisation:

Theorem 11 ([13]). *Let $\mathfrak{M}$ be a monster for a homogeneous class. Suppose there exists a dependence relation, written $\downarrow$, satisfying*

1. *(Local character) For all finite $\mathbf{a}$ and set C, there exists a finite $B \subseteq C$ such that $\mathbf{a} \downarrow_B C$.*
2. *(Monotonicity) If $B \subseteq B_1 \subseteq C_1 \subseteq C$ and $\mathbf{a} \downarrow_B C$, then $\mathbf{a} \downarrow_{B_1} C_1$.*
3. *(Extension) If $\mathbf{a} \downarrow_B C$ and $D \supseteq C$, then $\mathbf{a}' \downarrow_B D$ for some $\mathbf{a}' \models \mathrm{tp}(\mathbf{a}/CB)$.*
4. *(Transitivity) If $B \subseteq C \subseteq D$, then $\mathbf{a} \downarrow_B D$ if and only if $\mathbf{a} \downarrow_B C$ and $\mathbf{a} \downarrow_C D$.*
5. *(Symmetry) If $A \downarrow_B C$, then $C \downarrow_B A$.*
6. *(Finite character) If $\mathbf{a} \downarrow_B \mathbf{c}$ for all finite $\mathbf{a} \in A$ and $\mathbf{c} \in C$, then $A \downarrow_B C$.*
7. *(Invariance) $\downarrow$ is invariant under automorphisms.*
8. *(Bounded extensions) There exists a cardinal λ, such that for each $A \subseteq C$, the size of $\{\mathrm{tp}(\mathbf{a}/C) : \mathbf{a} \downarrow_A C\}$ is at most $\lambda + |A|$.*

Then $\mathfrak{M}$ is superstable and $A \downarrow_C B$ if and only if for all finite $\mathbf{a} \in A$ and $\mathbf{b} \in B$, $\mathrm{tp}(\mathbf{a}/C\mathbf{b})$ does not divide over C.

8.2 Excellent classes

An even more general framework than the homogeneous one is that of excellent classes. These were defined by Shelah [31, 32], where he studied the class of models of a sentence in $\mathscr{L}_{\omega_1\omega}$ for countable $\mathscr{L}$. Classifying these corresponds to classifying the class of atomic models of a countable first order theory. For these categoricity transfer can be proved if the class contains sufficiently homogeneous models. Excellence is a property that ensures such models exist.

Grossberg and Hart [11] managed to classify the excellent classes and prove a Main Gap theorem: If an excellent class has the dimensional order property (DOP), then it has the maximal number of models; if it does not have DOP, then all models can be decomposed as a tree of countable models.

Grossberg and Hart use nonsplitting as their starting point for independence and define $A \downarrow_B C$ if and only if for every $\mathbf{a} \in A$ there is $\mathbf{b} \in B$ so that $\mathrm{tp}(\mathbf{a}/BC)$ does not split over $\mathbf{b}$. This notion satisfies symmetry over models, as well as monotonicity and transitivity, but it lacks free extensions in general. However, extension holds over the so-called *good* sets and one of the consequences of excellence is the existence of these good sets.

8.3   Abstract elementary classes

Abstract elementary classes are the most general classes for which classification
theory has been actively developed. Their roots can be seen to go back to Jonsson
[17, 18] who axiomatised classes of models for which one could construct universal
homogeneous models. The most crucial generalisation abstract elementary classes
make to Jonsson's classes is in allowing different notions of *strong submodel* instead
of just studying the submodel relation. The definition of abstract elementary classes
was given by Shelah [33]. The work done in classifying these so far is gathered in
Shelah's two volumes [35, 36]. A very readable treatment on abstract elementary
classes is Baldwin's book [4].

The approach is syntax-free, and both types and forking-like notions have to be
defined via embeddings or automorphisms.

Definition 9. A pair $(\mathbb{K}, \prec_{\mathbb{K}})$, where $\mathbb{K}$ is a class of structures of a given vocabulary
τ and $\prec_{\mathbb{K}}$ is a partial order on the class, is an *abstract elementary class* (AEC) if the
following axioms are satisfied

1. Both $\mathbb{K}$ and $\prec_{\mathbb{K}}$ are closed under isomorphism.
2. If $M, N \in \mathbb{K}$ and $M \prec_{\mathbb{K}} N$, then $M \subseteq N$, i.e., M is a submodel of N.
3. If $M, M', N \in \mathbb{K}$ satisfy $M \prec_{\mathbb{K}} N$, $M' \prec_{\mathbb{K}} N$ and $M \subseteq M'$, then $M \prec_{\mathbb{K}} M'$.
4. There exists a cardinal $LS(\mathbb{K}) \geq |\tau| + \aleph_0$ such that for any $M \in \mathbb{K}$ and set $A \subset M$
   there is $N \in \mathbb{K}$ such that $A \subseteq N \prec_{\mathbb{K}} M$ and $|N| \leq |A| + LS(\mathbb{K})$.
5. a. If $(M_i)_{i<\mu} \subseteq \mathbb{K}$ is a $\prec_{\mathbb{K}}$-increasing chain, then $\bigcup_{i<\mu} M_i \in \mathbb{K}$ and $M_i \prec_{\mathbb{K}}$
      $\bigcup_{i<\mu} M_i$ for each $i < \mu$.
   b. If $(M_i)_{i<\mu} \subseteq \mathbb{K}$ is a $\prec_{\mathbb{K}}$-increasing chain and for some $N \in \mathbb{K}$, $M_i \prec_{\mathbb{K}} N$ for
      all $i < \mu$, then $\bigcup_{i<\mu} M_i \prec_{\mathbb{K}} N$.

The canonical example of an AEC is an elementary class together with the
elementary submodel relation. The idea is to extract the most crucial properties of
this relation, and demand them of the class under consideration, but not demand that
there be a logic defining the strong submodel relation one is interested in.

In this setting there are no syntactic types, so types will have to be replaced by
Galois-types. The formulation below is from [16].

Definition 10. 1. For $M, N_1, N_2 \in \mathbb{K}$ with $M \prec_{\mathbb{K}} N_i$, $\mathbf{a}_1 \in N_1$ and $\mathbf{a}_2 \in N_2$, write

$$(M, N_1, \mathbf{a}_1) E^* (M, N_2, \mathbf{a}_2)$$

if there are some $N \in \mathbb{K}$ and embeddings $f_1 : N_1 \longrightarrow N$ and $f_2 : N_2 \longrightarrow N$
satisfying $f_i(N_i) \prec_{\mathbb{K}} N$ such that $f_1(\mathbf{a}_1) = f_2(\mathbf{a}_2)$ and $f_1 \upharpoonright M = f_2 \upharpoonright M$.
2. Let E be the transitive closure of E^*. Then E is an equivalence relation.
3. The *Galois-type* of $\mathbf{a}$ in N over M is the E-equivalence class of $(M, N, \mathbf{a})$.

If the AEC satisfies amalgamation, already E^* is an equivalence relation, and
Galois-types are well behaved. For classification the basic axioms of an AEC are
often not enough and in many cases axioms for joint embedding, amalgamation and

no maximal model are assumed. This enables the construction of a monster model for the class, and then Galois-types will correspond to orbits under automorphisms fixing models (pointwise).

The most widely studied context for AECs, as far as independence is concerned, is that of *good frames*. It is designed to be the analog of superstable elementary classes. This is a context where one assumes the existence of a local independence notion for types over models of a fixed cardinality. One then studies, on one hand, which conditions would suffice for the existence of such an independence notion, on the other, what one could deduce from it, e.g., transferability of the good frames from one cardinality to another. Good frames consist of

- the collection $\mathbb{K}_\lambda$ of all λ-sized models of an AEC,
- for each $M \in \mathbb{K}_\lambda$ a well-behaved collection $S^{bs}(M)$ of types over M, called *basic*, and
- a notion "$p \in S^{bs}(N)$ does not fork over $M \prec_{\mathbb{K}} N$", satisfying some axioms for independence.

Boney [6] has studied good frames under the additional assumption of tameness (stating that differences in types can be detected over small enough models). An attempt to study independence without amalgamation has been done by Jarden and Shelah [16].

Recently Boney et al. [7] have proved a very general uniqueness result for independence notions in AECs extracting the minimum of requirements for a uniqueness proof. They work in the context of an abstract elementary class $(\mathbb{K}, \prec)$ that satisfies amalgamation, joint embedding and no maximal model.

In this setting they study the abstract notion of a general independence relation.

Definition 11 (Boney et al. [7]). An *independence relation* $\downarrow$ is a set of triples of the form (A, M, N) where A is a set, M, N are models (i.e. $N, N \in \mathbb{K}$), $M \prec N$. Write $A \downarrow_M N$ for $(A, M, N) \in \downarrow$. When $A = \{a\}$, we may write $a \downarrow_M N$ for $A \downarrow_M N$. The relation $\downarrow$ is required to satisfy

- (*I*) Invariance: Assume $(A, M, N) \equiv (A', M', N')$. Then $A \downarrow_M N$ if and only if $A' \downarrow_{M'} N'$.
- (*M*) Left and right monotonicity: If $A \downarrow_M N$, $A' \subseteq A$, $M \prec N' \prec N$, then $A' \downarrow_M N'$.
- (*B*) Base monotonicity: If $A \downarrow_M N$, and $M \prec M' \prec N$, then $A \downarrow_{M'} N$.

$\downarrow_M$ denotes $\downarrow$ restricted to the base set M.

Boney et al. consider the following additional properties of independence:

- $(C)_\kappa$ Continuity: If $A \not\downarrow_M N$, then there exists $A^- \subseteq A$, $B^- \subseteq N$ of size $< \kappa$ such that for all $N_0 \succ M$ containing B^-, we have $A^- \not\downarrow_M N_0$.
- (*T*) Left transitivity: If $M_1 \downarrow_{M_0} N$, and $M_2 \downarrow_{M_1} N$, with $M_0 \prec M_1 \prec M_2$, then $M_2 \downarrow_{M_0} N$.
- (T_*) Right transitivity: If $A \downarrow_{M_0} M_1$, and $A \downarrow_{M_1} M_2$, with $M_0 \prec M_1 \prec M_2$, then $A \downarrow_{M_0} M_2$.

- (S) Symmetry: If $A \downarrow_M N$, then there is $M' \succ M$ with $A \subseteq M'$ such that $N \downarrow_M M'$. If A is a model extending M, one can take $M' = A$.
- (U) Uniqueness: If $A \downarrow_M N$, $A' \downarrow_M N$, and $f : A \equiv_M A'$, then $g : A \equiv_N A'$ for some g so that $g \upharpoonright A = f \upharpoonright A$.
- (E) The following properties hold:
  - (E_0) Existence: for all A, $A \downarrow_M M$.
  - (E_1) Extension: Given a set A, and $M \prec N \prec N'$, if $A \downarrow_M N$, then there is $A' \equiv_N A$ such that $A' \downarrow_M N'$.
- (L) Local character: $\kappa_\alpha(\downarrow) < \infty$ for all α, where $\kappa_\alpha(\downarrow) := \min\{\lambda \in REG \cup \{\infty\} :$ for all $\mu = \mathrm{cf}\,\mu \geq \lambda$, all increasing, continuous chains $\langle M_i : i \leq \mu \rangle$ and all sets A of size α, there is some $i_0 < \mu$ so $A \downarrow_{M_0} M_\mu\}$.

When (P) is a property other than local character and M is a model , $\downarrow$ is said to have $(P)_M$ if $\downarrow_M$ has (P).

They prove the following:

Theorem 12 (Boney et al. [7]). *In an abstract elementary class satisfying joint embedding, amalgamation and no maximal model, if $\downarrow^1$ and $\downarrow^2$ are independence notions (i.e. satisfy invariance as well as left, right and base monotonicity) and*

- $\downarrow^1$ *satisfies* $(E)_M$ *and* $(U)_M$,
- $\downarrow^2$ *satisfies* $(E)_M$, $(U)_M$ *and* (L),

then $\downarrow^1_M = \downarrow^2_M$.

In particular there can be at most one independence relation satisfying existence, extension, uniqueness and local character.

9 Conclusion

I have described independence notions in various model theoretic frameworks, beginning with nonforking for first order theories and proceeding to different generalisations of it. The more general the setting, the more restrictive the applicability of independence gets, as one will lose the formula-wise information forking provides in first order theories and has to restrict attention to complete types, or even only complete types over (saturated enough) models. However, in the frameworks described, there is some form of 'independence calculus' that can be used for classification. The independence notions studied also demonstrate a common uniqueness property: whenever one can define a well-behaved independence notion, it tends to be unique.

Acknowledgements The author was partially supported by grant 251557 of the Academy of Finland. I also wish to express my gratitude to the anonymous referee for valuable comments.

References

1. Adler, H.: Introduction to theories without the independence property. Arch. Math. Log. (to appear)
2. Adler, H.: A geometric introduction to forking and thorn-forking. J. Math. Log. **9**(1), 1–20 (2009)
3. Baldwin, J.T.: Fundamentals of Stability Theory. Perspectives in Mathematical Logic. Springer, Berlin (1988)
4. Baldwin, J.T.: Categoricity. University Lecture Series, vol. 50. American Mathematical Society, Providence, RI (2009)
5. Berenstein, A., Buechler, S.: Simple stable homogeneous expansions of Hilbert spaces. Ann. Pure Appl. Logic **128**(1–3), 75–101 (2004)
6. Boney, W.: Tameness and extending frames. J. Math. Logic **14**(2) (2014)
7. Boney, W., Grossberg, R., Kolesnikov, A., Vasey, S.: Canonical forking in AECs. Submitted. arXiv:1404.1494
8. Dolich, A.: Forking and independence in o-minimal theories. J. Symb. Log. **69**(1), 215–240 (2004)
9. García, D., Onshuus, A., Usvyatsov, A.: Generic stability, forking, and thorn-forking. Trans. Am. Math. Soc. **365**(1), 1–22 (2013)
10. Grädel, E., Väänänen, J.: Dependence and independence. Stud. Logica **101**(2), 399–410 (2013)
11. Grossberg, R., Hart, B.: The classification of excellent classes. J. Symb. Log. **54**(4), 1359–1381 (1989)
12. Harnik, V., Harrington, L.: Fundamentals of forking. Ann. Pure Appl. Logic **26**(3), 245–286 (1984)
13. Hyttinen, T., Lessmann, O.: A rank for the class of elementary submodels of a superstable homogeneous model. J. Symb. Log. **67**(4), 1469–1482 (2002)
14. Hyttinen, T., Paolini, G.: Reduction of database independence to dividing in atomless Boolean algebras. Arch. Math. Logic (2016)
15. Hyttinen, T., Shelah, S.: Strong splitting in stable homogeneous models. Ann. Pure Appl. Log. **103**(1–3), 201–228 (2000)
16. Jarden, A., Shelah, S.: Non-forking frames in abstract elementary classes. Ann. Pure Appl. Log. **164**(3), 135–191 (2013)
17. Jónsson, B.: Universal relational systems. Math. Scand. **4**, 193–208 (1956)
18. Jónsson, B.: Homogeneous universal relational systems. Math. Scand. **8**, 137–142 (1960)
19. Kim, B.: Forking in simple unstable theories. J. Lond. Math. Soc. (2) **57**(2), 257–267 (1998)
20. Kim, B., Pillay, A.: Simple theories. Ann. Pure Appl. Log. **88**(2–3), 149–164 (1997). Joint AILA-KGS Model Theory Meeting (Florence, 1995)
21. Lascar, D.: Ranks and definability in superstable theories. Isr. J. Math. **23**(1), 53–87 (1976)
22. Lascar, D., Poizat, B.: An introduction to forking. J. Symb. Log. **44**(3), 330–350 (1979)
23. Morley, M.: Categoricity in power. Trans. Am. Math. Soc. **114**(2), 514–538 (1965)
24. Onshuus, A.: Properties and consequences of thorn-independence. J. Symb. Log. **71**(1), 1–21 (2006)
25. Pillay, A.: An introduction to stability theory. In: Oxford Logic Guides, vol. 8. Clarendon Press, Oxford University Press, New York (1983)
26. Poizat, B.: A course in model theory: an introduction to contemporary mathematical logic. In: Universitext. Springer, New York (2000). Translated from the French by Moses Klein and revised by the author
27. Shelah, S.: Finite diagrams stable in power. Ann. Math. Log. **2**(1), 69–118 (1970/1971)
28. Shelah, S.: Stability, the f.c.p., and superstability; model theoretic properties of formulas in first order theory. Ann. Math. Log. **3**(3), 271–362 (1971)
29. Shelah, S.: Classification Theory and the Number of Nonisomorphic Models. Studies in Logic and the Foundations of Mathematics, vol. 92. North-Holland, Amsterdam, New York (1978)
30. Shelah, S.: Simple unstable theories. Ann. Math. Log. **19**(3), 177–203 (1980)

31. Shelah, S.: Classification theory for nonelementary classes. I. The number of uncountable models of $\psi \in L_{\omega_1,\omega}$. Part A. Isr. J. Math. **46**(3), 212–240 (1983)
32. Shelah, S.: Classification theory for nonelementary classes. I. The number of uncountable models of $\psi \in L_{\omega_1,\omega}$. Part B. Isr. J. Math. **46**(4), 241–273 (1983)
33. Shelah, S.: Classification of nonelementary classes. II. Abstract elementary classes. In: Classification Theory (Chicago, IL, 1985). Lecture Notes in Mathematics, vol. 1292, pp. 419–497. Springer, Berlin (1987)
34. Shelah, S.: Classification theory and the number of nonisomorphic models. Studies in Logic and the Foundations of Mathematics, vol. 92, 2nd edn. North-Holland, Amsterdam (1990)
35. Shelah, S.: Classification Theory for Abstract Elementary Classes, vol. 2. Studies in Logic (London), vol. 20. College Publications, London (2009)
36. Shelah, S.: Classification Theory for Abstract Elementary Classes. Studies in Logic (London), vol. 18. College Publications, London (2009)
37. van der Waerden, B.L.: Moderne Algebra. Teil I. Julius Springer, Berlin (1930)
38. Whitney, H.: On the abstract properties of linear dependence. Am. J. Math. **57**(3), 509–533 (1935)

Dependency as Question Entailment

Ivano Ciardelli

Abstract In the past few years, a tight connection has emerged between logics of dependency and logics of questions. The aim of this paper is to show that this connection stems from a fundamental relation existing between dependency and questions. Once we expand our view on logic by bringing questions into the picture, dependency emerges as a facet of the fundamental logical notion of entailment, namely entailment among questions. Besides providing an insightful conceptual picture, this perspective yields a general and well-behaved way of dealing with dependency in logical systems.

1 Introduction

Recent years have witnessed the parallel development of two novel logical frameworks: Dependence Logic and Inquisitive Semantics.

At first sight, these frameworks seem to pursue radically different enterprises. The former line of work originates from Henkin's observation that certain natural patterns of quantification over individuals are not expressible in first-order logic. For instance, it is impossible to write a first-order formula expressing that for every x_1 and x_2, there exist a y_1 determined only by x_1 and a y_2 determined only by x_2, such that a certain formula $\varphi(x_1, x_2, y_1, y_2)$ holds. To provide the tools to express such patterns, Henkin [22] introduced the so-called *branching quantifiers*, and Hintikka and Sandu [24] later developed this work in the framework of Independence Friendly (IF) logic, which allows for quantified variables to be marked as independent of other variables. In recent years, Väänänen [33] proposed a new approach to the issue: he noticed that dependency and quantification may be separated out. In the resulting Dependence Logic, quantifiers have the standard form, but the language is enriched with a new kind of atomic formula $=(x_1, \ldots, x_n, y)$, expressing the fact that the value of y is determined by the values of $x_1, \ldots, x_n$. Thus, in Dependence Logic, the pattern of quantification mentioned above would be expressed as follows:

I. Ciardelli (✉)
ILLC, University of Amsterdam, Amsterdam, The Netherlands
e-mail: i.a.ciardelli@uva.nl

© Springer International Publishing Switzerland 2016
S. Abramsky et al. (eds.), *Dependence Logic*, DOI 10.1007/978-3-319-31803-5_8

$$\forall x_1 \forall x_2 \exists y_1 \exists y_2 (\; =(x_1, y_1) \; \wedge \; =(x_2, y_2) \; \wedge \; \varphi(x_1, x_2, y_1, y_2))$$

Clearly, in the standard semantic context consisting of a model and an assignment, every variable simply gets a specific value, and there is no way to make sense of the idea that some variables are determined by others. Instead, building on the work by Hodges [25, 26] on IF logic, Dependence Logic is interpreted with respect to a model and a *set* of assignments, called a *team*. Relative to a team X, we can say that the value of y is determined by the value of $x_1, \ldots, x_n$ in case the value that an assignment $g \in X$ assigns to y is fully determined by the values it assigns to $x_1, \ldots, x_n$. Dependence atoms are thus interpreted by means of the following clause:

$$M \models_X \; =(x_1, \ldots, x_n, y) \iff \forall g, g' \in X, \text{ if } g(x_i) = g'(x_i) \text{ for all } i, \text{ then } g(y) = g'(y)$$

Due to the similarity between assignments for individual variables in predicate logic and valuations for propositional letters in propositional logic, dependence atoms were later considered also in the setting of propositional and modal logic [34, 39]. There, an atom has the form $=(p_1, \ldots, p_n, q)$, and it is interpreted, relative to a set s of valuations, as expressing that the truth-value that a valuation $w \in s$ assigns to q is determined by the truth-values it assigns to $p_1, \ldots, p_n$.

Inquisitive Semantics stems from a different line of work, whose aim is to give a logical account of *information exchange* as a process of requesting and providing information. This enterprise requires a framework in which not only statements, which are used to *provide* information, but also questions, which are used to *request* information, may be interpreted. An early example of such a framework is the Logic of Interrogation (LoI) of Groenendijk [13], in which statements and questions are interpreted with respect to pairs $\langle w, w' \rangle$ of possible worlds: for a statement α, $\langle w, w' \rangle \models \alpha$ holds in case α is true at both worlds; for a question μ, $\langle w, w' \rangle \models \mu$ holds in case the true answer to μ is the same in the two worlds. In general, the set $[\varphi]$ of pairs satisfying φ is a symmetric and transitive relation. For a statement α, this relation has a unique connected component, which coincides with the set of worlds where α is true, i.e., with the proposition expressed by α in classical logic. For a question μ, the relation $[\mu]$ typically has multiple connected components, corresponding to the possible answers to μ. In this way, classical logic is extended in a conservative way to encompass questions.

However, this view is only suitable for a restricted class of questions, namely, *unique-answer* questions — questions that have a unique true answer at each world. The pursuit of greater generality has lead first to drop the transitivity requirement [14, 27, 36] and finally to abandon the relational setting altogether [4, 7, 15] in favor of what is now called Inquisitive Semantics. Like LoI, Inquisitive Semantics interprets statements and questions in a uniform way. Only, the points of interpretations are no longer *pairs* of possible worlds, but rather *sets* of possible worlds, regarded intuitively as *information states*. One way to read the fundamental relation $\models$ of the semantics is that $s \models \varphi$ holds in case the information state s *settles* φ. The situation is similar to that of LoI: statements α have just one "way" of being settled, corresponding to the set of all worlds where α is true, while questions have multiple "ways" of being settled, corresponding to their possible answers. Thus, like LoI, Inquisitive Semantics extends classical logic to encompass questions, but now without being confined to unique-answer questions.

Recently, Yang [39] remarked some striking, unexpected similarities between the frameworks of Dependence Logic and Inquisitive Semantics. Indeed, in propositional Inquisitive Semantics and in propositional Dependence Logic, meanings are the same kind of object, namely *downward closed sets of information states*, where information states are sets of propositional valuations; furthermore, many connectives and properties ended up being considered independently in both theories—so much so that some versions of these systems could be seen as variants of each other.

The purpose of this paper is to show that this convergence, far from being an accident, stems from a fundamental connection between questions and dependency: namely, dependency is nothing but *question entailment*. This implies that logics of questions and logics of dependency share the same object of investigation. In light of this, the convergence of these two historically independent lines of research looks like a welcome indication that these approaches are on the right track.

More specifically, in this paper we will look at a simple and natural way of extending classical logic to questions, and we will find that the relation of dependency emerges naturally from this approach as a case of entailment. We will see that, besides providing an insightful conceptual picture, this perspective also suggests a general and well-behaved way of handling dependencies in logical systems.

The paper is organized as follows. Section 2 discusses how the relation of dependency emerges as a particular kind of entailment once we broaden our perspective on logic so as to encompass questions. Section 3 defines a concrete extension of classical propositional logic with questions, and discusses dependence relations in this setting. Section 4 investigates the logic of this propositional system, providing an axiomatization and bringing out an interesting proof-theoretic side of the relation between questions and dependencies. Section 5 takes the discussion to the setting of predicate logic, defining a system which extends classical first-order logic with questions, and showing how, besides the standard dependencies of dependence logic, a wide variety of dependence relations can be recognized and expressed in this system. Section 6 wraps up and concludes.

2 Dependency is question entailment

This section is the conceptual core of the paper. We discuss the move from a truth-conditional semantics to a support-conditional semantics, which specifies when a sentence is *settled* in an information state rather than when it is *true* at a world. We show that this allows us to extend classical logic to questions, and that, in this setting, dependency emerges naturally as entailment between questions in context. Moreover, we discuss how a support-conditional semantics may be equipped with an operation of implication which allows us to internalize entailment in the language. Finally, we discuss the relation between dependency construed as a relation between questions, and dependency construed as a relation between variables.

2.1 A motivating example

Suppose a certain disease may give rise to two symptoms, S_1 and S_2, the latter being much more distressing than the former. Suppose the disease may be countered by means of a certain treatment, which however carries some associated risk. A hospital has the following protocol for dealing with the disease: if a patient presents symptom S_2, then the treatment is always administered. If the patient only presents symptom S_1, on the other hand, the treatment is only administered in case the patient is in good overall physical conditions; if not, the risks connected to the treatment are regarded as outweighing the benefits, and the treatment is not administered.

Given the hospital's protocol, whether or not a given patient should be administered the treatment is determined by two things: (i) which symptoms the patient presents and (ii) whether the patient is in good physical conditions.

In other words, in the given context, a certain relation holds between the question v of whether the patient should get the treatment, and the questions μ_1 of which symptoms the patient presents and μ_2 of whether the patient is in good physical conditions. This relation amounts to the following: any pair consisting of an answer α_1 to μ_1 and of an answer α_2 to μ_2 determines—that is, entails in the given context—a corresponding answer $\beta_{\alpha_1,\alpha_2}$ to v. For instance, the fact that the patient suffers only from S_1, together with the fact that the patient is not in good physical conditions, determines the fact that the patient should not be administered the treatment.

Another way to characterize this relation is the following: in the context of the given protocol, as soon as the questions μ_1 and μ_2 are settled, the question v is settled as well; that is, as soon as we settle the patient's symptoms and physical conditions, we have also settled whether or not the treatment should be administered.

We will refer to this relation between questions as *dependency*. In this section, we will look in detail at the logical status of this relation, and we will find out that it is nothing but a manifestation of the familiar notion of *entailment*, once logic is generalized beyond statements, to encompass also questions.[1]

---

[1] A terminological remark is in order here: this technical usage of the word *dependency* is not quite in line with the ordinary sense of the term, being weaker in one respect, and stronger in another. On the one hand, if v is already settled in a certain way in a context, independently of the answer to another question μ, then in our technical sense, v *does* depend on μ, although in ordinary language, we would say that it does not. In this sense, the technical notion of dependency is *weaker* than the ordinary notion, since dependencies are not required to be non-trivial. On the other hand, if in the given context the answer to v is partly determined by the answer to μ, and partly by other factors,

2.2   Support semantics for classical logic

Usually, logic is not concerned with questions, but rather with *statements*, a.k.a. *declarative sentences*. For statements, the default assumption—shared by logic and natural language semantics—is that meaning consists in truth-conditions: that is, the meaning of a statement α consists in the specification of what a state of affairs has to be like for α to be true.

We will refer to the formal representation of a state of affairs as a *possible world*.[2] Thus, in the standard approach, semantics consists in the specification of a relation $w \models \alpha$ between possible worlds w and statements α, which holds in case α is true in the possible world w. The fundamental logical relation of *entailment* is then defined in terms of preservation of truth: α entails β if the truth of α implies the truth of β.

$$\alpha \models \beta \iff \text{ for all possible worlds } w, \ w \models \alpha \text{ implies } w \models \beta$$

This is, in a nutshell, the standard semantic foundation of classical logic, in its many flavors: propositional, modal, first-order, etc.

Now, given this perspective on its fundamental notions, it would seem that the applicability of logic is inherently confined to statements: after all, it is not even clear what it would mean for a question—or an imperative, for that matter—to be *true* or *false* in a certain state of affairs.

The first step in the way that will lead us to bring questions within the purview of logic is to notice that it is possible to give an alternative semantic foundation for classical logic, which starts out from a more *information-oriented* perspective. Rather than saying that the meaning of a statement α is given by laying out in which circumstances α is true, we may say that the meaning of α is given by laying out *what information it takes to settle that α*. In this perspective, α is evaluated not with respect to states of affairs, but instead with respect to *states of information*.[3]

---

then in our technical sense, ν does *not* depend on μ, although in the ordinary sense, it does: in our example, for instance, we would ordinarily say that the treatment depends on the symptoms, even though it is not fully *determined* by the symptoms. In this sense, the technical notion of dependency is stronger than the usual notion. A better name for the relation that we are going to investigate would probably be *determinacy*; however, we will stick to the term *dependency* for the sake of consistency with the literature.

[2] The exact nature of possible worlds depends on the specific logical framework. Usually, a possible world may be identified with a model for the language at stake. In so-called *intensional* logics, which aim at representing a whole variety of states of affairs in a single model, possible worlds are internalized as particular entities within the model. In some cases, several choices are possible, depending on one's notion of a state of affairs. We will return to this issue in section 5.1.

[3] Information-oriented semantics are frequently considered in logic as a starting point for non-classical logics, such as intuitionistic logic or Veltman's [37] data logic. However, non-classicality is in no way an inherent consequence of the information-oriented perspective: as we will see, it is utterly possible to give an information-based semantics for classical logic. Since our goal here is to show how classical logic can be extended naturally to questions, and *not* to reform classical logic, we will take such a semantics as our starting point.

Following a tradition that goes back to Hintikka [23], we may model an information state by identifying it with a set s of possible worlds, namely those worlds which are compatible with the given information. In other words, if s is a set of possible worlds, then s encodes the information that the actual state of affairs corresponds to one of the possible worlds in s.

In the informational approach, the semantics will be given by a relation $s \models \alpha$, called *support*, between information states s and statements α, which holds just in case α is *settled* in the information state s. This semantic foundation brings along a corresponding notion of entailment as preservation of support: α entails β if settling that α implies settling that β.

$$\alpha \models \beta \iff \text{for all information states } s, \ s \models \alpha \text{ implies } s \models \beta$$

In spite of superficial differences, the truth-conditional approach and the informational approach are really two sides of the same coin. For, consider an information state s. When does s settle a statement α? Well, just in case it follows from the information available in s that α is true. That is, just in case s is only compatible with worlds in which α is true. But the worlds compatible with s are precisely those that belong to s. This means, then that the support conditions for a statement α are determined by its truth-conditions as follows:

$$s \models \alpha \iff \text{for all } w \in s, \ w \models \alpha$$

From this connection it also follows that, vice versa, truth-conditions are determined by support conditions. For, α is true at a world w just in case, were we to know that w is the actual world, α would be settled.

$$w \models \alpha \iff \{w\} \models \alpha$$

So, the support conditions for a statement are definable from its truth-conditions, and vice versa. More importantly, the truth-conditional notion of entailment and the support-conditional one coincide. To see this, suppose α truth-conditionally entails β, and consider any state $s \models \alpha$: this means that α is true everywhere in s, so β must be true everywhere in s too, and thus supported in s; this shows that α entails β in the support-conditional sense.

Conversely, suppose α entails β in the support-conditional sense, and consider a world w in which α is true: then, $\{w\}$ is a state which supports α, and since α entails β in the support-conditional sense, $\{w\}$ must support β as well, which means that β must be true at w; therefore, α entails β in the truth-conditional sense.

Thus, support-conditional semantics provides an alternative, information-oriented semantic foundation for classical logic.

2.3   Bringing questions into the picture

Given the equivalence between the truth-conditional and the support-conditional view of the semantics of statements, it is not surprising that the former perspective, based on simpler semantic objects, is taken as the standard one.

The support-conditional approach has an advantage, however: unlike the truth-conditional one, it naturally accommodates *questions*, besides statements. For, while it is not clear what it means for a question to be true or false at a world, it is perfectly clear what it means for a question to be *settled* in a certain information state. For instance, an information state s settles the polar question *whether it rains* just in case s settles that it rains, or it settles that it doesn't; that is, in case s consists uniquely of rain worlds, or uniquely of non-rain worlds. To give another example, a state settles the question *who is the culprit?* if it settles of some individual d that d is the culprit, that is, if it consists uniquely of worlds where d is the culprit.

Not only are support conditions defined for questions: there are also good reasons to regard them as a natural candidate for the role of question meaning. For, questions are used primarily (though not uniquely) in order to specify requests for information: it is therefore natural to expect that to know the meaning of a question is to know what information is requested by asking it, that is, what information state has to be brought about in order for the question to be settled. That is precisely what is encapsulated into the question's support conditions.

2.4   Entailment

We have now seen that the relation of support allows us to interpret questions on a par with statements. As a consequence, the notion of entailment defined as preservation of support is just as meaningful for questions as it is for statements. Let us look at this notion in some more detail: in general, entailment will be defined as a relation $\Phi \models \psi$ between a set Φ of sentences, which may include questions as well as statements, and a sentence ψ, which may be either a statement or a question.

$$\Phi \models \psi \iff \text{ for every information state } s, \text{ if } s \models \varphi \text{ for all } \varphi \in \Phi, \text{ then } s \models \psi$$

Focusing on the case of a single premise, we have four possible entailment patterns: statement-to-statement, statement-to-question, question-to-statement, and question-to-question. Let us examine briefly the significance of each case.

- Statement-to-statement. If α and β are statements, then $\alpha \models \beta$ expresses the fact that whenever we settle that α, we have also settled that β. As we have already discussed, this coincides with the familiar, truth-conditional notion of entailment: $\alpha \models \beta$ in case β is true whenever α is.
- Statement-to-question. If α is a statement and μ is a question, then $\alpha \models \mu$ expresses the fact that whenever we settle that α, we also settle μ. In other

words, establishing that α is true is always sufficient to reach a state where μ is resolved. Thus, we may regard $\alpha \models \mu$ as expressing the fact that α *logically resolves* μ. For example, the statement "The patient does not present symptom S_1 and is in good physical conditions" entails the question "Is the patient in good physical conditions?," but it does not entail the question "What symptoms does the patient present?," since it leaves it undetermined whether or not the patient presents symptom S_2.

- Question-to-statement. If μ is a question and α is a statement, then $\mu \models \alpha$ means that whenever we settle μ—in whatever way—we also settle that α. In other words, it is impossible to resolve μ unless we establish that α is the case. We may thus regard $\mu \models \alpha$ as expressing the fact that the question μ *presupposes* α.

  For instance, the question "When did Galileo discover Jupiter's moons?" entails the statement "Galileo discovered Jupiter's moons," since it is impossible to resolve the question without also establishing the statement as true.

- Question-to-question. If μ and ν are both questions, then $\mu \models \nu$ expresses the fact that whenever we settle μ, we also settle ν. This is precisely the relation of *dependency* among questions that we set out to examine, but now in its *purely logical* version, since *all* possible worlds—not just some contextually relevant ones—are taken into account. We may thus read $\mu \models \nu$ as expressing that the question μ *logically determines* the question ν. For instance, the question "When and where did Galileo discover Jupiter's moons?" entails the question "When did Galileo discover Jupiter's moons?', since any information that settles the former question also settles the latter question.

Thus, support-conditional semantics gives rise to an interesting general notion of entailment, which concerns questions as well as statements, and which unifies four natural logical notions: (i) a statement being a logical consequence of another; (ii) a statement logically resolving a question; (iii) a question logically presupposing a statement; and, finally, (iv) a question logically determining another.

2.5   Entailment in context

When we think about a statement being a consequence of another, it is rarely the purely *logical* notion of consequence that we are concerned with. Rather, we typically take many facts about the world for granted, and then assess whether *on that basis*, the truth of one statement implies the truth of the other. We say, for instance, that "Galileo was born in Italy" is a consequence of "Galileo was born in Pisa"; in doing so, we take for granted the fact that Pisa is located in Italy: worlds in which the location of Pisa is different from the actual one are simply not taken into account.

The same holds for questions: when we think about dependencies, it is rarely purely *logical* dependency that is at stake. Rather, we are usually concerned with the relations that one question bears to another, given certain facts about the world.

In our initial example of a dependency, for instance, it is the hospital's protocol that provides the context relative to which the dependency holds.

We can capture these relations by means of a notion of *entailment in context*, where a context is modeled as an information state s: in assessing entailment relative to s, one is taking for granted that the world is one of those in s; thus, only worlds in s, and states consisting of such worlds, are taken into account. Formally, we set

$$\Phi \models_s \psi \iff \text{ for every state } t \subseteq s, \text{ if } t \models \varphi \text{ for all } \varphi \in \Phi, \text{ then } t \models \psi$$

The notion of contextual entailment gives us a way to capture cases of consequence, resolution, presupposition, and dependency which hold not purely logically, but rather within a specific context.

Focusing on dependency, let us see how our hospital protocol example is indeed captured as an instance of entailment in context. Let us denote by four digits $ijkl$ a possible world in which:

- i is the truth-value of "the patient presents symptom S_1";
- j is the truth-value of "the patient presents symptom s_2";
- k is the truth-value of "the patient is in good physical conditions";
- l is the truth-value of "the patient should be administered the treatment."

Clearly, only eight of the sixteen possible worlds $ijkl$ are compatible with the hospital protocol, corresponding to the eight possible states for a patient. The context s determined by the protocol consists of the set s of these eight worlds, which are displayed in Figure 1(a). In the context s of the protocol, only these eight worlds are taken into account in assessing entailments.

Now, a state $t \subseteq s$ settles the question μ_1 of which symptoms the patient presents in case it settles whether the patient presents symptom S_1 and whether she presents symptom S_2. This holds just in case all the worlds $ijkl \in t$ have the same indices i and j. The maximal substates of s which settle μ_1 are the depicted in Figure 1(b).

Moreover, a state $t \subseteq s$ settles the question μ_2 of whether the patient is in good physical conditions just in case all of the worlds in $ijkl \in t$ have the same index k. The maximal substates of s which settle μ_2 are depicted in Figure 1(c).

Finally, a state $t \subseteq s$ settles the question ν of whether the treatment is prescribed just in case all the worlds in $ijkl \in t$ have the same index l. The maximal states which settle ν are depicted in Figure 1(d).

Now, clearly, relative to the context s, neither μ_1 nor μ_2 by itself entails ν. For instance, μ_1 is settled in the state $\{1011, 1000\}$, corresponding to the third row in the pictures, but ν is not. Similarly, μ_2 is settled in the state $\{1111, 0111, 1011, 0010\}$, corresponding to the left column, but ν is not. Hence, we have $\mu_1 \not\models_s \nu$ and $\mu_2 \not\models_s \nu$, that is, ν is not determined by either μ_1 or μ_2 in the given context.

At the same time, μ_1 and μ_2 together *do* entail ν relative to s. For, consider a state $t \subseteq s$ which settles both μ_1 and μ_2: since t settles μ_1, t must be included in one of the rows in the picture; and since t settles μ_2, t must be included in one of the columns. Thus, t must be included in a singleton; hence, trivially, all the worlds in $w_{ijkl} \in t$ must have the same index l, which means that t settles ν as well. This shows that we have $\mu_1, \mu_2 \models_s \nu$, that is, ν is jointly determined by μ_1 and μ_2 relative to s.

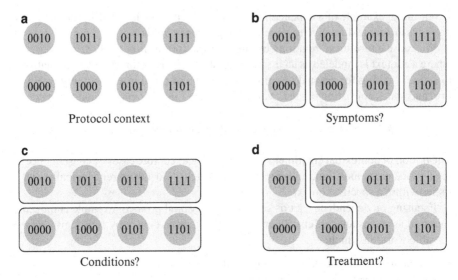

Fig. 1 The meanings of the three questions involved in the protocol example. For the sake of simplicity, only the *maximal* states in which each question is settled are displayed.

Thus, as we had anticipated, we have captured the dependence relation of our initial example as a case of entailment—more precisely, as a case of question entailment in context.

Quite naturally, contextual entailments can be turned into *logical* entailments by making the relevant contextual material into an explicit premise. Indeed, if Γ is a set of statements, and if $|\Gamma|$ is the set of worlds at which these statements are all true, we have the following connection:

$$\Phi \models_{|\Gamma|} \psi \iff \Gamma, \Phi \models \psi$$

That is, if a context s is describable by a set Γ of statements, contextual entailment relative to s amounts to logical entailment with the statements in Γ as additional premises. In our example, the context s may be expressed by a statement α describing the hospital's protocol, for instance: "the patient should be treated if and only if she presents symptom S_2, or she presents symptom S_1 and is in good physical conditions." Thus, the contextual entailment $\mu_1, \mu_2 \models_s \nu$ also has a purely logical counterpart $\alpha, \mu_1, \mu_2 \models \nu$ in which the hospital's protocol is made into an explicit premise.

To sum up our findings, just as the ordinary notion of consequence amounts to entailment between statements in context, so the notion of *dependency* amounts to entailment between *questions* in context. Far from being an exotic logical notion, thus, dependency turns out to be an incarnation of entailment, and therefore central to the concerns of logic.

2.6 Internalizing entailment

In a support-based semantics, the contexts to which entailment can be relativized are the same kind of object at which formulas are evaluated. This ensures that a support-based logical system can always be enriched with an operation of *implication* which internalizes in the object language the meta-language notion of entailment. In other words, any logic based on a support semantics may be equipped with a connective $\longrightarrow$ such that, for any sentences φ and ψ, the sentence $\varphi \longrightarrow \psi$ is settled in a state s iff φ entails ψ relative to s. In symbols:

$$s \models \varphi \longrightarrow \psi \iff \varphi \models_s \psi$$

In this precise sense, the operator $\longrightarrow$ reflects the entailment relation within the object language. Simply by making explicit what the condition $\varphi \models_s \psi$ amounts to, we get the inductive support clause governing this operation:

$$s \models \varphi \longrightarrow \psi \iff \text{for any state } t \subseteq s, \ t \models \varphi \text{ implies } t \models \psi$$

Interestingly, this is, *mutatis mutandis*, precisely the interpretation of implication that we find in most information-based semantics, such as Kripke and Beth semantics for intuitionistic logic and Veltman's data semantics.

If we apply this to statements, what we get is the standard material conditional of classical logic. To see this, suppose α and β are statements, and let us denote by $|\alpha|$ the set of worlds at which α is true, and by $|\beta|$ the set of worlds at which β is true. Recall that, for statements, support conditions are determined by truth-conditions in the following way: $s \models \alpha \iff s \subseteq |\alpha|$, and the same for β. Using this, we get

$$s \models \alpha \longrightarrow \beta \iff \forall t \subseteq s, \ t \models \alpha \text{ implies } t \models \beta$$
$$\iff \forall t \subseteq s, \ t \subseteq |\alpha| \text{ implies } t \subseteq |\beta|$$
$$\iff s \cap |\alpha| \subseteq |\beta| \iff s \subseteq \overline{|\alpha|} \cup |\beta|$$

where $\overline{|\alpha|}$ denotes the set-theoretic complement of $|\alpha|$, i.e., the set of worlds where α is false. So, the conditional $\alpha \longrightarrow \beta$ is supported in a state s just in case the corresponding material conditional is true everywhere in s. Thus, the standard material conditional of classical logic may be seen as arising precisely by internalizing within the language the relation of contextual entailment between statements.

The fact that a conditional is an object-language counterpart of the meta-language relation of entailment should not come too much as a surprise. Indeed, the relation between the conditional operator and the entailment relation is so tight that students of a first course in logic often have a hard time teasing the two notions apart.

What is more interesting from our perspective is that the clause given above defines an operation which *generalizes* the material conditional. For, we have seen that support semantics is suitable for interpreting questions, besides statements. If our language does indeed contain questions, implication between them is naturally defined: for any two questions μ and ν, this gives us a corresponding conditional $\mu \longrightarrow \nu$ with the property of being supported in a state s just in case $\mu \models_s \nu$, that is, just in case the question μ determines the question ν relative to s.

What this means is that, if a certain support-based logic has the resources to express questions, it can also be equipped in a canonical way with the resources to express dependencies between these questions.

Thus, in the support-conditional setting, the classical entailment relation generalizes to questions, capturing the notion of dependency, and in parallel, the classical material conditional generalizes to questions, providing the means to express these dependencies within the logical language.

2.7 Conditional dependencies

In our hospital protocol example, a patient's symptoms do not fully determine whether the treatment should be administered. However, suppose we know that the patient is in good physical conditions: then, her symptoms *do* determine whether the treatment should be administered. If g is the statement "the patient is in good physical conditions," we may say that the question μ_1 (symptoms?) determines the question ν (treatment?) *conditionally on g*.

In general, we will say that questions $\mu_1, \ldots, \mu_n$ determine another question ν *conditionally on α* in a context s if the dependency holds relative to the set of α-worlds in the context, that is, if we have

$$\mu_1, \ldots, \mu_n \models_{s \cap |\alpha|} \nu$$

where $|\alpha|$ is the set of worlds where α is true. We will refer to such a relation as a *conditional dependency* in s. Now, given the relation between entailment and support for implication, and given that support for a statement boils down to truth at each world, we have the following equivalence.

$$\mu_1, \ldots, \mu_n \models_{s \cap |\alpha|} \nu \iff s \cap |\alpha| \models \mu_1 \wedge \cdots \wedge \mu_n \longrightarrow \nu$$
$$\iff \text{for all } t \subseteq s, \ t \models \alpha \text{ implies } t \models \mu_1 \wedge \cdots \wedge \mu_n \longrightarrow \nu$$
$$\iff s \models \alpha \longrightarrow (\mu_1 \wedge \cdots \wedge \mu_n \longrightarrow \nu)$$

This brings out another remarkable feature of the conditional operator $\longrightarrow$. So far, we saw that when applied to two statements, this operator gives us the standard material conditional, while when applied to two questions, it yields a formula which expresses a dependency. We can now see that, in addition, the same operator provides a perspicuous way of expressing *conditional* dependencies, simply by conditioning the formula $\mu_1 \wedge \cdots \wedge \mu_n \longrightarrow \nu$ expressing the dependency with the formula α expressing the condition. This shows that, in a support-based logic equipped with the operator $\longrightarrow$, a formula for conditional dependency is nothing but the conditional version of a formula for dependency.

2.8 Relation between questions or relation between variables?

Before concluding the present section, one important issue remains to be discussed. We have characterized dependency as a relation which holds between *questions*. This is not the only conceivable view. In Dependence Logic, dependency is regarded as a relation that holds not between questions, but between *variables*. Thus, in our example, we may say that, given the hospital's protocol, the value of the Boolean variable z denoting the truth-value of "the patient should get the treatment" is determined by the value of the variable x denoting the patient's symptoms together with the value of the Boolean variable y denoting the truth-value of "the patient is in good conditions." Clearly, these two views are related, but in what way exactly?

In one direction, the correspondence is smooth: given a variable x, we can associate it with the identity question λ_x of what the value of x is, that is, the question which is settled in a state just in case it is established that x has a certain value d. To say that the value of y is determined by the value of x in s is simply to say that, relative to s, the identity question λ_y is determined by the identity question λ_x. Thus, any case of dependency construed as a relation between variables can be safely re-conceptualized as a relation between the corresponding identity questions.[4]

What about the converse? Well, suppose μ is a question which has a unique true answer at each world; then we can associate μ with a variable x_μ, whose value at a world w is the unique true answer to μ at w. To say that the question ν is determined by the question μ in s is simply to say that, relative to s, what the answer to ν is determined by what the answer to μ is; and this in turn means that, relative to s, the value of x_ν is determined by the value of x_μ. Thus, at least for a broad class of questions, the question-view and the variable-view are inter-translatable.[5]

One difference arises with the fact that not every question has a unique answer at every world. A notorious counterexample is given by questions which are resolved merely by mentioning an instance, so-called *mention-some* questions: What is a typical French dish? What is an example of a continuous function? Where can I

---

[4] For the reader interested in the formal details, we will make this precise in Section 5.5.

[5] However, a complication for the variable approach should be noted here. In order for the translation from question-dependencies to variable-dependencies to work, we need the domain of our model to include more than individuals. For, depending on what our view on the notion of answer is, the unique answer to μ at a world will either be a set of n-tuples for some $n \geq 0$, as in categorial theories of questions [20, 29, 31], or it will be a set of worlds, as in the partition theory [13, 17]. Thus, we need to work with domains which include either arbitrary n-tuples of individuals or sets of worlds. While obviously possible, this seems to make the logical setup unnecessarily cumbersome.

buy an Italian newspaper in Amsterdam? For such questions μ, we cannot simply let the variable x_μ denote at any world w the true answer to μ at w: for, in general, there is no unique true answer. Since variables are bound to have a unique value at any world, it seems that dependencies involving such questions cannot be construed as relations involving variables.[6] Thus, one advantage of regarding dependency as a relation between questions is the extra generality that it affords, allowing us to cover also dependencies involving non-unique answer questions; we will discuss examples of this kind in sections 3.3 and 5.5.

However, there is a more important conceptual and practical advantage to be gained from construing dependency as involving questions, rather than variables. This advantage is precisely the principled view of *dependency as entailment* developed in this section: once we think of dependency as a relation between questions, we can recognize it as a facet of entailment—of the very same notion of entailment with which we are acquainted in classical logic, only applied to a different class of sentences. Besides giving rise to a simple and elegant conceptual picture, this allows us to treat dependency by means of familiar tools. The possibility of expressing dependencies by means of implication is a prime example of this, and more examples will be encountered later on in the paper, such as the connection discussed in Section 4.4 between proofs involving questions and programs for computing dependencies.

2.9 Summing up

We have seen that classical logic can be given an alternative, information-oriented semantics in terms of *support conditions*, which determines when a sentence is *settled* in an information state, rather than when it is true at a world. This semantics brings along a notion of entailment, which can be internalized in the language by means of an implication operation.

Unlike truth-conditional semantics, this support-conditional semantics is applicable to *questions* in a natural way. This renders the notion of entailment meaningful for questions as well as statements. While entailment between statements amounts to the familiar notion, entailment between questions captures precisely the notion of dependency that we set out to investigate. Given that entailment may be internalized in the language by means of the implication operation, dependencies can then be expressed in a support-based logic as implications between questions.

In order to make our points as simply and generally as possible, our presentation has so far abstracted away from the details of a specific logical system, focusing

---

[6]One may think of letting x_μ denote at a world w the set of all true answers to μ at w. A moment's reflection reveals that this will not work: saying that ν is determined by μ does not amount to saying that the set of answers to ν is determined by the set of answers to μ, but merely to saying that any answer to μ determines some answer to ν.

instead on the fundamental ideas underlying the support-based semantic approach, and how they allow us to extend classical logic in a conservative way to deal with questions and dependencies. In the rest of the paper we will see these general ideas in action in two particular formal incarnations.

3 Propositional logic

In this section, the ideas discussed abstractly so far will be made concrete in the simplest possible setting, that of propositional logic. We will describe a system that enriches classical propositional logic with the tools to express questions and dependencies. This system may be seen as an extension of propositional Inquisitive Semantics [4, 7, 16] with the tensor disjunction connective adopted in Dependence Logic; alternatively, it can be seen as an extension of propositional Dependence Logic [39] with the inquisitive connectives $\lor\!\!\lor$ and $\longrightarrow$. It is part of the thesis of this paper that there is no principled boundary, other than historical, that allows us to label a system a version of Inquisitive Semantics rather than a version of Dependence Logic, or vice versa; therefore, we will simply refer to this system as *propositional logic of Questions and Dependencies*, and denote it $\mathsf{QD_P}$.

Let $\mathscr{P}$ denote a set of propositional letters. In this setting a *state of affairs* is determined simply by a propositional valuation, a function $w : \mathscr{P} \longrightarrow \{0, 1\}$ which specifies which atomic sentences are true and which are false. Thus, *possible worlds* are simply propositional valuations, and *information states* are sets of valuations. The set consisting of *all* valuations will be called the *global state* and denoted ω.

3.1 Support semantics for classical propositional logic

Let us first look at how we may define a support semantics for classical propositional logic. The set $\mathscr{L}_c$ of *classical formulas* of our language consists of propositional formulas built up from atoms and the *falsum* constant $\bot$ by means of the binary connectives $\land$, $\longrightarrow$, and $\lor$, that we refer to as *conjunction, implication*, and *tensor disjunction*.[7] *Negation* is defined as implication to falsum, that is, we write $\neg\varphi$ as an abbreviation for $\varphi \longrightarrow \bot$.

The relation of *support* between information states $s \subseteq \omega$ and sentences $\varphi \in \mathscr{L}_c$ is defined by the following clauses.

---

[7]We refer to $\lor$ as *tensor* disjunction to distinguish it from *inquisitive* disjunction $\lor\!\!\lor$, the question-forming operation that we will introduce below.

Definition 1 (Support).

- $s \models p \iff w(p) = 1$ for all $w \in s$
- $s \models \bot \iff s = \emptyset$
- $s \models \varphi \wedge \varphi \iff s \models \varphi$ and $s \models \psi$
- $s \models \varphi \vee \psi \iff s = t \cup t'$ for some t, t' such that $t \models \varphi$ and $t' \models \psi$
- $s \models \varphi \longrightarrow \psi \iff$ for all $t \subseteq s, t \models \varphi$ implies $t \models \psi$

The clauses may be read as follows. An atom p is settled in s in case the information in s implies that p is true. The falsum constant $\bot$ is only settled in the empty state, which represents the state of inconsistent information, in which all possible worlds have been discarded. A conjunction is settled in s in case both conjuncts are. A tensor disjunction is settled in s if one can divide the worlds in s into two subsets—two "cases"—each establishing one of the disjuncts. Finally, an implication is settled in s in case enhancing s so as to settle the antecedent is guaranteed to lead to a state in which the consequent is settled as well. Notice that the clauses for implication and falsum determine the following clause for negation.

Remark 1 (Derived clause for negation).

- $s \models \neg\varphi \iff$ for all $t \subseteq s$, $t \models \varphi$ implies $t = \emptyset$

That is, $\neg\varphi$ is settled in s in case enhancing s so as to settle φ would lead to the state of inconsistent information. It is easy to verify inductively that the given support-based semantics for classical formulas satisfies three fundamental properties.

Proposition 1. *For any $\varphi \in \mathscr{L}_c$, the following properties hold.*

Persistency property: if $s \models \varphi$ and $t \subseteq s$, then $t \models \varphi$
Empty state property: $\emptyset \models \varphi$
Regularity property: if $s \models \varphi$ for every $s \in S$, then $\bigcup S \models \varphi$

We will refer to the set $[\varphi] = \{s \mid s \models \varphi\}$ of states supporting a formula φ as the *meaning* of φ, and to the maximal states in $[\varphi]$ as the *alternatives* for φ.

So far, our semantics is simply a support-based formulation of classical propositional logic. On the one hand, we can recover truth by defining it as support with respect to singleton states.

Definition 2 (Truth). We say a formula φ is *true* at a world w, and write $w \models \varphi$, in case $\{w\} \models \varphi$.

The *truth-set* of φ, denoted $|\varphi|$, is the set of all the worlds where φ is true.

The following proposition ensures that this notion indeed coincides with truth in classical propositional logic.

Proposition 2 (Truth is classical). *For any world w and any $\varphi \in \mathscr{L}_c$: $w \models \varphi \iff \varphi$ is classically true at w.*

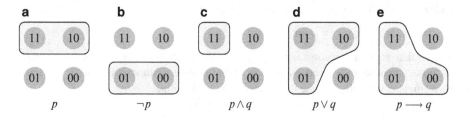

Fig. 2 The meaning of some classical formulas. In world 11 p and q are true, in world 10 p is true and q is false, etc. Only *alternatives*—i.e., *maximal* supporting states—are displayed. As implied by Proposition 3, each formula has a unique alternative, i.e., the set of all worlds where it is true.

On the other hand, let us say that a formula φ is *truth-conditional* in case support for it simply amounts to truth at each world.[8]

Definition 3 (Truth-conditionality). A formula φ is truth-conditional if for any state s: $s \models \varphi \iff w \models \varphi$ for all $w \in s$

It is then easy to prove inductively the following proposition.

Proposition 3. *Every classical formula is truth-conditional.*

This proposition implies that every classical formula φ has a unique maximal supporting state—a unique alternative—that coincides with the formula's truth-set, $|\varphi|$. This is illustrated by the examples in figure 2.

Thus, the meaning of a classical formula in $\mathsf{QD_P}$ can be seen as an alternative representation of its standard truth-conditional meaning. And indeed, as we will see, Proposition 3 implies that entailment between classical formulas is just entailment in classical propositional logic.

3.2 Enriching propositional logic with questions

Now that we have reconstructed classical propositional logic in a support-based fashion, we can start taking advantage of the extra richness of the new semantic context by enriching our language with formulas that stand for *questions*, rather than statements. We will do this by extending our repertoire of logical constants with a new connective $\mathbin{\backslash\!\backslash}$, called *inquisitive disjunction*. Intuitively, $\alpha \mathbin{\backslash\!\backslash} \beta$ will stand for an alternative question *whether α or β*, which is settled just in case one among α and β is settled. Thus, the support conditions for $\mathbin{\backslash\!\backslash}$ will be stricter than those of the tensor disjunction $\vee$: in order to settle $\alpha \mathbin{\backslash\!\backslash} \beta$, it is not sufficient to establish *that* one or the other must hold; we really must be able to establish one of the two.[9]

---

[8]In the dependence logic literature, the truth-conditionality property is referred to as *flatness*.

[9]It seems quite possible that, in natural language, an alternative question *whether α or β* is only settled if we establish which one of α and β holds, to the exclusion of the other. If so, such a

Definition 4 (Support for inquisitive disjunction).

- $s \models \varphi \lor\kern-0.8ex\lor \psi \iff s \models \varphi$ or $s \models \psi$

Notice that a polar question *whether* φ, which is settled in case either φ or $\neg\varphi$ is settled, can be expressed by means of inquisitive disjunction as $\varphi \lor\kern-0.8ex\lor \neg\varphi$. For convenience, we will thus use the following abbreviation:

- $?\varphi := \varphi \lor\kern-0.8ex\lor \neg\varphi$.

Also, notice that occurrences or inquisitive disjunctions can be nested. It is clear from the given support conditions that inquisitive disjunction is associative, so need not keep track of the particular bracketing. We can thus form multiple inquisitive disjunctions such as $\varphi_1 \lor\kern-0.8ex\lor \ldots \lor\kern-0.8ex\lor \varphi_n$, which stand for the question which is resolved as soon as we settle at least one of $\varphi_1, \ldots, \varphi_n$.

Thus, the full language $\mathscr{L}$ of our system QD$_P$ is the propositional language generated from atoms in $\mathscr{P}$ and $\bot$ by means of the connectives $\land, \lor, \longrightarrow$, and $\lor\kern-0.8ex\lor$. For this extended language, the *persistency property* of Proposition 1 still holds: if a formula is settled in a state, it remains settled in any extension of the state. The *empty state property* holds as well: in the inconsistent information state, everything is settled. However, the *regularity* property fails, in general, for formulas containing inquisitive disjunctions; that is, support is not preserved under unions. To see this, consider the polar question $?p := p \lor\kern-0.8ex\lor \neg p$ and consider the state $|p|$ consisting of all the worlds at which p is true, and the state $|\neg p|$ consisting of all the worlds at which p is false: clearly, $?p$ is supported at $|p|$ and also at $|\neg p|$, but not at their union $|p| \cup |\neg p|$, which amounts to the set ω of all worlds.

Now that we can form questions by means of $\lor\kern-0.8ex\lor$, our semantics becomes more than a mere reformulation of classical logic. Truth is still defined, according to Definition 2, even for formulas containing inquisitive disjunction. Moreover, truth-conditions are still classical: that is, Proposition 2 holds for the full language $\mathscr{L}$, when both disjunctions $\lor$ and $\lor\kern-0.8ex\lor$ are assigned the usual disjunctive truth-conditions. However, in general, formulas containing inquisitive disjunction are *not* truth-conditional, that is, Proposition 3 fails for the full language $\mathscr{L}$: it can very well be the case that a formula φ is true at all the worlds in a state s, yet φ fails to be supported at s. An example is again the polar question $?p$, which is true at every world, but is not supported by a state s in case s contains both p-worlds and $\neg p$-worlds.[10]

---

question should be translated in our formal language not as $\alpha \lor\kern-0.8ex\lor \beta$, but as an *exclusive* inquisitive disjunction $\alpha \underline{\lor\kern-0.8ex\lor} \beta := (\alpha \lor\kern-0.8ex\lor \beta) \land \neg(\alpha \land \beta)$. Nothing important in this paper hinges on this empirical issue. What matters is that, however construed, such questions—and therefore also dependencies between them—can be represented and reasoned about in the system.

[10]One may wonder what our formal notion of truth even means for questions, at an intuitive level. To answer this question, notice that, by persistence, we have $w \models \mu \iff w \in s$ for some $s \models \mu$. Thus, μ is true at w just in case there exists some information state that is true at w and which settles μ; in other words, just in case μ has a true answer at w. This converges with the proposal made by Belnap at the end of his paper *Questions, answers, and presuppositions* [2]: "I should like

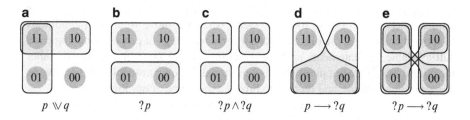

Fig. 3 The alternatives for some questions in the system QD$_P$.

Moreover, it no longer holds in general that formulas have a unique alternative. For instance, as illustrated in Figure 3, the question $?p$ has *two* alternatives, namely the states $|p|$ and $|\neg p|$. Indeed, the following proposition ensures that the formulas having a unique alternative are precisely the truth-conditional ones.

Proposition 4. *For any* $\varphi \in \mathscr{L}$, φ *is truth-conditional* $\Longleftrightarrow$ *there is a unique alternative for* φ

Proof. If φ is truth-conditional, then obviously $|\varphi|$ is the unique alternative for φ. To establish the converse, we need to make use of the fact that any supporting state is included in one alternative: the standard proof for inquisitive logic also works in the present setting: see [7], Proposition 2.10. Now, suppose φ has a unique alternative a. Suppose φ is true at any world in a state s. Then, for any $w \in s$, $\{w\}$ supports φ, and thus it must be included in one alternative for φ; since a is the unique alternative, we must have $\{w\} \subseteq a$. This implies that $s \subseteq a$, and since φ is supported at a, by persistency also $s \models \varphi$. This shows that φ is supported at a state s as soon as it is true at all the worlds in s, i.e., φ is truth-conditional.   □

Intuitively, we may regard the alternatives for a sentence as the different *ways* in which a sentence may be settled. It seems natural to regard formulas which may be settled in just one way—by establishing that they are true—as standing for *statements*, and formulas which may be settled in several different ways as standing for *questions*. Since in the first-order case the use of alternatives is problematic (see [5, 9] for discussion), we will use *truth-conditionality*, or lack thereof, as the defining feature of the classes of statements and questions.

Definition 5 (Statements and Questions). We call a formula φ a *statement* if it is truth-conditional, and a *question* if it isn't. We will use α, β, γ as variables for statements, and μ, ν, λ as variables for questions.[11]

---

in conclusion to propose the following linguistic reform: that we all start calling a question 'true' just when some direct answer thereto is true."

[11]This is not the terminology standardly used in inquisitive semantics. Usually (e.g., in [8]) sentences with a unique alternative are called *assertions*, and sentences with several alternatives are called *inquisitive*; due to a particular view of the effect of uttering a sentence, the term *question* is reserved for formulas whose alternatives cover the whole set of possible worlds.

We can now see how the general story discussed in the previous section is implemented in a concrete formal system: when we move from truth-conditions to support conditions, the old, truth-conditional meanings of propositional logic are still available for statements in the new setting. At the same time, however, other meanings, which are not purely truth-conditional, become available. This allows us to represent *questions*, and thereby to broaden the scope of our logic.

To familiarize ourselves with the sort of questions which become expressible by means of inquisitive disjunction, let us take a look at figure 3. Figure 3(a) represents the alternative question $p \lor q$, which is settled in case one of p and q is settled; this question has two alternatives, corresponding to the statements p and q. Figure 3(b) represents the polar question $?p$, which is settled in case either of p and $\neg p$ is established: this question has two alternatives, corresponding to the statements p and $\neg p$. Figure 3(c) represents the conjunctive question $?p \land ?q$, which is settled just in case both polar questions $?p$ and $?q$ are settled; this question has four alternatives, corresponding to the statements $p \land q$, $p \land \neg q$, $\neg p \land q$, and $\neg p \land \neg q$. Figure 3(d) represents the conditional question $p \longrightarrow ?q$, which is settled in a state s just in case enhancing s with the assumption that p leads to a state that settles $?q$: this question has two alternatives, corresponding to the statements $p \longrightarrow q$ and $p \longrightarrow \neg q$.

Finally, considering Figure 3(e) leads us to the topic of the next sub-section: dependencies among propositional questions, and how they may be expressed dependencies within the language.

3.3 Propositional dependencies

We have seen in Section 2 that, in a support-based logic equipped with questions, the fact that a question ν is determined by a question μ can be captured by means of the conditional $\mu \longrightarrow \nu$. In particular, this is then true for the system QD$_P$.

As an example, consider the formula $?p \longrightarrow ?q$, whose meaning is represented in Figure 3(e). By definition, this formula is supported in a state s just in case $?p \models_s ?q$, that is, in case the question $?p$ determines the question $?q$ relative to s. The four alternatives for $?p \longrightarrow ?q$, depicted in Figure 3(e), correspond to the following four statements, each expressing one particular way for the dependency to obtain:

1. $(p \longrightarrow q) \land (\neg p \longrightarrow q) \equiv q$
2. $(p \longrightarrow q) \land (\neg p \longrightarrow \neg q) \equiv q \leftrightarrow p$
3. $(p \longrightarrow \neg q) \land (\neg p \longrightarrow q) \equiv q \leftrightarrow \neg p$
4. $(p \longrightarrow \neg q) \land (\neg p \longrightarrow \neg q) \equiv \neg q$

We prefer the term *statement* over *assertion*, since the latter is normally used to refer to a *speech act*, rather than to a sentence. As for the term *question*, we want to have a more liberal notion, since we also want to consider questions that can only be resolved in *some* worlds. This does not commit us to the view that questions are *informative* in the sense of [8], since we make no assumptions about the effect of uttering sentences in context. If we wanted to describe such an effect, we could do it along the lines of the system InqD$_\pi$ in [10].

Notice that $?p \longrightarrow ?q$ is settled in a state s just in case one of the four statements above is established in s, i.e., just in case we have information that allows us to infer some answer to $?q$ from each answer to $?p$. This insight will be made precise in the next section via the notion of *resolutions* for an implication.

In standard propositional dependence logic PD [39], which does not have implication, dependencies are introduced by means of a logical primitive, called *dependence atom*. For $p_1, \ldots, p_n, q \in \mathscr{P}$, PD is equipped with a corresponding formula $=(p_1, \ldots, p_n, q)$, which is interpreted by means of the following clause:

$$s \models\, =(p_1, \ldots, p_n, q) \iff \forall w, w' \in s\text{: if } w(p_i) = w'(p_i) \text{ for all } i, \text{ then } w(q) = w'(q)$$

It is easy to verify that this truth-based clause is equivalent to the following support-based one:

$$s \models\, =(p_1, \ldots, p_n, q) \iff \forall t \subseteq s : \text{ if } t \models ?p_i \text{ for all } i, \text{ then } t \models ?q$$

Now, in our terminology, what the right-hand side expresses is precisely the fact that the question $?q$ is determined by the questions $?p_1, \ldots, ?p_n$ relative to s:

$$s \models\, =(p_1, \ldots, p_n, q) \iff ?p_1, \ldots, ?p_n \models_s ?q$$

Thus, what dependence atoms capture is, indeed, a particular case of dependency in our sense—a particular case of contextual question entailment.

Moreover, since question entailment is internalized by implication, it follows that a dependence atom $=(p_1, \ldots, p_n, q)$ expresses exactly the same meaning as the implication $?p_1 \wedge \cdots \wedge ?p_n \longrightarrow ?q$ in QDP.[12] Incidentally, this means that we may regard the system PD of propositional dependence logic as a fragment of QDP simply by making the following abbreviation:

$$=(p_1, \ldots, p_n, q) := ?p_1 \wedge \cdots \wedge ?p_n \longrightarrow ?q$$

At the same time, what our perspective brings out is that dependence atoms are only a special case of a more general pattern. What dependence atoms capture are dependencies among *atomic polar questions*, i.e., questions of the form $?p$ for an atom p. In general, however, dependencies may concern all sorts of questions. Expressing dependencies by means of implication is completely general in this respect: for any questions $\mu_1, \ldots, \mu_n, \nu$ expressible in QDP, the fact that $\mu_1, \ldots, \mu_n$ determine ν can be captured in QDP by means of the implication $\mu_1 \wedge \cdots \wedge \mu_n \longrightarrow \nu$.

To make this point concrete, let us consider a few instances of the dependence pattern that involve questions other than atomic polar questions. First, we may consider dependencies involving polar questions which are not atomic. For instance,

---

[12]In particular, for $n = 0$ we have that the constancy atom $=(p)$ of propositional dependence logic is equivalent to the polar question $?p$.

we may want to express that whether John will go to the party depends on whether at least one of Bill and Sue will go there. This may be expressed by a conditional of the form:

$$?(b \lor s) \longrightarrow ?j$$

Secondly, we may want to express dependencies between questions which are not polar questions at all. For instance, we may want to express that whether *John* or *Bill* is in the office (read with stress on the italicized words) depends on whether it is *Monday*, *Tuesday*, or *Wednesday*. This dependency involves two alternative questions, the question of whether it is Monday, Tuesday, or Wednesday ($m \lor\!\!\!\lor t \lor\!\!\!\lor w$), and the question whether John or Bill is in the office ($j \lor\!\!\!\lor b$). Thus, it may be expressed by means of the following conditional:

$$m \lor\!\!\!\lor t \lor\!\!\!\lor w \longrightarrow j \lor\!\!\!\lor b$$

The alternatives for this conditional correspond to statements such as the following, which expresses that on Monday and Tuesday John is in the office, and on Wednesday Bill is.

$$(m \longrightarrow j) \land (t \longrightarrow j) \land (w \longrightarrow b)$$

It is worth remarking that, while the truth-based clause for dependency used in dependence logic can be adapted in a simple way to handle dependencies between polar questions which are not atomic—such as the one in the previous example—a truth-conditional clause would not be able to cope with this example: in this case, the dependency cannot be seen as a relation between the truth-values of two statements; rather, it is crucial to regard it as a relation that involves two different *questions*.

For another interesting example, suppose we want to express the following: whether Mary will dance with John *if he asks her* depends on whether she is in a good mood. This dependency involves a polar question, *whether Mary is in a good mood* ($?g$), and a conditional question, *whether Mary will dance if John asks her* ($a \longrightarrow ?d$), whose meaning has the form depicted in Figure 3(d). Thus, the dependency may be expressed by means of the following conditional:

$$?g \longrightarrow (a \longrightarrow ?d)$$

The various alternatives for the above conditional correspond to statements such as, e.g., the following one, expressing that if Mary is in a good mood, then she will dance with John if he asks her, but if she is not is a good mood, she won't.

$$(g \longrightarrow (a \longrightarrow d)) \land (\neg g \longrightarrow (a \longrightarrow \neg d))$$

Notice that $?g \longrightarrow (a \longrightarrow ?d) \equiv a \longrightarrow (?g \longrightarrow ?d)$: this means that our dependency of a conditional question $a \longrightarrow ?d$ on a polar question $?g$ can be equivalently regarded as a dependency among polar questions which holds conditionally on the statement a. That is, our situation can be equivalently described as one where, conditionally on John asking Mary to dance, whether she will dance depends on whether she is in a good mood. Notice that $?g \longrightarrow ?d$ is nothing but the dependence atom $=(g, d)$: so, this is an example in which a basic dependency holds, not *simpliciter*, but rather conditionally on a statement.

Summing up, in QD$_P$ implication gives us a uniform way to express dependencies between arbitrary propositional questions. The dependence atoms of propositional dependence logic are a special case, namely dependencies between atomic polar questions. However, QD$_P$ can also handle in a natural way many other cases of propositional dependencies involving, for instance, non-atomic polar questions, alternative questions, and conditional questions. In addition, *conditional* propositional dependencies are also captured straightforwardly in QD$_P$.

3.4   Higher-order dependencies and the embedding problem

If we consider the operator $=(\cdot, \cdot)$ of propositional dependence logic as a propositional connective, we will immediately notice one rather odd feature: its application is restricted to a particular kind of formulas, namely propositional atoms. This restriction is syntactic in nature, so it is natural to consider how it may be lifted. In fact, there is an obvious way to make the semantic clause for $=(\cdot, \cdot)$ applicable to arbitrary formulas. Since truth at a world is defined for all formulas in the language, we can give the following definition:

$$s \models =(\varphi_1, \ldots, \varphi_n, \psi) \iff \forall w, w' \in s : \text{if } (w \models \varphi_i \iff w' \models \varphi_i) \text{ for all } i$$
$$\text{then } w \models \psi \iff w' \models \psi$$

Such a generalization of the dependence operator has indeed been considered in the dependence logic literature, in particular in the context of modal dependence logic, where it enhances the expressive power of the logic (see, e.g., [11, 21]). Even in this work, however, the application of the operator $=(\cdot, \cdot)$ is taken to be restricted to classical formulas, i.e., to formulas which do not in turn contain occurrences of $=(\cdot, \cdot)$.

This is because, the moment the above clause for the operator $=(\cdot, \cdot)$ is applied to formulas which themselves contain occurrences of this operator, the results are no longer satisfactory. For instance, notice that a formula $=(\varphi_1, \ldots, \varphi_n, \psi)$ is necessarily supported at any singleton state, and thus, by definition, true at any world. But then, the above clause predicts that a higher-order dependence formula such as

$$=(=(\varphi_1, \ldots, \varphi_n, \psi) , =(\chi_1, \ldots, \chi_m, \xi))$$

is a tautology, no matter what the formulas $\varphi_i, \psi, \chi_i, \xi$ are. This shows that, in propositional and modal dependence logic, it is hard to make sense of higher-order dependencies—dependencies among dependence formulas. This issue is sometimes referred to as the *embedding problem* for dependence atoms.

Now let us take a look at this problem from the perspective of our system $\mathsf{QD_P}$. We have seen that, in $\mathsf{QD_P}$, a dependency is expressed by a conditional $\mu \longrightarrow \nu$ involving two questions as antecedent and consequent. As Figure 3(e) shows, such a conditional is itself typically a question: $\mu \longrightarrow \nu$ can be settled in several ways, corresponding to the different ways in which the dependency may obtain.[13] For instance, we have seen that to settle $?p \longrightarrow ?q$ is to establish one of the following four statements, each corresponding to a specific way for $?p$ to determine $?q$.[14]

1. $(p \longrightarrow q) \wedge (\neg p \longrightarrow q) \equiv q$
2. $(p \longrightarrow q) \wedge (\neg p \longrightarrow \neg q) \equiv q \leftrightarrow p$
3. $(p \longrightarrow \neg q) \wedge (\neg p \longrightarrow q) \equiv q \leftrightarrow \neg p$
4. $(p \longrightarrow \neg q) \wedge (\neg p \longrightarrow \neg q) \equiv \neg q$

Being a question, a conditional $\mu \longrightarrow \nu$ can itself stand in the relation of dependency with other questions. For instance, given any questions μ, μ', ν, ν', we may consider the following contextual entailment:

$$\mu \longrightarrow \nu \models_s \mu' \longrightarrow \nu'$$

What this captures is that, in s, settling the question $\mu \longrightarrow \nu$ implies settling $\mu' \longrightarrow \nu'$. Now, we have just seen that to settle $\mu \longrightarrow \nu$ is to establish a specific way for μ to determine ν. Similarly, to settle $\mu' \longrightarrow \nu'$ is to establish a way for μ' to determine ν'. Thus, what the above dependency amounts to is that, in the context s, any specific way for ν to depend on μ yields a corresponding way for ν' to depend on μ'.

As a concrete example, suppose in our state s is it settled that $q \leftrightarrow \neg r$. Then, as soon as we settle a way for $?q$ to depend on $?p$, this also determines a way for $?r$ to depend on $?p$. For instance, if we learn that $q \leftrightarrow p$, then we know that $r \leftrightarrow \neg p$; if we learn that $q \leftrightarrow \neg p$, then we know that $r \leftrightarrow p$; and so on. This means that the following higher-order dependency holds in the given context:

---

[13]One may be tempted to read $\mu \longrightarrow \nu$ as "ν is determined by μ" and thus to think that $\mu \longrightarrow \nu$ should qualify as a statement. That is not quite right, as witnessed by the fact that $?p \longrightarrow ?q$ is true at all worlds, and that its negation is a contradiction. If we wanted to have a formal counterpart of the statement "ν is determined by μ," we should proceed as follows. We should associate to every world w a "modal base," that is, an information state $\sigma(w)$. We would then introduce a binary operator D which produces for any φ and ψ a statement with the following truth-conditions: $w \models D(\varphi, \psi) \iff \sigma(w) \models \varphi \longrightarrow \psi$. Since $D(\varphi, \psi)$ is a statement, this determines the associated support conditions. In this way, we obtain a formula $D(\mu, \nu)$ capable of being true or false at a world w depending on whether ν is determined by μ at the associated information state $\sigma(w)$.

[14]The idea of a "way for μ to determine ν" will be made precise in the next section by means of the notion of *resolutions* of a formula. We will see that the resolutions of an implication correspond to functions from resolutions of the antecedent to resolutions of the consequent. This is reminiscent of the Brouwer-Heyting-Kolmogorov interpretation of intuitionistic logic (see, e.g., [32]), where a proof of an implication is defined as a function that turns any proof of the antecedent into a corresponding proof of the consequent.

$$?p \longrightarrow ?q \models_s ?p \longrightarrow ?r$$

For a concrete and more interesting example of a higher-order dependency, we refer to section 5.5, where we discuss such dependencies in the setting of first-order logic.

For the time being, let us simply point out that, if dependencies are regarded as relations between questions and captured by means of implications between questions, no embedding problem arises. In QD$_P$, all the connectives apply unrestrictedly to all sorts of formulas. Moreover, we can make good sense of the relation of dependency among two (or more) dependence implications $\mu \longrightarrow \nu$ and $\mu' \longrightarrow \nu'$: this simply means that any way for ν to depend on μ yields a corresponding way for ν' to depend on μ'. Since nothing prevents us from nesting implications, such a higher-order dependency can then be expressed straightforwardly as $(\mu \longrightarrow \nu) \longrightarrow (\mu' \longrightarrow \nu')$.

To end this section, it is worth stopping to ask why the embedding problem arises for dependence atoms. Our perspective suggests the following diagnosis. Dependency is a relation between questions. Question meaning is captured by support conditions: hence, to check whether a dependency holds, we must in general look at the support conditions for the questions involved. However, if our dependency concerns *polar* questions $?\alpha_1, \ldots, ?\alpha_n, ?\beta$, then dependency boils down to a relation among the declaratives $\alpha_1, \ldots, \alpha_n, \beta$, which involves only their *truth*-conditions. It is this truth-conditional relation which is detected by the dependence atom. But this is only a special case: as we have seen, not all questions are polar questions. In particular, a dependence implication $\mu \longrightarrow \nu$ is not a *polar* question, as Figure 3(e) clearly shows. In order to account for the relation of dependency between such questions, therefore, a support-sensitive clause for dependency seems crucial.

4   Reasoning with questions and dependencies

In this section, we look at the logic of the system QD$_P$. We will provide a sound and complete proof system for this logic, and we will find that the connection between questions and dependencies has an interesting proof-theoretic side to it: namely like entailments among questions capture dependencies, so proofs of these entailments describe methods whereby these dependencies may be computed.

4.1   Properties of entailment

As expected, entailment in QD$_P$ is defined simply as preservation of support.

Definition 6 (Entailment). $\Phi \models \psi \iff$ for any information state $s \subseteq \omega$, $s \models \Phi$ implies $s \models \psi$.

As discussed abstractly in Section 2.4, this support-based notion of entailment extends the classical, truth-conditional notion of entailment to encompass *questions*, besides statements. This extension is conservative: indeed, it follows easily from Proposition 3 that entailment towards statements is always truth-conditional.

Proposition 5 (Entailment towards statements is truth-conditional). *If α is a statement, then $\Phi \models \alpha \iff$ for all worlds $w \in \omega, w \models \Phi$ implies $w \models \alpha$.*

Since Proposition 2 ensures that our semantics assigns the standard truth-conditions to all classical formulas, it follows that entailment restricted to $\mathscr{L}_c$ coincides with entailment in classical logic.

Proposition 6 (Conservativity on classical propositional logic). *Let $\Gamma \cup \{\alpha\} \subseteq \mathscr{L}_c$. Then $\Gamma \models \alpha \iff \Gamma$ entails α in classical propositional logic.*

At the same time, as we discussed in Section 2.4, entailment now assumes a more general significance: in particular, if Λ is a set of questions and ν is a question, then the entailment $\Lambda \models \nu$ captures the fact that ν is logically dependent on the questions in Λ. Moreover, we have seen in Section 2.5 that adding a set Γ of statements as assumptions, the entailment $\Gamma, \Lambda \models \nu$ captures the fact that ν is determined by Λ given Γ, that is, relative to the context $|\Gamma|$:[15]

$$\Gamma, \Lambda \models \nu \iff \Lambda \models_{|\Gamma|} \nu$$

To familiarize with this more general notion of entailment, let us examine how our initial example of a dependency can be formalized as a case of entailment in the system $\mathsf{QD_P}$. We will make use of four propositional atoms:

- s_1: the patient presents symptom S_1;
- s_2: the patient presents symptom S_2;
- g : the patient is in good physical conditions;
- t : the treatment should be administered.

The protocol of our hospital is encoded by the following classical formula, expressing that the treatment should be administered just in case the patient has the heavy symptom S_2, or has the mild symptom S_1 and is in good physical conditions.

$$\gamma := t \leftrightarrow s_2 \vee (s_1 \wedge g)$$

Our dependency amounts to the fact that, in the context of the protocol γ, the question $?t$ of whether the patient should get the treatment is determined by (i) the question $?s_1 \wedge ?s_2$ of what symptoms the patient has and (ii) by the question $?g$ of whether the patient is in good physical conditions. This corresponds to the validity of the contextual entailment $?s_1 \wedge ?s_s, ?g \models_{|\gamma|} ?t$, and thus, to the validity of the following *logical* entailment:

---

[15]The state $|\Gamma|$ is the set of worlds where all formulas in Γ are true, $\{w \mid w \models \alpha$ for all $\alpha \in \Gamma\}$.

$$\gamma, \ ?s_1 \wedge ?s_2, \ ?g \ \models \ ?t$$

An alternative way to look at things is the following: it follows from our discussion so far that the implication $?s_1 \wedge ?s_2 \wedge ?g \longrightarrow ?t$ captures the fact that a patient's symptoms and conditions determine whether or not the treatment should be administered. Now, in any state in which our protocol γ is in force, such a dependency does indeed hold. This is captured by the validity of the following entailment:

$$\gamma \ \models \ ?s_1 \wedge ?s_2 \wedge ?g \longrightarrow ?t$$

Let us now turn to examine the fundamental logical features of $\mathsf{QD_P}$. First, and not surprisingly, $\mathsf{QD_P}$ enjoys compactness; the proof given in [4] for inquisitive semantics (Theorem 3.1.10) carries over unmodified to the present setting.

Proposition 7 (Compactness). *If $\Phi \models \psi$, then there exists $\varphi_1, \ldots, \varphi_n \in \Phi$ such that $\varphi_1, \ldots, \varphi_n \models \psi$.*

This ensures that we do not need to worry about infinite sets of assumptions. Moreover, the deduction theorem holds, which witnesses once more the tight connection between implication and entailment.

Proposition 8 (Deduction theorem). *For any set of formulas Φ, and formulas ψ and χ: $\Phi, \psi \models \chi \iff \Phi \models \psi \longrightarrow \chi$*

Next, recall that Proposition 3 ensures that all classical formulas are statements, i.e., truth-conditional. We have seen that the introduction of inquisitive disjunction expands the expressive power of our system, allowing it to express not only statements, but also questions. However, as far as statements go, inquisitive disjunction does not add any expressive power: that is, any statement in $\mathsf{QD_P}$ is equivalent to a classical formula. To see this, let us first associate to any formula $\varphi \in \mathscr{L}$ a classical formula φ^{cl} having exactly the same truth-conditions as φ.

Definition 7 (Classical variant of a formula). The classical variant of a formula $\varphi \in \mathscr{L}$ is the formula φ^{cl} obtained by replacing all occurrences of $\lor\!\!\lor$ in φ by $\lor$.[16]

It is immediate from the definition that $\varphi^{cl} \in \mathscr{L}_c$. Moreover, a straightforward inductive proof suffices to establish that φ^{cl} has the same truth-conditions as φ.

Proposition 9. *For all $\varphi \in \mathscr{L}$, $|\varphi| = |\varphi^{cl}|$.*

As a consequence of this, we obtain the following proposition, which characterizes statements as being those formulas which are equivalent to some classical formula.

---

[16] In dependence logic, the analogue of the operation $(\cdot)^{cl}$ is the operation of *flattening*, in which the dependence atoms occurring in a formula are replaced by $\top$. Like the classical variant operation, flattening always yields a classical formula whose support conditions with respect to singletons coincide with those of the original formula.

Proposition 10. *For any $\varphi \in \mathcal{L}$, φ is a statement $\Longleftrightarrow$ $\varphi \equiv \alpha$ for some $\alpha \in \mathcal{L}_c$.*

Proof. If φ is a statement, by definition its meaning is completely determined by its truth-conditions. Now, φ^{cl} is a classical formula, and thus by Proposition 3, it is a statement as well. Since the previous proposition ensures that φ and φ^{cl} have the same truth-conditions, it follows $\varphi \equiv \varphi^{cl}$, and so, φ is equivalent to a classical formula. Conversely, if $\varphi \equiv \alpha$ for some classical formula α, then since classical formulas are statements, φ must be a statement as well. $\qquad\square$

The previous proposition tells us that questions are not equivalent to any single classical formula. However, any formula in $\mathsf{QD_P}$ is equivalent to an *inquisitive disjunction* of classical formulas. To see this, let us first associate to any formula φ a set $\mathcal{R}(\varphi)$ of classical formulas, that we will refer to as the *resolutions* of φ.

Definition 8 (Resolutions). The set $\mathcal{R}(\varphi)$ of resolutions of a formula is defined recursively as follows:

- $\mathcal{R}(p) = \{p\}$
- $\mathcal{R}(\bot) = \{\bot\}$
- $\mathcal{R}(\varphi \wedge \psi) = \{\alpha \wedge \beta \mid \alpha \in \mathcal{R}(\varphi) \text{ and } \beta \in \mathcal{R}(\psi)\}$
- $\mathcal{R}(\varphi \vee \psi) = \{\alpha \vee \beta \mid \alpha \in \mathcal{R}(\varphi) \text{ and } \beta \in \mathcal{R}(\psi)\}$
- $\mathcal{R}(\varphi \longrightarrow \psi) = \{\bigwedge_{\alpha \in \mathcal{R}(\varphi)}(\alpha \longrightarrow f(\alpha)) \mid f : \mathcal{R}(\varphi) \longrightarrow \mathcal{R}(\psi)\}$
- $\mathcal{R}(\varphi \mathbin{\mathpalette\@vee\relax} \psi) = \mathcal{R}(\varphi) \cup \mathcal{R}(\psi)$

Notice that, by definition, resolutions never contain inquisitive disjunction. That is, $\mathcal{R}(\varphi)$ is always a set of *classical* formulas. If α is itself a classical formula, then $\mathcal{R}(\alpha) = \{\alpha\}$. On the other hand, if μ is a question, then we may regard the resolutions of μ as playing the role of syntactically generated answers to μ. For instance, we have $\mathcal{R}(?p) = \{p, \neg p\}$.

Now, an easy inductive proof suffices to show that any $\varphi \in \mathcal{L}$ is equivalent to the inquisitive disjunction of its resolutions.

Proposition 11 (Inquisitive normal form). *For any $\varphi \in \mathcal{L}$, $\varphi \equiv \bigvee\mathcal{R}(\varphi)$*

Incidentally, notice that, as we anticipated in the previous section, the resolutions of an implication $\varphi \longrightarrow \psi$ correspond to functions from resolutions of φ to resolutions of ψ. The normal form result tells us that to settle an implication $\varphi \longrightarrow \psi$ is to have a function $f : \mathcal{R}(\varphi) \longrightarrow \mathcal{R}(\psi)$ such that $\alpha \longrightarrow f(\alpha)$ is settled for all $\alpha \in \mathcal{R}(\varphi)$.

Proposition 12. *Let $\varphi, \psi \in \mathcal{L}$ and $s \subseteq \omega$. $s \models \varphi \longrightarrow \psi \Longleftrightarrow$ there is a function $f : \mathcal{R}(\varphi) \longrightarrow \mathcal{R}(\psi)$ such that for all $\alpha \in \mathcal{R}(\varphi)$, $s \models \alpha \longrightarrow f(\alpha)$.*

Resolutions allow us to give an insightful characterization of entailment in $\mathsf{QD_P}$. To see this, let us write $\overline{\varphi}$ for a sequence $\varphi_1, \ldots, \varphi_n$ of formulas, and let us write $\overline{\alpha} \in \mathcal{R}(\overline{\varphi})$ to mean that $\overline{\alpha}$ is a sequence $\alpha_1, \ldots, \alpha_n$ such that $\alpha_i \in \mathcal{R}(\varphi_i)$ for $1 \leq i \leq n$.

It is easy to verify that our normal form result implies the following proposition, which states that an entailment holds just in case any particular way of resolving the assumptions entails some corresponding way of resolving the conclusion.

Proposition 13. $\overline{\varphi} \models \psi \iff$ *for all* $\overline{\alpha} \in \mathscr{R}(\overline{\varphi})$ *there is some* $\beta \in \mathscr{R}(\psi)$ *s.t.* $\overline{\alpha} \models \beta$.

This proposition grounds entailment among arbitrary formulas of our logic in entailment among classical formulas, and thus, by Proposition 6, into classical logic.

Now, suppose our assumptions $\overline{\alpha}$ are classical formulas. Then, each α_i is a resolution of itself, and the previous proposition ensures $\overline{\alpha} \models \beta$ for some $\beta \in \mathscr{R}(\psi)$. So, we have the following proposition.

Proposition 14 (Split property for resolutions). *Let* $\overline{\alpha}$ *be a sequence of classical formulas. If* $\overline{\alpha} \models \psi$, *then* $\overline{\alpha} \models \beta$ *for a* $\beta \in \mathscr{R}(\psi)$.

From this, the following property of QDₚ follows easily.

Proposition 15 (Split property for $\lor\!\!\!\lor$). *Let* $\overline{\alpha}$ *be a sequence of classical formulas. If* $\overline{\alpha} \models \varphi \lor\!\!\!\lor \psi$, *then* $\overline{\alpha} \models \varphi$ *or* $\overline{\alpha} \models \psi$.

Proof. Suppose $\overline{\alpha} \models \varphi \lor\!\!\!\lor \psi$. By the previous proposition, this implies $\overline{\alpha} \models \beta$ for some $\beta \in \mathscr{R}(\varphi \lor\!\!\!\lor \psi) = \mathscr{R}(\varphi) \cup \mathscr{R}(\psi)$. Suppose $\beta \in \mathscr{R}(\varphi)$. Then by Proposition 11 we have $\beta \models \varphi$, and thus $\overline{\alpha} \models \varphi$. Similarly, if $\beta \in \mathscr{R}(\psi)$, we have $\overline{\alpha} \models \psi$. □

Notice that, simply by taking the sequence $\overline{\alpha}$ to be empty in the previous proposition, we obtain that our logic has the *disjunction property* for inquisitive disjunction.

Proposition 16 (Disjunction Property for $\lor\!\!\!\lor$). *If* $\models \varphi \lor\!\!\!\lor \psi$, *then* $\models \varphi$ *or* $\models \psi$.

In fact, we will see that the split property does not just hold for purely *logical* entailment, but also for entailment relative to an arbitrary context s.

Proposition 17 (Local split property for $\lor\!\!\!\lor$). *Let* α *be a sequence of classical formulas, and let* $s \subseteq \omega$ *be any state. If* $\overline{\alpha} \models_s \varphi \lor\!\!\!\lor \psi$, *then* $\overline{\alpha} \models_s \varphi$ *or* $\overline{\alpha} \models_s \psi$.

Proof. By Proposition 3, we know that there is a unique maximal substate of s at which all the formulas in $\overline{\alpha}$ are supported, namely $s \cap |\overline{\alpha}| := s \cap |\alpha_1| \cap \cdots \cap |\alpha_n|$. If $\overline{\alpha} \models_s \varphi \lor\!\!\!\lor \psi$, then since $s \cap |\overline{\alpha}|$ is a subset of s which supports all formulas in $\overline{\alpha}$, we must have $s \cap |\overline{\alpha}| \models \varphi \lor\!\!\!\lor \psi$. By the support conditions for $\lor\!\!\!\lor$, this implies that either $s \cap |\overline{\alpha}| \models \varphi$ or $s \cap |\overline{\alpha}| \models \psi$. By persistency, in the former case we have that $\overline{\alpha} \models_s \varphi$, in the latter case that $\overline{\alpha} \models_s \psi$. □

Since entailment in context can be internalized by means of implication, it is easy to verify that the local split property for $\lor\!\!\!\lor$ amounts to the validity of the following entailment pattern, which plays a key role in the axiomatization of QDₚ.

$(\lor\!\!\!\lor\text{-split}) \quad \alpha \longrightarrow \varphi \lor\!\!\!\lor \psi \models (\alpha \longrightarrow \varphi) \lor\!\!\!\lor (\alpha \longrightarrow \psi) \quad$ where $\alpha \in \mathscr{L}_c$

Finally, an important aspect of QDₚ is that validity is *not* preserved under uniform substitution. This is due to the fact that, in QDₚ—and indeed in all systems of inquisitive and dependence logics—atoms do not stand for arbitrary sentences; they only stand for arbitrary *statements*. This means that their semantics is truth-conditional—as Proposition 3 ensures. As a consequence, atoms—and statements more generally—obey all laws of classical logic. On the other hand, questions do not obey classical logic, as the following proposition shows.

Proposition 18 (The double negation law characterizes truth-conditionality).
$\varphi \equiv \neg\neg\varphi \iff \varphi$ *is a statement.*

Proof. The proof relies on two observations about negation in $\mathsf{QD_P}$: first, the semantic clause for negation implies that a negation $\neg\varphi$ is always a statement, i.e., it is always truth-conditional. Second, if we compute the support conditions for negation on singleton states, we find that they give rise to the standard truth-conditions: $w \models \neg\varphi \iff w \not\models \varphi$; in particular, this means that φ and $\neg\neg\varphi$ always have the same truth-conditions. Now, if φ is truth-conditional, then since $\neg\neg\varphi$ is truth-conditional as well, we have $s \models \varphi \iff$ for all $w \in s, w \models \varphi \iff$ for all $w \in s, w \models \neg\neg\varphi \iff s \models \neg\neg\varphi$, which shows that $\varphi \equiv \neg\neg\varphi$.

On the other hand, if φ is not a statement, then since $\neg\neg\varphi$ *is* a statement, we must have $\varphi \not\equiv \neg\neg\varphi$. $\qquad\qquad\square$

4.2 Proof system

We now turn to the task of providing a proof system for our logic. The system we will present here is based on the axiomatization of standard propositional inquisitive logic in [7] and on the rules for tensor disjunction given in [39]. Unlike the standard proof system for inquisitive logic, which is Hilbert-style, the present system is given in natural-deduction style. This allows for more concise and more perspicuous formal proofs, which in turn will enable us to get a grasp of the role played by questions in logical proofs.

The rules of the system are displayed in Figure 4. Notice that φ, ψ, and χ are intended to range over all formulas, while α ranges over classical formulas only.

Let us comment briefly on each of these rules: first, notice that conjunction, implication, and the falsum constant are handled by their standard rules. This applies even when these connectives apply to questions. This is interesting, as it shows that, somehow unexpectedly, we can reason with conjunctive and conditional questions just as we normally reason with conjunctions and conditionals. In particular, notice that formulas expressing dependencies, which are implications between questions, can be handled by simple hypothetical reasoning and modus ponens.

Also, since we are defining $\neg\varphi := \varphi \longrightarrow \bot$, the usual intuitionistic rules for $\neg$ are particular cases of the rules for $\longrightarrow$. For the introduction, we have: if from φ you can derive $\bot$, infer $\neg\varphi$. For the elimination, we have: from φ and $\neg\varphi$, infer $\bot$.

For tensor disjunction, the situation is more tricky. While the standard introduction rule is still sound, the standard elimination rule is only sound provided the conclusion is a statement. To see that it may fail when the conclusion is a question, notice that we have both $p \models ?p$ and $\neg p \models ?p$, since either one of p and $\neg p$ logically resolves $?p$. Yet, we have $p \vee \neg p \not\models ?p$: $p \vee \neg p$ is a plain tautology, which certainly does not logically resolve the polar question $?p$.

$$\frac{\varphi \quad \psi}{\varphi \wedge \psi} \ (\wedge i) \qquad\qquad\qquad \frac{\varphi \wedge \psi}{\varphi} \ (\wedge e_1) \quad \frac{\varphi \wedge \psi}{\psi} \ (\wedge e_2)$$

$$\begin{array}{c} [\varphi] \\ \vdots \\ \psi \\ \hline \varphi \longrightarrow \psi \end{array} \ (\longrightarrow i) \qquad\qquad\qquad \frac{\varphi \quad \varphi \longrightarrow \psi}{\psi} \ (\longrightarrow e)$$

$$\frac{\varphi}{\varphi \vee \psi} \ (\vee i_1) \quad \frac{\psi}{\varphi \vee \psi} \ (\vee i_2) \qquad\qquad \begin{array}{c} [\varphi] \quad [\psi] \\ \vdots \quad \vdots \\ \alpha \quad \alpha \quad \varphi \vee \psi \\ \hline \alpha \end{array} \ (\vee e)$$

$$\begin{array}{c} [\varphi] \quad [\psi] \\ \vdots \quad \vdots \\ \varphi' \quad \psi' \quad \varphi \vee \psi \\ \hline \varphi' \vee \psi' \end{array} \ (\vee r) \qquad\qquad\qquad\qquad \frac{\bot}{\varphi} \ (\bot e)$$

$$\frac{\varphi}{\varphi \vvee \psi} \ (\vvee i_1) \quad \frac{\psi}{\varphi \vvee \psi} \ (\vvee i_2) \qquad\qquad \begin{array}{c} [\varphi] \quad [\psi] \\ \vdots \quad \vdots \\ \chi \quad \chi \quad \varphi \vvee \chi \\ \hline \chi \end{array} \ (\vvee e)$$

$$\frac{\varphi \vee (\psi \vee \chi)}{(\varphi \vee \psi) \vee \chi} \ (\vee a) \qquad \frac{\varphi \vee \psi}{\psi \vee \varphi} \ (\vee c) \qquad \frac{\varphi \vee (\psi \vvee \chi)}{(\varphi \vee \psi) \vvee (\varphi \vee \chi)} \ (\vee d)$$

$$\frac{\alpha \longrightarrow (\varphi \vvee \psi)}{(\alpha \longrightarrow \varphi) \vvee (\alpha \longrightarrow \psi)} \ (\mathsf{Split}) \qquad\qquad \frac{\neg\neg \alpha}{\alpha} \ (\neg\neg e)$$

Fig. 4 A natural-deduction system for QD$_\mathsf{P}$. The variable α is restricted to classical formulas.

Thus, we need to restrict the application of the rule ($\vee$e); since Proposition 10 guarantees that classical formulas are representative of all statements, we restrict the application of the rule to classical formulas. However, the consequence of restricting the $\vee$–elimination rule in this way is that we need additional rules to characterize what *non-classical* formulas are consequences of a certain tensor disjunction. This is the role of the rule ($\vee$r) of tensor replacement, which allows us to replace each disjunct of a tensor by a consequence of it, and of the rules of tensor associativity ($\vee$a), tensor commutativity ($\vee$c), and distributivity of $\vee$ over $\mathbb{\vee}$ ($\vee$d).

As opposed to tensor disjunction, *inquisitive* disjunction $\mathbb{\vee}$ is simply handled by the standard inference rules for disjunction. This proof-theoretic simplicity is linked to the fact that it is $\mathbb{\vee}$, and not $\vee$, that performs the operation of *join* in the algebra of meanings of our logic [1, 28]. Intuitively, the introduction rule for $\mathbb{\vee}$ states that settling a disjunct of a question is *sufficient* to settle the question, while the elimination rule states that it is also *necessary*.

The role of the (Split) rule is to capture the local split property for $\mathbb{\vee}$ of Proposition 17, which we have seen to be equivalent with the entailment pattern $\alpha \longrightarrow (\varphi \mathbb{\vee} \psi) \models (\alpha \longrightarrow \varphi) \mathbb{\vee} (\alpha \longrightarrow \psi)$. Essentially, what the (Split) rule captures is that, if a statement resolves a question in context, it must do so by contextually entailing some specific answer to it.

Finally, Proposition 18 shows that the double negation law is characteristic of statements. Thus, the rule ($\neg\neg$e) of double negation elimination for classical formulas can be seen as characterizing classical formulas as being statements. Notice that, as a consequence, our system includes a complete natural-deduction system for classical propositional logic, as given, e.g., in [12].

4.3   Completeness

We will write $P : \Phi \vdash \psi$ to mean that P is a proof whose conclusion is ψ and whose set of undischarged assumptions is included in Φ. We will then use the familiar notation $\Phi \vdash \psi$ to mean that a proof $P : \Phi \vdash \psi$ exists. Moreover, we will write $\varphi \dashv\vdash \psi$ to mean that φ and ψ are interderivable, that is, $\varphi \vdash \psi$ and $\psi \vdash \varphi$.

As usual, proving that the rules are sound is a lengthy but straightforward matter.

Proposition 19 (Soundness). $\Phi \vdash \psi$ *implies* $\Phi \models \psi$

As we have already observed, our system includes all the rules of a standard natural-deduction system for classical logic. Since entailment for classical formulas *is* just entailment in classical logic, this means that we then have the following proposition.

Proposition 20 (Completeness for classical formulas). *If* $\Phi \cup \{\psi\} \subseteq \mathscr{L}_c$, *then* $\Phi \models \psi$, *implies* $\Phi \vdash \psi$.

Furthermore, our system also contains all the rules of a system for *intuitionistic* logic, when $\mathbb{\vee}$, instead of $\vee$, is taken to play the role of intuitionistic disjunction. To state this, let us make the following notational convention: if $\varphi \in \mathscr{L}_c$, then φ^{inq}

is the formula obtained from φ by replacing all occurrences of $\vee$ by $\mathbb{W}$. Then, we have the following proposition.

Proposition 21 (Instances of intuitionistic validities are provable). *Suppose* $\varphi(p_1, \ldots, p_n) \in \mathscr{L}_c$ *is a valid formula of intuitionistic propositional logic, and let* $\chi_1, \ldots, \chi_n \in \mathscr{L}$ *be any formulas of our language. Then* $\vdash \varphi^{inq}(\chi_1, \ldots, \chi_n)$.

Proof. Since $\varphi(p_1, \ldots, p_n)$ is intuitionistically valid, there is a proof of it using only the standard intuitionistic rules for $\wedge, \vee, \longrightarrow$, and $\bot$. In our system, we can simply replicate the same proof with $\mathbb{W}$ in place of $\vee$, and with $\chi_1, \ldots, \chi_n$ in place of the atoms. $\square$

Indeed, if we remove from our system the rules dealing with tensor disjunction, what we get is a complete system for standard inquisitive logic, which is known to be a non-substitution closed intermediate logic.[17]

As in the completeness proofs given in [7] and [39], a crucial aspect of the system is that it allows us to prove that every formula φ is equivalent to its normal form $\mathbb{W}\mathscr{R}(\varphi)$. The inductive proof of this result is given in the Appendix.

Proposition 22 (Provability of normal form). *For any* φ, $\varphi \dashv\vdash \mathbb{W}\mathscr{R}(\varphi)$

Using Propositions 20 and 22, we can speedily prove the completeness theorem.

Theorem 1 (Completeness). $\models \varphi$ *implies* $\vdash \psi$.

Proof. Suppose $\models \varphi$. By Proposition 11, this implies $\models \mathbb{W}\mathscr{R}(\varphi)$, and thus by the disjunction property (Proposition 16), $\models \alpha$ for some $\alpha \in \mathscr{R}(\varphi)$. Since α is a classical formula, the completeness of our system for classical propositional logic (Proposition 20) gives $\vdash \alpha$, whence by an application of ($\mathbb{W}$i) we obtain $\vdash \mathbb{W}\mathscr{R}(\varphi)$. Finally, by Proposition 22 we can conclude $\vdash \varphi$. $\square$

Notice that, by the compactness of the logic and the semantic deduction theorem, completeness implies *strong* completeness.

Corollary 1 (Strong completeness). $\Phi \models \psi$ *implies* $\Phi \vdash \psi$.

Proof. Suppose $\Phi \models \psi$. By compactness (Proposition 7) we have $\varphi_1, \ldots, \varphi_n \models \psi$ for some $\varphi_1, \ldots, \varphi_n \in \Phi$. By the semantic deduction theorem (Proposition 8), this means that $\models \varphi_1 \wedge \cdots \wedge \varphi_n \longrightarrow \psi$, which by completeness implies $\vdash \varphi_1 \wedge \cdots \wedge \varphi_n \longrightarrow \psi$. Finally, since $\varphi_1 \wedge \cdots \wedge \varphi_n$ is provable from Φ, we obtain $\Phi \vdash \psi$. $\square$

---

[17]For a completeness proof and for details on the tight connections of inquisitive logic to intuitionistic logic and other intermediate logics, see [4, 7].

4.4 Computational content of proofs involving questions

Now that we have a complete proof system for our logic, let us take a look at what proofs involving questions look like. Let us consider once again our initial example of a dependency. We have seen that, in propositional logic, our dependency corresponds to the validity of the following entailment relation:[18]

$$\gamma, \ ?s_1, \ ?s_2, \ ?g \ \models \ ?t$$

where γ stands for the protocol description $t \leftrightarrow s_2 \vee (s_1 \wedge g)$. The following is a proof of the validity of this entailment in our system, where sub-proofs involving only inferences in classical logic have been omitted and denoted by $(C_1), (C_2), (C_3), (C_4)$.

$$
\cfrac{
\cfrac{\cfrac{\gamma \ [s_2]}{\cfrac{t}{-} \ (C_1)}}{\cfrac{}{?t} \ (\vee i)} \quad ?s_1 \quad
\cfrac{\cfrac{\cfrac{\gamma \ [\neg s_1] \ [\neg s_2]}{-} \ (C_2)}{\cfrac{\neg t}{?t} \ (\vee i)} \quad ?g \quad \cfrac{\cfrac{\cfrac{\gamma \ [s_1] \ [g]}{\cfrac{t}{-} \ (C_3)}}{\cfrac{}{?t} \ (\vee i)} \quad \cfrac{\cfrac{\gamma \ [\neg s_2] \ [\neg g]}{\cfrac{\neg t}{-} \ (C_4)}}{\cfrac{}{?t} \ (\vee i)}}{?t} \ (\vee e)}{?t} \ (\vee e)
}{?t}
$$

The undischarged assumptions of this proof are the protocol γ and the determining questions $?s_1$, $?s_2$, and $?g$, while the conclusion is the determined question $?t$. Now, what is the reasoning encoded by this proof? Is there a way to see how it establishes that, in the context of γ, the given dependency holds?

Well, we can look at our proof as spelling out the following argument. First, assume we are given information as to whether the patient has symptom S_2. That means that either we have the information that the patient have S_2 or we have information that she doesn't. These two cases correspond to the two main branches of the proof. In the former case, from the information that the patient has S_2 and from the protocol we can infer that the treatment ought to be administered. Thus, in this case we have some information as to $?t$. In the latter case, corresponding to the right branch of the proof, we must make use of the assumption that we are given information as to whether the patient has symptom S_1. Again, this means that we consider two cases: if the information we have is that the patient doesn't suffer from S_1, then from our assumptions and the protocol we can infer that no treatment is needed; thus, again, we have some information as to $?t$. Finally, if the information we are given is that the patient *does* suffer from S_1, we need to make use of the assumption that we are given information as to whether she is in good physical conditions: again, each answer to this question, together with the protocol

---

[18]In order to simplify the proof, here we have replaced the conjunctive question $?s_1 \wedge ?s_2$ (what are the symptoms?) by two distinct polar questions. This change is merely cosmetic, and dispensable.

and our prior assumptions s_1 and $\neg s_2$, yields a corresponding deliberation about the treatment. Thus, the proof ensures that in all cases, as soon as we settle the questions $?s_1$, $?s_2$, and $?g$, on the basis of γ we will be in a position to settle $?t$.

Interestingly, the paraphrase we have just given shows that this proof does more than just *witnessing* that the patient's symptoms and conditions determine whether the treatment should be administered: it actually describes *how* to use information about the patient's symptoms and conditions to make a corresponding deliberation about the treatment. For instance, we can read off from the proof that if the patient has only symptom S_1 and is not in good physical conditions (the right-most branch of the proof), then the treatment should not be administered.

In other words, our proof has *constructive content*: it does not just witness that a certain dependency holds, but actually describes a method to *compute* it, that is, a method for turning any given resolutions of the assumptions into a corresponding resolution of the conclusion that they entail, on the basis of the protocol.

It turns out that this is not a peculiarity of this particular proof, but a general feature of proofs in QD$_P$ and related systems. This is made precise by the following result. The proof is first given in [6] for a system that can be identified with a fragment of QD$_P$, but the extension to our richer language is straightforward.

Theorem 2 (Resolution Algorithm). *Let* $P : \overline{\varphi} \vdash \psi$ *and let* $\overline{\alpha} \in \mathscr{R}(\overline{\varphi})$. *There is a procedure which, inductively on* P, *constructs a proof* $Q : \overline{\alpha} \vdash \beta$ *having as conclusion a resolution* $\beta \in \mathscr{R}(\psi)$.

What the theorem shows is that there is a systematic way of reading a proof in our system as an algorithm that takes as input specific resolutions of the assumptions, and outputs a corresponding resolution of the conclusion—together with a proof ensuring that the output resolution is indeed entailed by the input resolutions. This is illustrated in Fig. 5.

Thus, while proofs involving only classical formulas are essentially just proofs in classical logic, proofs involving questions have a specific kind of constructive content, reminiscent of the proofs-as-programs interpretation of intuitionistic logic:

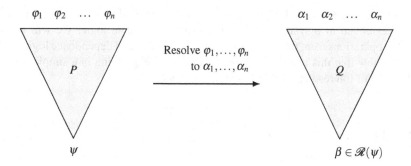

Fig. 5 An illustration of the resolution algorithm: given a proof $P : \overline{\varphi} \vdash \psi$ and resolutions $\overline{\alpha}$ of $\overline{\varphi}$, the algorithm builds a proof $Q : \overline{\alpha} \vdash \beta$ of a corresponding resolution β of ψ.

a proof having some question assumptions and a question conclusion encodes an actual procedure for computing a certain dependency. Thus, the connection discussed in Section 2 between questions and dependency is not confined to the semantic side of logic, but also has an interesting proof-theoretic side to it.

4.5 Summing up

In this section, we have seen that $\mathsf{QD_P}$ may be axiomatized by extending a system for classical logic with rules dealing with the question-forming operation $\vee\!\!\vee$ and with its interaction with the other connectives. While being a conservative extension of classical logic, $\mathsf{QD_P}$ exhibits constructive features when it comes to questions: indeed, the disjunction property holds for $\vee\!\!\vee$, and both the double negation law and the law of excluded middle are not generally valid for questions. This constructive aspect of the logic is reflected in the fact that proofs involving questions have an interesting computational interpretation: they encode methods for computing dependencies.

In reasoning with dependencies, an important asset of $\mathsf{QD_P}$ is that dependencies are not expressed by means of a primitive logical constant, as in propositional dependence logic, but decomposed into more basic operations, such as $\vee\!\!\vee$, $\wedge$, and $\longrightarrow$. As shown in [1] and [28], these operations correspond to fundamental operations on the space of meanings of our logic. The proof-theoretic consequence of this is that each of these connectives can be handled by means of simple, and indeed essentially standard, inference rules, thus allowing for a natural and insightful proof system. By contrast, no such natural rules are available if dependencies are expressed by means of a primitive operator.

5   Predicate logic

In this section, we look at how the propositional system $\mathsf{QD_P}$ defined in Section 3 can be extended to a system $\mathsf{QD_{FO}}$ of first-order predicate logic. We will bring together operations from first-order inquisitive semantics and dependence logic, and we will show that this results in a system capable of expressing in a simple way a wide range of interesting questions and dependencies.

5.1 What semantic setting?

In extending our logic from the propositional to the first-order setting, we immediately face an important choice: what objects should play the role of *possible worlds* in this setting?

The first candidate that comes to mind are first-order models for the given language. After all, the standard view is that a first-order model is to a first-order language what a valuation is to a propositional language: a specification of a complete state of affairs. Indeed, this is the view taken in the first-order implementations of Inquisitive Semantics developed so far ([4, 8], a.o.): worlds are identified with first-order models, and states with sets of models, which are assumed for simplicity to share the same domain. In this view, an information state reflects uncertainty as to which first-order model represents the actual world. By evaluating sentences with respect to such states, we interpret questions concerning the features of the model, as well as dependencies among them.

As we will see, many interesting dependence patterns do indeed concern features of the model, and could naturally be handled in this semantic framework. However, there is an important class of dependencies that can*not* be interpreted in this framework. These are precisely the dependencies between first-order variables which are the crucial concern of first-order Dependence Logic. As discussed in Section 2.8, dependencies between variables can be re-conceptualized as dependencies between questions about the variables' value. For instance, to say that y is determined by x is simply to say that λ_y, the question of what the value of y is, is determined by λ_x, the question of what the value of x is. However, the questions λ_y and λ_x do not concern features of the *model*, but rather features of the *assignment*, namely the value it assigns to y and x, respectively. If we want to give support conditions for these questions, our states must reflect not only uncertainty about the model, but also uncertainty about the assignment.

Curiously, in Dependence Logic, the situation is opposite to that of Inquisitive Semantics: it is *only* assignment-uncertainty which is taken into account. That is, whereas in first-order Inquisitive Semantics formulas are evaluated with respect to a set of models and a single assignment, in first-order Dependence Logic formulas are evaluated with respect to a single model and a set of assignments. However, this confines us to dealing *only* with questions and dependencies about the values of variables. One odd consequence of this choice is that things that were expressible in *propositional* dependence logic are no longer expressible in *first-order* dependence logic. For instance, as we have seen, in the propositional case we can express the fact that whether q is determined by whether p. However, the moment the atoms p and q are represented in first-order logic, say, as two sentences Pa and Qb, we can no longer express the same dependency. In a fixed model, both Pa and Qb just have a definite truth-value, and we can no longer make sense of the notion of whether Qb being determined by whether Pa.

To overcome the limitations of each of the two frameworks, we will set up our semantics in such a way that the information encoded in a state may be partial in both respects, with regard to the features of the model, and with regard to the values of variables. We will take a *possible world* w to be a pair, consisting of a first-order model M and an assignment g, and we will think of an information state as a set of such worlds. For simplicity, we will assume here that all the worlds in a state share the same domain, that is, that there is no uncertainty about what the relevant individuals are, but only about what properties these individuals have, and which individual each variable denotes.

Definition 9 (Worlds and states). Let $\mathscr{S}$ be a first-order signature.

- A *possible world* is a pair $w = \langle M_w, g_w \rangle$, where:

  - M_w is a first-order model for $\mathscr{S}$
  - g_w is an assignment into the domain of M_w

- An *information state* is a set s of possible worlds whose models are based on the same domain, denoted D_s, or simply D, when no confusion arises.

This more general framework encompasses both frameworks discussed above as particular cases: the single-assignment framework of standard inquisitive semantics is the case in which all worlds w in our state s have the same assignment component g_w, differing only in their model component M_w; conversely, the single-model framework of standard dependence logic is the case in which all the worlds have the same model component M_w, differing only in the assignment component g_w.[19,20]

5.2   Quantification in support semantics

Let us now turn to the definition of the system $\mathsf{QD_{FO}}$. The quantifier-free fragment of the language will be a simple adaptation of the propositional logic $\mathsf{QD_P}$ discussed above. That is, quantifier-free formulas are built up from atomic sentences of the

---

[19]An anonymous reviewer pointed out that an equivalent semantic setup has been recently advocated by Väänänen [35] with a somewhat different motivation in mind. Väänänen's goal is to develop a logic capable of expressing interesting properties of a set-theoretic *multiverse*, i.e., a structure containing a multitude of distinct models of set theory. In his system, formulas are evaluated with respect to a multiset of first-order models and to a function mapping each of these models to an assignment into the corresponding domain. While seemingly more complex, this setup is essentially equivalent to our setup based on sets of model-assignment pairs.

[20]Interestingly, the move we are making here has an almost exact parallel in the history of dynamic semantics. The fundamental idea of dynamic semantics is that the meaning of a sentence lies in its potential to bring about a change in an information state (either the conversational context or the hearer's information state). Formally, what this means is that meanings are taken to be functions from information states to information states. In the Dynamic Predicate Logic (DPL) of [18], this approach was used to deal with *anaphora*, that is, with the process whereby pronouns get their referent. In this system, pronouns are modeled semantically as free variables. Accordingly, information states are taken to be sets of assignment, and a fixed underlying model is assumed. Later on, however, the dynamic approach was used by [38] to deal with modals and default reasoning; for these applications, it is crucial for an information state to contain different *models*, while assignments do not play a role. Finally, in [19], these lines of research were brought together in a unified framework, where states are taken to be sets of world-assignment pairs, just as in the present paper.

signature $\mathscr{S}$ and the falsum constant $\bot$ by means of the connectives $\wedge, \longrightarrow, \vee,$ and $\vvv$. The clause for an atomic sentence $R\bar{t}$ simply states that $R\bar{t}$ is supported at a state in case it is true at all the worlds in the state.

- $s \models R\bar{t} \iff$ for all $w \in s$: $\bar{t}_w \in R_w$

where $\bar{t}_w$ is the interpretation of the tuple $\bar{t}$ of terms in world w, defined in the obvious way, and R_w is the interpretation of the relation symbol R in the model M_w. The clauses for $\bot$ and the connectives are the same as in the propositional case.

Let us now look at quantification. In this support-based setting, two different ways to deal with the standard first-order quantifiers suggest themselves. To state the relevant clauses, we first need to introduce some notational convention.

If w is a world, x a variable, and d an individual, let $w[x \mapsto d] := \langle M_w, g_w[x \mapsto d] \rangle$. We can now introduce some operations on states which are useful for quantification. First, if s is a state and d an element in its domain, then $s[x \mapsto d]$ is the state that results from replacing each world w by $w[x \mapsto d]$. Second, if $f : s \longrightarrow D$ is a function from worlds in the state s to individuals, then $s[x \mapsto f]$ is the state that results from replacing each world w by $w[x \mapsto f(w)]$. Notice that $s[x \mapsto d]$ is nothing but $s[x \mapsto c_d]$ where $c_d : s \longrightarrow D$ is the constant function that assigns the individual d to every world in s. Finally, if A is a set of individuals, then $s[x \mapsto A]$ is the state that results from replacing each world $w \in s$ with a number of variants of it, namely all the worlds obtained as $w[x \mapsto d]$ for some $d \in A$. More formally, we have the following.

Definition 10. Let s be a state and let x be a variable. Then:

- if $d \in D$, then $s[x \mapsto d] = \{w[x \mapsto d] \mid w \in s\}$
- if $f : s \longrightarrow D$, then $s[x \mapsto f] = \{w[x \mapsto f(w)] \mid w \in s\}$
- if $A \subseteq D$, then $s[x \mapsto A] = \{w[x \mapsto d] \mid w \in s, d \in A\}$

Using these operations, we can equip our system with the standard dependence logic quantifiers, which we will denote by $\exists^d$ and $\forall^d$.

- $s \models \exists^d x \varphi \iff s[x \mapsto f] \models \varphi$ for some $f : s \longrightarrow D$
- $s \models \forall^d x \varphi \iff s[x \mapsto D] \models \varphi$

The first clause says that the formula $\exists^d x \varphi$ is supported at a state if we can set the value of x at each world in a suitable way so that the resulting state supports φ. The second clause says that the formula $\forall^d x \varphi$ is supported at a state if expanding the state by letting x take all the possible values in D leads to a state which supports φ. The crucial feature of $\forall^d$ is that, in the process of interpretation, it creates a *range*—with one possible world for each element in the domain—relative to which other operators may then be interpreted. For instance, in the process of interpreting $\forall^d x \varphi$ relative to a singleton state $s = \{w\}$, the state is expanded to a new state $s[x \mapsto D] = \{w[x \mapsto d] \mid d \in D\}$, whose elements correspond to the different values for x; the formula φ is then interpreted relative to this state which reflects the variability in the possible value for x.[21]

Besides these quantifiers, we will also equip our system with the standard inquisitive semantics quantifiers, which we denote by $\exists^i$ and $\forall^i$. The clauses for these quantifiers are more similar to the familiar ones.[22]

- $s \models \exists^i x \varphi \iff s[x \mapsto d] \models \varphi$ for some $d \in D$
- $s \models \forall^i x \varphi \iff s[x \mapsto d] \models \varphi$ for all $d \in D$

Setting the value of a variable x to a certain individual d throughout the state—that is, looking at the state $s[x \mapsto d]$—allows us to check what the state s settles about the individual d, rather than about the variable x. In the subsequent evaluation of φ, x will be interpreted *rigidly*, as denoting the same individual d throughout all the possible worlds. Thus, the clause for $\exists^i$ (respectively, $\forall^i$) says that $\exists^i x \varphi$ is supported in s just in case s settles φ of some (respectively, every) individual d.

Notice that $\exists^i$ and $\forall^i$ are the quantifier counterparts of $\lor\!\!\!\lor$ and $\land$. To see this, suppose a state s has a finite domain D_s whose elements are rigidly denoted by constants $a_1, \ldots, a_n$. That is, suppose the value of $(a_i)_w$ is the same at any world $w \in s$, and $D = \{(a_1)_w, \ldots, (a_n)_w\}$ for a world $w \in s$. Then, we have the following:

- $s \models \exists^i x \varphi(x) \iff s \models \varphi(a_1) \lor\!\!\!\lor \ldots \lor\!\!\!\lor \varphi(a_n)$
- $s \models \forall^i x \varphi(x) \iff s \models \varphi(a_1) \land \cdots \land \varphi(a_n)$

In a similar way, the dependence existential quantifier is the quantifier counterpart of tensor disjunction. That is, in a state like the one we just described, we have

- $s \models \exists^d x \varphi(x) \iff s \models \varphi(a_1) \lor \cdots \lor \varphi(a_n)$

Notice that $\exists^i$ is stronger than $\exists^d$: if $\exists^i x \varphi$ is supported at s, then there exists some $d \in D$ such that $s[x \mapsto d] \models \varphi$; as we remarked above, $s[x \longrightarrow d] = s[x \mapsto c_d]$ where c_d is the constant function mapping each world to d; thus, we also have $s \models \exists^d x \varphi$. Dually, $\forall^d$ is stronger than $\forall^i$: for, observe that for all $d \in D$, $s[x \mapsto d] \subseteq s[x \mapsto D]$. As we will see in a moment, $\mathsf{QD_{FO}}$ still satisfies the persistence property. Hence, from $s \models \forall^d x \varphi$ it follows that $s \models \forall^i x \varphi$.

From different perspectives, both sets of quantifiers are natural choices. On the one hand, from an algebraic perspective, it is the quantifiers $\forall^i$ and $\exists^i$ that correspond to generalized meet and join operations, thus behaving abstractly like the standard quantifiers of classical logic. On the other hand, it is shown in [1] that the quantifiers $\forall^d$ and $\exists^d$ are the multi-assignment counterpart of the standard quantifiers in a precise category-theoretical sense.

---

[21]This feature of $\forall^d$ has been exploited in linguistics by Brasoveanu and Farkas [3] to provide a solution to the long standing issue of how indefinites like *a man* take their scope.

[22]These quantifiers have also been considered in the dependence logic literature, where they are denoted $\exists^1$ and $\forall^1$.

5.3 A glance at the system

Let us now take a quick look at the features of the resulting system. The fundamental properties of the propositional system QD_P carry over to the first-order setting. First of all, the semantics still satisfies the persistency and empty state properties.

Proposition 23. *For any formula φ, the following holds.*

- *Persistence property: if $s \models \varphi$ and $t \subseteq s$, $t \models \varphi$*
- *Empty state property: $\emptyset \models \varphi$*

As in the propositional case, we can define truth as support with respect to singleton states. The following proposition, easily verified by induction, ensures that the truth-conditions determined in this way are the classical ones.

Proposition 24 (Truth is classical). *Let φ be a formula and let $w = \langle M, g \rangle$ be a world. Let φ^* be the first-order formula obtained by replacing every operator in φ by its classical counterpart. Then:*

$$w \models \varphi \iff M \models_g \varphi^* \text{ in classical logic}$$

Recall that we call a formula *truth-conditional* in case support for it simply amounts to truth at every world. The next proposition, which may be proven by induction, states that inquisitive disjunction and the inquisitive existential quantifier are the only sources of non truth-conditional meanings in the system.

Proposition 25. *Any $\vee\!\!\!\vee, \exists^i$–free formula is truth-conditional.*

Let $\mathscr{L}_c^Q$ denote the language generated from atomic formulas and $\perp$ by means of the connectives $\wedge, \longrightarrow, \vee$, and the quantifiers $\exists^d$ and $\forall^d$. We will refer to formulas in $\mathscr{L}_c^Q$ as *classical* first-order formulas. The previous proposition implies that the classical fragment of QD_{FO}—the fragment consisting of classical formulas—may be regarded as a support-based implementation of classical first-order logic, and that entailment among classical formulas coincides with entailment in first-order logic. Thus, the full system QD_{FO} can be regarded as a *conservative extension* of classical first-order logic with questions.

Notice that, by Proposition 25, we could just as well have chosen $\forall^i$ instead of $\forall^d$ as support-based counterpart of the classical universal quantifier. This illustrates an important fact: in general, a classical logical constant may be extended in several non-equivalent ways beyond the truth-conditional realm. In this case, the clauses for $\forall^d$ and $\forall^i$ give the same results when restricted to truth-conditional formulas—to statements—but come apart in the outcome they yield when applied to questions.[23]

---

[23]For instance, it is easy to check that $\forall^i x \exists^i y(x = y)$ is a valid formula, but $\forall^d x \exists^i y(x = y)$ is not.

5.4 First-order questions

In $\mathsf{QD_{FO}}$, we have two basic question-forming operators: inquisitive disjunction $\mathbb{W}$, and the inquisitive existential quantifier $\exists^i$. We have already seen in the propositional setting how the former allows us to express a wide range of questions, including polar questions ($?\alpha$), alternative questions ($\alpha \mathbb{W} \beta$), and conditional questions (e.g., $\alpha \longrightarrow ?\beta$). Let us now take a look at the new sorts of questions which become expressible in the first-order setting.

Consider a formula φ: relative to a variable x, φ determines an intensional property φ^x, i.e., a function that maps each world w to a set of entities, namely:

$$\varphi^x(w) := \{d \in D \mid w[x \mapsto d] \models \varphi\}$$

Now let α be a statement. According to the clause for $\exists^i$, the formula $\exists^i x\alpha$ is settled in s just in case s establishes of some specific individual $d \in D$ that it is included in the denotation of α^x. More formally:

$$s \models \exists^i x\alpha \iff \text{for some } d \in D: \ d \in \alpha^x(w) \text{ for all } w \in s$$

Thus, $\exists^i x\alpha$ can be seen as expressing a question which is settled by providing an instance of the property α^x. Thus, if α stands for the statement "x is a European capital," then $\exists^i x\alpha$ stands for a question such as (1), which is settled precisely in case we establish of something that it is a European capital. Thus, $\exists^i$ allows us to express the so-called *mention-some* questions.

(1) What is one European capital?

Now suppose we want to express, instead, the *mention-all* question which is only settled in case we lay out exactly what the extension of α^x is. This question is settled just in case of each individual it is settled whether or not it has property α^x. This suggests to express this question by means of the formula $\forall^i x?\alpha$—that is, spelling out the question mark, $\forall^i x(\alpha \mathbb{W} \neg\alpha)$. And, indeed, we have the following:

$$
\begin{aligned}
s \models \forall^i x?\alpha \iff & \text{ for all } d \in D, \quad s[x \mapsto d] \models ?\alpha \\
\iff & \text{ for all } d \in D, \quad s[x \mapsto d] \models \alpha \text{ or } s[x \mapsto d] \models \neg\alpha \\
\iff & \text{ for all } d \in D, \quad \text{for all } w \in s, \ w[x \mapsto d] \models \alpha \text{ or} \\
& \qquad\qquad\qquad\quad \text{for all } w \in s, \ w[x \mapsto d] \not\models \alpha \\
\iff & \text{ for all } d \in D, \quad \text{for all } w \in s, \ d \in \alpha^x(w) \text{ or} \\
& \qquad\qquad\qquad\quad \text{for all } w \in s, \ d \notin \alpha^x(w) \\
\iff & \text{ for all } d \in D, \quad \text{for all } w, w' \in s, \ d \in \alpha^x(w) \iff d \in \alpha^x(w') \\
\iff & \text{ for all } w, w' \in s, \ \text{for all } d \in D, \ d \in \alpha^x(w) \iff d \in \alpha^x(w') \\
\iff & \text{ for all } w, w' \in s, \ \alpha^x(w) = \alpha^x(w')
\end{aligned}
$$

This shows that $s \models \forall^i x?\alpha$ holds just in case in s it is settled what the extension of α^x is. For instance, if we suppose again that α stands for the statement "x is a European capital," then $\forall^i x?\alpha$ stands for a question such as (1).[24]

(2) What are all the European capitals?

As an important particular case, let us see how we can express within our language the identity question concerning the value of a certain term t. Consider the formula $x = t$, where x is a variable not occurring in t. Then, $(x = t)^x$ expresses is the property of being identical to t, whose denotation at a world w consists of exactly one element, namely the referent t_w of t in the world w. Thus, if we consider the formula $\exists^i x(x = t)$, we have

$$s \models \exists^i x(x = t) \iff \text{for some } d \in D : \ d \in (x = t)^x(w) \text{ for all } w \in s$$
$$\iff \text{for some } d \in D : \ d \in \{t_w\} \text{ for all } w \in s$$
$$\iff \text{for some } d \in D : \ d = t_w \text{ for all } w \in s$$
$$\iff \text{for all } w, w' \in s : \ t_w = t_{w'}$$

That is, $\exists^i x(x = t)$ is settled in a state s just in case it is established in s what the value of the term t is. In line with the notation used in our informal discussion in Section 2.8, we will make the following notational convention:

• $\lambda_t := \exists^i x(x = t)$, where x is an arbitrary variable not occurring in t

We will refer to λ_t as the *identity question* about t. Incidentally, notice that, since $(x = t)^x$ denotes a singleton at any possible world, the distinction between mention-some and mention-all evaporates: establishing of some object that it has the property of being identical to t is the same thing as establishing what is the denotation of the property of being identical to t. Hence, λ_t could also have been defined equivalently as $\forall^i x?(x = t)$.

5.5 First-order dependencies

Defining the dependence atom
In standard first-order dependence logic, dependencies are expressed by means of a dedicated kind of atomic formulas, called *dependence atoms*, similar to the ones we discussed in the propositional setting, but having individual variables rather than

---

[24]Incidentally, this means that Groenendijk's *Logic of Interrogation* [13, 30] may be identified with the fragment of $\mathsf{QD_{FO}}$ consisting of classical formulas plus formulas of the form $\forall^i \bar{x}?\alpha$, where α is classical and $\bar{x}$ is a possibly empty sequence of variables.

propositional variables as constituents. For any variables $x_1, \ldots, x_n, y$, the language is equipped with a corresponding atom $=(x_1, \ldots, x_n, y)$, which is interpreted by means of the following clause:

$$s \models \ =(x_1, \ldots, x_n, y) \iff \forall w, w' \in s, \ g_w(x_i) = g_{w'}(x_i) \text{ for all } i, \text{ then } g_w(y) = g_{w'}(y)$$

As in the propositional case, this truth-based clause can be seen to be equivalent with the following, support-based one.

$$s \models \ =(x_1, \ldots, x_n, y) \iff \text{ for all } t \subseteq s, \text{ if } t \models \lambda_{x_i} \text{ for all } i, \text{ then } t \models \lambda_y$$

Now, notice that the right-hand side amounts to the fact that the question λ_y is entailed by the questions $\lambda_{x_1}, \ldots, \lambda_{x_n}$ in the context s. So, we have

$$s \models \ =(x_1, \ldots, x_n, y) \iff \lambda_{x_1}, \ldots, \lambda_{x_n} \models_s \lambda_y$$

This shows that, just like the propositional dependence atom, the first-order dependence atom captures a particular case of question entailment in context, namely entailment between identity questions. This provides a fully precise ground to our informal discussion, in section 2.8, of how dependencies between variables can be re-conceptualized as dependencies between questions. Moreover, due to the connection between entailment in context and implication, this means that dependence atoms are definable in $\mathsf{QD_{FO}}$ as implications involving identity questions.[25]

$$=(x_1, \ldots, x_n, y) \quad := \quad \lambda_{x_1} \wedge \cdots \wedge \lambda_{x_n} \longrightarrow \lambda_y$$

Thus, for instance, the Henkin pattern of quantification that we used as an example in the introduction section may be expressed in $\mathsf{QD_{FO}}$ as follows:

$$\forall^d x_1 \forall^d x_2 \exists^d y_1 \exists^d y_2 ((\lambda_{x_1} \longrightarrow \lambda_{y_1}) \wedge (\lambda_{x_2} \longrightarrow \lambda_{y_2}) \wedge \varphi(x_1, x_2, y_1, y_2))$$

The fact that the dependence atom is expressible by means of implication was already remarked, in a slightly different form, in [1]. However, what our perspective brings out is that, as in the propositional case, dependence atoms are only a special case of a much more general pattern. In this section, we will take a look at a number of other interesting kinds of dependencies that become expressible by combining implication and the wide range of questions available in $\mathsf{QD_{FO}}$.

Dependencies among features of the assignment
Suppose we are working with the standard signature of arithmetic, and suppose all the worlds in our state s are based on the same model, namely the standard model of the natural numbers. Now let even(x) be the statement that x is even. Then, the

---

[25]Notice in particular that, for $n = 0$, we have that the constancy atom $=(x)$ of standard dependence logic is equivalent to the identity question λ_x.

polar question $?\text{even}(x)$ captures the question about the parity of x. Clearly, this question is much weaker than λ_x: in order to have $s \models \lambda_x$, all the assignments in s have to agree about their value on x; in order to have $s \models ?\text{even}(x)$, on the other hand, it suffices that they agree on the *parity* of x—though they may well disagree on their precise value on x.

In more visual terms, we can think of both the question λ_x and the question $?\text{even}(x)$ as inducing partitions of the space of assignments: the partition induced by λ_x has infinitely many blocks, corresponding to the infinitely many possible values for x; the partition induced by $\text{even}(x)$ has only two blocks, corresponding to the two possible parities for x.

Now imagine that, in a certain state s, the value of y is not completely determined by the value of x, yet the parity of y *is* determined by the value of x. In this state, an interesting dependency relation between x and y holds, but one which is much weaker than the relation expressed by the dependence atom $=(x, y)$. In QD_{FO}, the relevant dependency can be expressed in a simple way by the following conditional:

$$\lambda_x \longrightarrow ?\,\text{even}(y)$$

Conversely, it may be that, in a certain state s, it is not even necessary to know the exact value of x in order to know the value of y: it suffices to know the parity of x. That is, in order to determine y we only need access to some partial information about x. In this state, too, an interesting dependency relation between x and y holds, but one which is much stronger than the relation expressed by the dependence atom $=(x, y)$. In QD_{FO}, this relation can be expressed by means of the following conditional:

$$?\,\text{even}(x) \longrightarrow \lambda_y$$

Finally, there are also cases of dependencies between variables that are simply incomparable to the standard ones. For instance, it may be that the parity of y in a state is determined by the parity of x, which is expressed by the following conditional.

$$?\,\text{even}(x) \longrightarrow ?\,\text{even}(y)$$

Of course, there is nothing special about the use of a *polar* question $?\text{even}(x)$ in these examples. To see that analogous phenomena arise with different kinds of questions, suppose $\text{mod}_n(x)$ denotes the remainder of the division of x by n. Then, the question $\lambda_{\text{mod}_n(x)}$ asks for the value of this remainder, that is, it asks for the equivalence class of x modulo n. Intuitively, the question $\lambda_{\text{mod}_n(x)}$ partitions the set of all assignments into n cells, each one corresponding to one equivalence class modulo n.

Then, the following implication expresses the fact that the value of y is fully determined by the value of $\mathsf{mod}_n(x)$, that is, by the equivalence class of x modulo n:

$$\lambda_{\mathsf{mod}_n(x)} \longrightarrow \lambda_y$$

Conversely, the following formula expresses that the equivalence class of y modulo n is determined by x.

$$\lambda_x \longrightarrow \lambda_{\mathsf{mod}_n(y)}$$

Of course, such conditionals may then be embedded in a sentence which contains quantifiers binding the variables x and y. In this way we can express, for instance, that for any x there exists a corresponding y, depending only on the parity of x, which stands in the relation φ to x.

$$\forall^d x \exists^d y((?\,\mathsf{even}(x) \longrightarrow \lambda_y) \wedge \varphi(x,y))$$

As another example, we may express that for all x and y there is a corresponding z, whose parity is fully determined by x and whose equivalence class modulo 3 is fully determined by y, which stands in the relation φ to x and y.

$$\forall^d x \forall^d y \exists^d z((\lambda_x \longrightarrow ?\,\mathsf{even}(z)) \wedge (\lambda_y \longrightarrow \lambda_{\mathsf{mod}_3(z)}) \wedge \varphi(x,y,z))$$

Thus, regarding dependency as a relation between questions leads us to a broader perspective on this notion, which makes it possible to recognize and express a variety of different dependence relations, some of which are weaker than the standard ones, some stronger, and some simply incomparable.

Dependencies among features of the model

In standard dependence logic, the semantics is based on a fixed underlying first-order model. By contrast, in $\mathsf{QD_{FO}}$ information states also reflect uncertainty as to the features of the model: this makes it possible to capture dependence relations that concern features of the model, rather than features of the assignment. Let us briefly consider a few basic examples.

First of all, all the kinds of dependencies expressible in the propositional system $\mathsf{QD_P}$ are also expressible in $\mathsf{QD_{FO}}$: this includes dependencies between polar questions, alternative questions, and conditional questions, of the kind discussed in section 3.3. Thus, $\mathsf{QD_{FO}}$ avoids the puzzling situation found in standard dependence logic, where dependencies expressible in the propositional system are no longer expressible in the first-order system.

Additionally, we can now express dependencies between properly first-order questions, such as the mention-some and mention-all questions discussed in the previous sub-section. As a first example, consider the following implication between two mention-some questions: this implication expresses the fact that any instance of the property α^x determines a corresponding instance of the property β^x.

$$\exists^i x \alpha \longrightarrow \exists^i x \beta$$

For a concrete illustration, suppose $L(x)$ stands for the statement "x is a leap year," and suppose we are in a state s in which we do not know exactly which years are leap years, but we do know that there are never two leap years in a row. In this state the dependency $\exists^i x L(x) \longrightarrow \exists^i x \neg L(x)$ holds, since from an instance of a leap year we can derive an instance of a common (non-leap) year: if we enhance s by settling that a certain year x is a leap year, then we have also settled that $x + 1$ is a common year. Vice versa, the converse dependency $\exists^i x \neg L(x) \longrightarrow \exists^i x L(x)$ does not hold in s, since from an instance of a common year we cannot without additional information obtain an instance of a leap year.

As a further example of first-order dependency, the following implication between mention-all question allows us to express the fact that the extension of the property β^x is completely determined by the extension of the property α^x.

$$\forall^i x ? \alpha \longrightarrow \forall^i x ? \beta$$

For an illustration, suppose $E(x)$ is the statement "x is an EU country" and $B(x)$ is the statement "x borders an EU country." Suppose we are in an information state s in which we know nothing about which countries belong to the EU, but we know exactly what the world map looks like, and so we know which countries share a border. The dependency $\forall^i x ? E(x) \longrightarrow \forall^i x ? B(x)$ holds in this state, since once we enhance our state by settling what the EU countries are, we also settle which countries border an EU country. On the other hand, the converse dependency $\forall^i x ? B(x) \longrightarrow \forall^i x ? E(x)$ does not hold: in the given state, settling which countries border an EU country is not enough to determine what the EU countries are.

Higher-order dependencies

Let us conclude this quick tour of the variety of dependencies expressible in $\mathsf{QD_{FO}}$ by showing a concrete example of the higher-order dependencies discussed in section 3.3. Suppose our language contains three individual constants, $t, p,$ and v, which stand, respectively, for the *temperature*, the *pressure*, and the *volume* of a certain gas.

Now, the implication $\lambda_t \longrightarrow \lambda_p$ is supported in a state s just in case, relative to s, the pressure of the gas is determined by the temperature in a certain way. As discussed already in section 3.3, this formula is itself a question, which may be described as asking for a way to determine the pressure from the temperature. This *way* may be construed as a function: indeed, $s \models \lambda_t \longrightarrow \lambda_p$ holds just in case there is a function $f : D \longrightarrow D$ such that $p_w = f(t_w)$ for all $w \in s$—that is, in case there is a function f of which it is established in s that applying it to the value of t yields the value of p.

The situation is analogous for the formula $\lambda_t \longrightarrow \lambda_v$, which is a question asking for a way to determine the volume from the temperature. Now consider the following higher-order dependence formula:

$$(\lambda_t \longrightarrow \lambda_p) \longrightarrow (\lambda_t \longrightarrow \lambda_v)$$

What the formula expresses is that any way for the pressure to depend on the temperature determines a corresponding way for the volume to depend on the temperature.

For instance, suppose in s it is established that our gas behaves as a perfect gas, satisfying the law $pv = kt$, where k is a constant whose value is established in s. Then, the above higher-order dependency holds. For, were we to establish that $p = f(t)$ for some function f, it would follow that $v = \frac{kt}{f(t)}$, so we would have a function g (namely the function $t \mapsto \frac{kt}{f(t)}$) of which it is established that it yields v from t. Hence, relative to s, any dependency of p on t yields a corresponding dependency of v on t. This is precisely what is required for the higher-order dependency above to hold.

This example is meant to illustrate that higher-order dependencies are meaningful and potentially important features of an informational scenario. In standard dependence logic, it is hard to see how such dependencies can be represented, since dependence atoms can only be applied to variables, not to other dependence atoms. In QD_{FO}, on the other hand, implications may be nested, and higher-order dependencies can be represented in a simple and elegant way.

5.6   Summing up

In this section we have introduced a first-order system QD_{FO} for questions and dependencies. This system is based on information states construed as sets of model-assignment pairs, thus reflecting partial information about the features of the model as well as the values of variables. This allows us to interpret in a natural way a broad range of questions concerning both these aspects, as well as their interaction. Besides the propositional questions familiar from Section 3, notable kinds of questions which are expressible in the system include identity questions, which ask for the value of a term, mention-some questions, which ask for an instance of a property, and mention-all questions, which ask for the extension of a property. Thanks to the availability of the implication operation, all dependencies between such questions may then be expressed in the language. The dependence atoms of standard dependence logic emerge as a particular case—the case in which the relevant questions are identity questions about the value of variables. However, our new perspective allows us to recognize and express many other dependence patterns; for instance, we may express that the extension (or an instance) of a property is determined by the extension (or an instance) of another. Or, we may express that some feature of a variable is determined by some feature of another—the standard dependencies constituting the limit case in which the relevant features are the variables' values.

6 Conclusion

In this paper, we have investigated the relation between questions and dependency. We started out by showing that we can take a new, informational perspective on classical logic, and that this move makes it possible to bring questions within the scope of logic. Once this is done, dependency emerges as a facet of the relation of entailment, namely entailment between questions in context. Moreover, in this semantic framework, an implication operation is naturally defined which allows us to express entailment in context within the language. Thus, in such a semantics, dependencies are generally expressible as implications between questions.

We made this discussion concrete by looking at a system of propositional logic which encompasses both propositional inquisitive semantics and dependence logic. We saw that this system is capable of expressing not only dependencies between atomic polar questions—corresponding to the dependence atoms in propositional dependence logic—but also other natural examples of dependence relations involving, e.g., complex polar questions, alternative questions, and conditional questions.

We investigated the features of the associated propositional logic, providing a complete proof system and showing that questions and dependencies can be manipulated in reasoning by means of simple and familiar logical rules. We found that the connection between questions and dependencies also shows up at the level of proofs: proofs involving questions do not just witness the existence of certain dependencies, but actually encode *methods for computing* these dependencies.

Finally, we took the discussion to the first-order level. We saw that, for a general account of first-order dependencies, our setup needs to be more general than the one adopted in first-order inquisitive semantics, which uses a single assignment, and also than the one adopted in dependence logic, which uses a single model: we proposed to evaluate formulas relative to information states construed as sets of world-assignment pairs. We defined a system of first-order logic which brings together operations from dependence logic and inquisitive semantics, and we saw that this system is capable of expressing not only dependencies between identity questions—corresponding to the dependence atoms of first-order dependence logic—but also many other interesting types of dependencies, such as dependencies between the extension of properties, and dependencies between features of different variables.

If the ideas presented here are sound, logics of questions and logics of dependency share the same object of investigation. Clearly, this calls for a transfer, or rather a *merge*, of insights and results that have been obtained within the two traditions. Interestingly, these are often complementary: work in Inquisitive Semantics has focused mostly on the logico-philosophical foundation of the semantics, on propositional connectives, axiomatization results, and on applications in linguistics; by contrast, work in Dependence Logic has focused mostly on quantification, on complexity results, and on applications in computer science. It thus seems that there is a great deal to be gained by bringing together these two lines of work. Along with [39] and [6], the present paper is a first step in this direction.

Acknowledgements I am indebted to Lucas Champollion, Jeroen Groenendijk, Lauri Hella, Rosalie Iemhoff, Juha Kontinen, Vít Punčochář, Floris Roelofsen, and Fan Yang for stimulating discussion of the ideas presented here. A special thanks is owed to Jouko Väänänen for bringing together in two separate occasions the communities working on Dependence Logic and Inquisitive Semantics. The present paper has grown out of ideas sparked by those meetings. Financial support from the Netherlands Organization of Scientific Research (NWO) is gratefully acknowledged.

Appendix

Proof of Proposition 22. We first observe one basic property of our system.

Replacement of equivalents: for all formulas $\varphi, \psi, \chi \in \mathscr{L}$, $\psi \dashv\vdash \chi$ implies $\varphi[\psi/p] \dashv\vdash \varphi[\chi/p]$.

This claim can be established straightforwardly by induction on φ. Equipped with this observation, we will proceed to prove Proposition 22 by induction on φ. If φ is an atom or $\bot$, then $\mathscr{R}(\varphi) = \{\varphi\}$ and the claim is trivially true. Let us now assume the claim holds for φ and ψ, that is, let us assume $\varphi \dashv\vdash \bigvee\mathscr{R}(\varphi)$ and $\psi \dashv\vdash \bigvee\mathscr{R}(\psi)$, and let us proceed to prove that it also holds for $\varphi \wedge \psi$, $\varphi \longrightarrow \psi$, $\varphi \vee \psi$, and $\varphi \bigvee \psi$.

- $\varphi \wedge \psi$. By induction hypothesis and replacement of equivalents, we have

$$\varphi \wedge \psi \dashv\vdash \bigvee\mathscr{R}(\varphi) \wedge \bigvee\mathscr{R}(\psi)$$

Proposition 21 ensures that any entailment pattern which is valid in intuitionistic logic is also valid in $\mathsf{QD_{FO}}$, when $\bigvee$ is interpreted as intuitionistic disjunction. In particular, this ensures that the distributivity of $\wedge$ over $\bigvee$ is provable. We thus have

$$\bigvee\mathscr{R}(\varphi) \wedge \bigvee\mathscr{R}(\psi) \dashv\vdash \bigvee_{\alpha \in \mathscr{R}(\varphi), \beta \in \mathscr{R}(\psi)}(\alpha \wedge \beta)$$

Given that $\mathscr{R}(\varphi \wedge \psi) = \{\alpha \wedge \beta \mid \alpha \in \mathscr{R}(\varphi) \text{ and } \beta \in \mathscr{R}(\psi)\}$, the latter formula is nothing but $\bigvee\mathscr{R}(\varphi \wedge \psi)$.

- $\varphi \longrightarrow \psi$. By induction hypothesis and replacement of equivalents, we have

$$\varphi \longrightarrow \psi \dashv\vdash \bigvee\mathscr{R}(\varphi) \longrightarrow \bigvee\mathscr{R}(\psi)$$

By reasoning intuitionistically with $\longrightarrow$, $\wedge$ and $\bigvee$ we then have

$$\bigvee\mathscr{R}(\varphi) \longrightarrow \bigvee\mathscr{R}(\psi) \dashv\vdash \bigwedge_{\alpha \in \mathscr{R}(\varphi)}(\alpha \longrightarrow \bigvee\mathscr{R}(\psi))$$

Now, for any $\alpha \in \mathscr{R}(\varphi)$, we have $\alpha \longrightarrow \bigvee\mathscr{R}(\psi) \dashv\vdash \bigvee_{\beta \in \mathscr{R}(\psi)}(\alpha \longrightarrow \beta)$. The right-to-left direction holds by simple intuitionistic reasoning, while the left-to-right direction simply amounts to the Split rule of our system (recall that resolutions are classical formulas, so this rule is indeed applicable). By replacement of equivalents, we then have the following:

$$\bigwedge_{\alpha \in \mathscr{R}(\varphi)} (\alpha \longrightarrow \bigvee \mathscr{R}(\psi)) \dashv\vdash \bigwedge_{\alpha \in \mathscr{R}(\varphi)} \bigvee_{\beta \in \mathscr{R}(\psi)} (\alpha \longrightarrow \beta)$$

Finally, using again the distributivity of $\wedge$ over $\bigvee$ we get that the right-hand side is provably equivalent with the following formula:

$$\bigvee_{f : \mathscr{R}(\varphi) \longrightarrow \mathscr{R}(\psi)} \bigwedge_{\alpha \in \mathscr{R}(\varphi)} (\alpha \longrightarrow f(\alpha))$$

Given that $\mathscr{R}(\varphi \longrightarrow \psi) = \{\bigwedge_{\alpha \in \mathscr{R}(\varphi)} \alpha \longrightarrow f(\alpha) \mid f : \mathscr{R}(\varphi) \longrightarrow \mathscr{R}(\psi)\}$, the latter formula is nothing but $\bigvee \mathscr{R}(\varphi \longrightarrow \psi)$.

- $\varphi \vee \psi$. By induction hypothesis and replacement of equivalents, we have

$$\varphi \vee \psi \dashv\vdash \bigvee \mathscr{R}(\varphi) \vee \bigvee \mathscr{R}(\psi)$$

Notice that we have $\varphi \vee (\psi \bigvee \chi) \dashv\vdash (\varphi \vee \psi) \bigvee (\varphi \vee \chi)$: the left-to-right direction of this equivalence is obtained by an application of the rule ($\vee$d) of our system, while the right-to-left direction is easily proved using the rules ($\bigvee$e), ($\bigvee$i) ($\vee$r).

Also, notice that by using the rule ($\vee$c), we can make sure that distributivity holds not only on the right side of a tensor, but also on the left side. Then, multiple applications of distributivity on both sides of the tensor yield

$$\bigvee \mathscr{R}(\varphi) \vee \bigvee \mathscr{R}(\psi) \dashv\vdash \bigvee_{\alpha \in \mathscr{R}(\varphi), \beta \in \mathscr{R}(\psi)} \alpha \vee \beta$$

Given that $\mathscr{R}(\varphi \vee \psi) = \{\alpha \vee \beta \mid \alpha \in \mathscr{R}(\varphi) \text{ and } \beta \in \mathscr{R}(\psi)\}$, the formula on the right is nothing but $\bigvee \mathscr{R}(\varphi \vee \psi)$.

- $\varphi \bigvee \psi$. By induction hypothesis and replacement of equivalents we have

$$\varphi \bigvee \psi \dashv\vdash \bigvee \mathscr{R}(\varphi) \bigvee \bigvee \mathscr{R}(\psi)$$

Given that $\mathscr{R}(\varphi \bigvee \psi) = \mathscr{R}(\varphi) \cup \mathscr{R}(\psi)$, the above formula is simply $\bigvee \mathscr{R}(\varphi \bigvee \psi)$.

References

1. Abramsky, S., Väänänen, J.: From IF to BI. Synthese **167**(2), 207–230 (2009)
2. Belnap, N.: Questions, answers, and presuppositions. J. Philos. **63**(20), 609–611 (1966)
3. Brasoveanu, A., Farkas, D.F.: How indefinites choose their scope. Linguist. Philos. **34**, 1–55 (2011)

4. Ciardelli, I.: Inquisitive semantics and intermediate logics. M.Sc. Thesis, University of Amsterdam (2009)
5. Ciardelli, I.: A first-order inquisitive semantics. In: Aloni, M., Bastiaanse, H., de Jager, T., Schulz, K. (eds.) Logic, Language, and Meaning: Selected Papers from the Seventeenth Amsterdam Colloquium, pp. 234–243. Springer, Berlin (2010)
6. Ciardelli, I.: Interrogative dependencies and the constructive content of inquisitive proofs. In: Kohlenbach, U., Barceló, P., de Queiroz, R. (eds.) Logic, Language, Information and Computation - 21st International Workshop. WoLLIC 2014. Lecture Notes in Computer Science, pp. 109–123. Springer, Berlin (2014)
7. Ciardelli, I., Roelofsen, F.: Inquisitive logic. J. Philos. Log. **40**(1), 55–94 (2011)
8. Ciardelli, I., Groenendijk, J., Roelofsen, F.: Inquisitive semantics (2012). NASSLLI Lecture notes for a course at the North Americal School of Logic, Language, and Information. Unpublished, available at www.illc.uva.nl/inquisitivesemantics
9. Ciardelli, I., Groenendijk, J., Roelofsen, F.: Towards a logic of information exchange: an inquisitive witness semantics. In: Bezhanishvili, G., Marra, V., Löbner, S., Richter, F. (eds.) Logic, Language, and Computation: Revised Selected Papers from the Ninth International Tbilisi Symposium on Logic, Language, and Computation, pp. 51–72. Springer, Berlin (2013)
10. Ciardelli, I., Groenendijk, J., Roelofsen, F.: On the semantics and logic of declaratives and interrogatives. Synthese **192**(6), 1689–1728 (2015)
11. Ebbing, J., Hella, L., Meier, A., Müller, J.S., Virtema, J., Vollmer, H.: Extended modal dependence logic. In: Libkin, L., Köhlenbach, U., de Queiroz, R. (eds.) Logic, Language, Information and Computation - 20th International Workshop. WoLLIC 2013. Lecture Notes in Computer Science, pp. 126–137. Springer, Berlin (2013)
12. Gamut, L.: Language, Logic and Meaning. Chicago University Press, Chicago, IL (1991)
13. Groenendijk, J.: The logic of interrogation. In: Matthews, T., Strolovitch, D. (eds.) Semantics and Linguistic Theory, pp. 109–126. Cornell University Press, Ithaca, NY (1999)
14. Groenendijk, J.: Inquisitive semantics: two possibilities for disjunction. In: Bosch, P., Gabelaia, D., Lang, J. (eds.) Seventh International Tbilisi Symposium on Language, Logic, and Computation. Springer, Berlin (2009)
15. Groenendijk, J.: Erotetic languages and the inquisitive hierarchy. In a Festschrift for Martin Stokhof (2011). http://dare.uva.nl/document/487828.
16. Groenendijk, J., Roelofsen, F.: Inquisitive semantics and pragmatics. Presented at the Workshop on Language, Communication, and Rational Agency at Stanford (2009). www.illc.uva.nl/inquisitivesemantics.
17. Groenendijk, J., Stokhof, M.: Studies on the semantics of questions and the pragmatics of answers. Ph.D. thesis, University of Amsterdam (1984)
18. Groenendijk, J., Stokhof, M.: Dynamic predicate logic. Linguist. Philos. **14**, 39–100 (1991)
19. Groenendijk, J., Stokhof, M., Veltman, F.: Coreference and modality. In: Lappin, S. (ed.) Handbook of Contemporary Semantic Theory, pp. 179–216. Blackwell, Oxford (1996)
20. Hausser, R., Zaefferer, D.: Questions and answers in a context-dependent Montague grammar. In: Guenthner, F., Schmidt, S.J. (eds.) Formal Semantics and Pragmatics for Natural Languages, pp. 339–358. Reidel, Dordrecht (1978)
21. Hella, L., Luosto, K., Sano, K., Virtema, J.: The expressive power of modal dependence logic. In: Kooi, R.G.B., Kurucz, A. (eds.) Advances in Modal Logic (AIML), pp. 294–312. College Publications, London (2014)
22. Henkin, L.: Some remarks on infinitely long formulas. In: Infinitistic Methods. Proceedings of the Symposium on the Foundations of Mathematics, pp. 167–183. Pergamon, New York (1961)
23. Hintikka, J.: Knowledge and belief: an introduction to the logic of the two notions. Cornell University Press, 1962.
24. Hintikka, J., Sandu, G.: Game-theoretical semantics. In: Handbook of Logic and Language, pp. 361–410. Elsevier, Amsterdam (1997)
25. Hodges, W.: Compositional semantics for a language of imperfect information. Log. J. IGPL **5**(4), 539–563 (1997)

26. Hodges, W.: Some strange quantifiers. In: Structures in Logic and Computer Science. Lecture Notes in Computer Science, pp. 51–65. Springer, New York (1997)
27. Mascarenhas, S.: Inquisitive semantics and logic. Master Thesis, University of Amsterdam (2009)
28. Roelofsen, F.: Algebraic foundations for the semantic treatment of inquisitive content. Synthese **190**(1), 79–102 (2013)
29. Scha, R.: Logical foundations for question answering. Ph.D. thesis, University of Groningen (1983)
30. ten Cate, B., Shan, C.C.: Axiomatizing Groenendijk's logic of interrogation. In: Aloni, M., Butler, A., Dekker, P. (eds.) Questions in Dynamic Semantics, pp. 63–82. Elsevier, New York (2007)
31. Tichy, P.: Questions, answers, and logic. Am. Philos. Q. **15**, 275–284 (1978)
32. Troelstra, A., van Dalen, D.: Constructivism in Mathematics, An Introduction, vol. 1. Studies in Logic and the Foundations of Mathematics, vol. 121. North-Holland, Amsterdam (1988)
33. Väänänen, J.: Dependence Logic: A New Approach to Independence Friendly Logic. Cambridge University Press, Cambridge (2007)
34. Väänänen, J.: Modal dependence logic. In: Apt, K., van Rooij, R. (eds.) New Perspectives on Games and Interaction. Amsterdam University Press, Amsterdam (2008)
35. Väänänen, J.: Multiverse set theory and absolutely undecidable propositions. In: Kennedy, J. (ed.) Interpreting Gödel: Critical Essays, Cambridge University Press, p. 180 (2014)
36. Velissaratou, S.: Conditional questions and which-interrogatives. M.Sc. Thesis, University of Amsterdam (2000)
37. Veltman, F.: Data semantics. In: Groenendijk, J., Janssen, T., Stokhof, M. (eds.) Formal Methods in the Study of Language. Mathematical Centre, Amsterdam (1981)
38. Veltman, F.: Defaults in update semantics. J. Philos. Log. **25**(3), 221–261 (1996)
39. Yang, F.: On extensions and variants of dependence logic: a study of intuitionistic connectives in the team semantics setting. Ph.D. thesis, University of Helsinki (2014)

Approximation Logics for Subclasses of Probabilistic Conditional Independence and Hierarchical Dependence on Incomplete Data

Sebastian Link

Abstract Probabilistic conditional independence constitutes a principled approach to handle knowledge and uncertainty in artificial intelligence, and is fundamental in probability theory and multivariate statistics. Similarly, first-order hierarchical dependence provides an expressive framework to capture the semantics of an application domain within a database system, and is essential for the design of databases. For complete data it is well known that the implication problem associated with probabilistic conditional independence is not axiomatizable by a finite set of Horn rules (Studený, Conditional independence relations have no finite complete characterization. In: Kubik, S., Visek, J. (eds.) Transactions of the 11th Prague Conference on Information Theory, Statistical Decision Functions and Random Processes, pp. 377–396. Kluwer, Dordrecht, 1992), and the implication problem for first-order hierarchical dependence is undecidable (Herrmann, Inf. Comput. 122(2):221–235, 1995). Moreover, both implication problems do not coincide (Studený, Conditional independence relations have no finite complete characterization. In: Kubik, S., Visek, J. (eds.) Transactions of the 11th Prague Conference on Information Theory, Statistical Decision Functions and Random Processes, pp. 377–396. Kluwer, Dordrecht, 1992) and neither of them is equivalent to the implication problem of some fragment of Boolean propositional logic (Sagiv et al., J. ACM 28(3):435–453, 1981). In this article, generalized saturated conditional independence and full first-order hierarchical dependence over incomplete data are investigated as expressive subclasses of probabilistic conditional independence and first-order hierarchical dependence, respectively. The associated implication problems are axiomatized by a finite set of Horn rules, and both shown to coincide with that of a propositional fragment under interpretations in the well-known approximation logic $\mathscr{S}$-3. Here, the propositional variables in the set $\mathscr{S}$ are interpreted classically, and correspond to random variables as well as attributes on which incomplete data is not permitted to occur.

S. Link (✉)
The University of Auckland, Auckland, New Zealand
e-mail: s.link@auckland.ac.nz

© Springer International Publishing Switzerland 2016
S. Abramsky et al. (eds.), *Dependence Logic*, DOI 10.1007/978-3-319-31803-5_9

1 Introduction

The concept of conditional independence is important for capturing structural aspects of probability distributions, for dealing with knowledge and uncertainty in artificial intelligence, and for learning and reasoning in intelligent systems [24, 50]. A conditional independence (CI) statement $I(Y, Z \mid X)$ represents the independence of two sets of random variables relative to a third: given three mutually disjoint subsets X, Y, and Z of a set S of random variables, if we have knowledge about the state of X, then knowledge about the state of Y does not provide additional evidence for the state of Z and vice versa. A fundamental problem is the implication problem, which is to decide for an arbitrary finite set S, and an arbitrary set $\Sigma \cup \{\varphi\}$ of CI statements over S, whether every probability model that satisfies every CI statement in Σ also satisfies φ. The significance of this problem is due to its relevance for building Bayesian networks [50]. The implication problem for CI statements is not axiomatizable by a finite set of Horn rules [56]. An important subclass of CI statements are saturated conditional independence (SCI) statements. These are CI statements $I(Y, Z \mid X)$ over S that satisfy $XYZ = S$, that is, the set union XYZ of X, Y, and Z is S. Geiger and Pearl have established an axiomatization for the implication problem of SCI statements by a finite set of Horn rules [21].

The notion of saturated conditional independence $I(Y, Z \mid X)$ over S is closely related to that of a multivalued dependency (MVD) $X \twoheadrightarrow Y|Z$ over S, studied in the framework of relational databases [5, 6, 15, 18, 25, 39, 40, 52]. Here, a set X of attributes is used to denote the X-value of a tuple over S, i.e., those tuple components that appear in the columns associated with X. Indeed, $X \twoheadrightarrow Y|Z$ expresses the fact that an X-value uniquely determines the set of associated Y-values independently of joint associations with Z-values where $Z = S - XY$. Thus, given a specific occurrence of an X-value within a tuple, so far not knowing the specific association with a Y-value and Z-value within this tuple, and then learning about the specific associated Y-value does not provide any information about the specific associated Z-value. Previous research has established an equivalence between the implication problem for SCI statements and that for MVDs [61]. In addition it is known that the implication problem of MVDs is equivalent to that of formulae in a Boolean propositional fragment $\mathfrak{F}'$ [52], even in nested databases with finite list, and record constructors [26]. Indeed, Sagiv et al. showed that it suffices to consider two-tuple relations in order to decide the implication problem of MVDs [52]. This enabled them to define truth assignments from two-tuple relations, and vice versa, in such a way that the two-tuple relation satisfies an MVD if and only if the truth assignment is a model for the $\mathfrak{F}'$-formula that corresponds to the MVD. It follows from these results that the implication of SCI statements is equivalent to that of $\mathfrak{F}'$-formulae.

Contribution. The purpose of this article is to summarize recent insight into the relationships between implication problems for fragments of conditional independencies, database dependencies, and propositional logic. The classical equivalences described above are extended in two directions. Firstly, extensions of saturated CI statements, multivalued dependencies, and the propositional fragment $\mathfrak{F}'$ are considered. These extensions include generalized saturated conditional independence (GSCI) statements $I(Y_1, \ldots, Y_k \mid X)$, which declare the independence between any finite number k of sets $Y_1, \ldots, Y_k$ of random variables, given X; as well as Delobel's class of full first-order hierarchical dependencies $X : [Y_1 \mid \ldots \mid Y_k]$ as an extension of MVDs $X \twoheadrightarrow Y_i | R - XY_i$ for $i = 1, \ldots, k$. Secondly, these extensions are handled in the presence of incomplete data. For the probabilistic framework this means that incomplete data can be present in some random variables, and for the database framework this means that null markers can be present in some attribute columns. As a mechanism to control the degree of incomplete data we permit random variables to be specified as complete, that is, incomplete data cannot be assigned to them. Similarly, attributes can be specified as NOT NULL to disallow occurrences of null markers in these columns. In fact, the industry standard SQL for defining and querying data permits attributes to be specified as NOT NULL [10]. As a main contribution we establish axiomatizations, by a finite set of Horn rules, for the implication problems of i) generalized saturated conditional independencies in the presence of an arbitrary finite set C of complete random variables, and ii) full first-order hierarchical dependencies in the presence of an arbitrary finite set R_s of attributes declared NOT NULL. It is shown that both implication problems coincide with the implication problem of a propositional fragment $\mathfrak{F}$ under interpretations by the well-known approximation logic $\mathcal{S}$-3. Indeed, the propositional variables in the set $\mathcal{S}$ correspond to the complete random variables in C as well as the NOT NULL attributes in R_s. The main proof arguments are based on special probability models that assign probability one half to two distinct assignments, and on two-tuple relations, since these allow us to define corresponding $\mathcal{S}$-3 truth assignments. The established equivalences are rather special, since any duality between two of these three frameworks fails already for general CI statements, embedded multivalued dependencies, and any Boolean propositional fragment over complete data. The equivalences are illustrated in Figure 1. In particular, they should be understood as strong drivers for the advanced treatment of (in)dependence statements as first-class citizens in some uniform framework for reasoning, such as dependence and independence logic [2, 13, 19, 22, 45, 58, 59].

Organization. Generalized conditional independence statements and complete random variables are defined in Section 2. Their combined implication problem is axiomatized in Section 3. In Section 4 we prove the equivalence between the C-implication of GSCI statements and $\mathcal{S}$-3 implication of the propositional

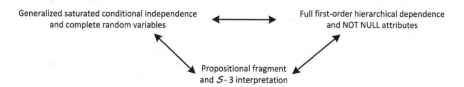

Fig. 1 Summary of Equivalences between Implication Problems

fragment $\mathfrak{F}$. In Section 5 this equivalence is extended to include Delobel's class of full first-order hierarchical dependencies and NOT NULL attributes. Related work is outlined in Section 6. We conclude in Section 7.

2   Generalized Conditional Independence under Incomplete Data

We use the framework of Geiger and Pearl [21]. We denote by S a finite set of distinct symbols $\{v_1, \ldots, v_n\}$, called *random variables*. A *domain mapping* is a mapping that associates a set, $dom(v)$, with each random variable v. The set $dom(v)$ is called the *domain* of v and each of its elements is called a *data value* of v. For $X \subseteq S$ we say that $\mathbf{x}$ is an assignment of X, if $\mathbf{x} \in \prod_{v \in X} dom(v)$. For an assignment $\mathbf{x} = (\mathbf{v}_1, \ldots, \mathbf{v}_k)$ of X with $\mathbf{v}_i \in dom(v_i)$, we write $\mathbf{x}(v_i)$ for the data value $\mathbf{v}_i$ of v_i. For some $Y \subseteq X$ we write $\mathbf{x}(Y)$ for the projection of $\mathbf{x}$ onto Y, that is, $\mathbf{x}(Y)$ denotes the restriction of the assignment $\mathbf{x}$ to the random variables in Y.

2.1   Complete Random Variables

In theory one can assume that the data values of assignments always exist and are known. In practice, these assumptions fail frequently. Indeed, it can happen in most samples that some data values do not exist, or that some existing data values are currently unknown. In statistics and machine learning, one speaks commonly of structural zeros in the first case, and of sampling zeros in the second case [17, 55]. In databases, one speaks of inapplicable nulls in the first case, and of unknown nulls in the second case [8, 9, 63]. In practice, it is often difficult to tell whether some data value does not exist, or exists but is currently unknown.

We use the notation $\mathbf{x}(v) = \mu$ to denote that no information is currently available about the data value $\mathbf{x}(v)$ of the random variable v assigned to $\mathbf{x}$. The interpretation of the marker μ as no information means that a data value does either not exist or a data value exists but is currently unknown.

It is an advantage to gain control over the occurrences of incomplete data values. For this purpose we introduce *complete random variables*. A random variable is defined to be *complete* if and only if $\mu \notin dom(v)$. Although we include μ in domains of random variables that are not complete, we prefer to think of $\boldsymbol{\mu}$ as a marker and not as a data value. In what follows we use C to denote the subset of complete random variables. It is a goal of this article to investigate the properties of generalized saturated conditional probabilistic independence in the presence of an arbitrarily chosen set C of complete random variables. Indeed, complete random variables are shown to provide an effective means to control the degree of uncertainty and to soundly approximate classical reasoning.

2.2 Conditional Independence under Complete Random Variables

A *probability model* over $(S = \{v_1, \ldots, v_n\}, C)$ is a pair (dom, P) where *dom* is a domain mapping that maps each v_i to a finite domain $dom(v_i)$, and $P : dom(v_1) \times \cdots \times dom(v_n) \rightarrow [0, 1]$ is a probability distribution having the Cartesian product of these domains as its sample space. Note that $\mu \notin dom(v_i)$ if and only if $v_i \in C$. An assignment $\mathbf{x}$ of $X \subseteq S$ is *complete* if and only if $\mathbf{x}(v) \neq \mu$ holds for all $v \in X$. As usual, for an assignment $\mathbf{x}$ of X, $P(\mathbf{x})$ denotes the marginal probability $P(X = \mathbf{x})$.

Definition 1. The expression $I(Y_1, \ldots, Y_k \mid X)$, where k is a non-negative integer, and $X, Y_1, \ldots, Y_k$ are mutually disjoint subsets of S, is called a *generalized conditional independence* (CI) statement over S. If $XY_1 \cdots Y_k = S$, we call $I(Y_1, \ldots, Y_k \mid X)$ a *generalized saturated conditional independence* (GSCI) statement. Let (dom, P) be a probability model over (S, C). A generalized CI statement $I(Y_1, \ldots, Y_k \mid X)$ is said to *hold for* (dom, P) if for all complete assignments $\mathbf{x}$ of X, and for all assignments $\mathbf{y}_i$ of Y_i for $i = 1, \ldots, k$,

$$P(\mathbf{y}_1, \ldots, \mathbf{y}_k, \mathbf{x}) \cdot P(\mathbf{x})^{k-1} = P(\mathbf{y}_1, \mathbf{x}) \cdot \ldots \cdot P(\mathbf{y}_k, \mathbf{x}) \tag{1}$$

Equivalently, (dom, P) is said to *satisfy* $I(Y_1, \ldots, Y_k \mid X)$.

Remark 1. The expressions $I(Y_1, \ldots, Y_k \mid X)$ are generalized in the sense that they cover CI statements as the special case where $k = 2$. We assume w.l.o.g. that the sets Y_i are non-empty. Indeed, for all positive k we have the property that a probability distribution satisfies $I(\emptyset, Y_2, \ldots, Y_k \mid X)$ if and only if the probability distribution satisfies $I(Y_2, \ldots, Y_k \mid X)$. In particular, for $k = 1$, the CI statement $I(Y \mid X)$ is always satisfied. One may now define an equivalence relation over the set of generalized CI statements over some fixed set S of random variables. Indeed, two such generalized CI statements are equivalent whenever they are satisfied by the same probability distributions over S. However, our inference rules do not need to be applied to such equivalence classes, as Remark 5 shows. For the sake of simplicity, we assume that in GSCI statements $I(Y_1, \ldots, Y_k \mid X)$ the sets Y_i are non-empty.

Remark 2. The satisfaction of generalized CI statements $I(Y_1, \ldots, Y_k \mid X)$ requires equation (1) to hold for *complete* assignments $\mathbf{x}$ of X only. The reason is that the mutual independence between the sets Y_i is conditional on X. That is, assignments that have *no information* about some random variable in X are not taken into account when judging the independence between distinct Y_i.

Remark 3. If every random variable is declared to be complete, that is, when $C = S$, and $k = 2$, then Definition 1 reduces to the standard definition of CI statements [21, 50].

We now introduce the running example of this article.

Example 1. Let $\{m(ovie), a(ctor), r(ole), c(rew), f(eature), l(anguage), s(ubtitle)\}$ denote the set S of random variables, that captures properties of blu-rays we want to model. Let $C = \{m, a, r, c, s\}$ denote the set of complete random variables, and let Σ consist of the GSCI statements $I(sar, c, fl \mid m)$ and $I(sc, ar \mid mfl)$, and let φ be $I(s, ar, flc \mid m)$. We may define the following probability model (dom, P) over (S, C):

- $dom(m) = \{$Rashomon, The Seven Samurai$\}$,
- $dom(a) = \{$T. Mifune, M. Kyo$\}$,
- $dom(r) = \{$Tajomaru, Masako$\}$,
- $dom(c) = \{$Kurosawa, Hashimoto$\}$,
- $dom(f) = \{$Tailer, Comments, $\mu\}$,
- $dom(l) = \{$Japanese, Maori, $\mu\}$,

and define P by assigning the probability one half to each of the following two assignments of (S, C):

movie	actor	role	crew	feature	language	subtitle
Rashomon	T. Mifune	Tajomaru	Kurosawa	μ	μ	Suomi
Rashomon	M. Kyo	Masako	Kurosawa	μ	μ	Deutsch

It follows that (dom, P) satisfies Σ, but violates φ.

For the remainder of the article we will be interested in GSCI statements. Let $\Sigma \cup \{\varphi\}$ be a set of GSCI statements over S. We say that Σ *C-implies* φ, denoted by $\Sigma \models_C \varphi$, if every probability model over (S, C) that satisfies every GSCI statement in Σ also satisfies the GSCI statement φ. The *implication problem for GSCI statements and complete r.v.* is defined as the following problem.

PROBLEM:	Implication problem of GSCI statements and complete r.v.
INPUT:	Pair (S, C) with set S of random variables and
	subset $C \subseteq S$ of complete random variables
	Set $\Sigma \cup \{\varphi\}$ of GSCI statements over S
OUTPUT:	Yes, if $\Sigma \models_C \varphi$; No, otherwise

Example 2. For $S = \{m, a, r, c, f, l, s\}$, $\Sigma = \{I(sar, c, fl \mid m), I(sc, ar \mid mfl)\}$ does not *C*-imply $\varphi = I(s, ar, cfl \mid m)$ for $C = \{m, a, r, c, s\}$, but Σ does *C′*-imply φ for $C' = \{f, l\}$. A proof of the former is given by the probability model over (S, C) in Example 1, which satisfies Σ, but violates φ. Intuitively, for φ to be implied by Σ one needs to specify f and l to be complete.

For Σ we let $\Sigma_C^* = \{\varphi \mid \Sigma \models_C \varphi\}$ be the *semantic closure* of Σ, i.e., the set of all GSCI statements *C*-implied by Σ. In order to characterize the implication problem of GSCI statements and complete r.v. we use a syntactic approach by applying inference rules. These inference rules have the form

$$\frac{\text{premise}}{\text{conclusion}}\text{condition}$$

and inference rules without any premises are called axioms. The premise consists of a finite set of GSCI statements, and the conclusion is a singleton GSCI statement. The condition of the rule is simple in the sense that it stipulates a simple syntactic restriction on the application of the rule. An inference rule is called *sound*, if every probability model over (S, C) that satisfies every GSCI statement in the premise of the rule also satisfies the GSCI statement in the conclusion of the rule, given that the condition is satisfied. We write $\Sigma \vdash_{\mathfrak{R}} \varphi$ if and only if there is some *inference* of φ from Σ by the set $\mathfrak{R}$ of inference rules. That is, there is some sequence $\gamma = [\sigma_1, \dots, \sigma_n]$ of GSCI statements such that $\sigma_n = \varphi$ and every σ_i is an element of Σ or results from an application of an inference rule in $\mathfrak{R}$ to some elements in $\{\sigma_1, \dots, \sigma_{i-1}\}$. For Σ, let $\Sigma_{\mathfrak{R}}^+ = \{\varphi \mid \Sigma \vdash_{\mathfrak{R}} \varphi\}$ be its *syntactic closure* under inferences by $\mathfrak{R}$. A set $\mathfrak{R}$ of inference rules is said to be *sound* (*complete*) for the implication of GSCI statements and complete r.v., if for every S, every $C \subseteq S$ and for every set Σ of GSCI statements over (S, C) we have $\Sigma_{\mathfrak{R}}^+ \subseteq \Sigma_C^*$ ($\Sigma_C^* \subseteq \Sigma_{\mathfrak{R}}^+$). The (finite) set $\mathfrak{R}$ is said to be a (finite) *axiomatization* for the implication problem of GSCI statements and complete r.v., if $\mathfrak{R}$ is both sound and complete.

Theorem 1 (Geiger and Pearl 1993). *The set $\mathfrak{G} = \{\mathscr{T}', \mathscr{S}', \mathscr{C}', \mathscr{W}'\}$ from Table 1 forms a finite axiomatization for the implication problem of SCI statements, that is, the special case of the implication problem for GSCI statements and complete r.v. where all GSCI statements are of the form $I(Y_1, Y_2 \mid X)$ and where all random variables are complete.* □

Table 1 Axiomatization $\mathfrak{G} = \{\mathscr{T}', \mathscr{S}', \mathscr{C}', \mathscr{W}'\}$ of SCI statements when every r.v. is complete

$\overline{I(S, \emptyset \mid \emptyset)}$	$\dfrac{I(Y_1, Y_2 \mid X)}{I(Y_2, Y_1 \mid X)}$
(saturated trivial independence, $\mathscr{T}'$)	(symmetry, $\mathscr{S}'$)
$\dfrac{I(Z, Y \mid X) \qquad I(Z_1, Z_2 \mid XY)}{I(Z_1, Z_2 Y \mid X)}$	$\dfrac{I(Y_1, Y_2 Z \mid X)}{I(Y_1, Y_2 \mid XZ)}$
(weak contraction, $\mathscr{C}'$)	(weak union, $\mathscr{W}'$)

Remark 4. Studený [56] showed that, in the special case where $C = S$ and $k = 2$, the implication problem of CI statements, i.e., to decide for any given set S of

random variables and any given set $\Sigma \cup \{\varphi\}$ of CI statements over S of the kind $I(Y_1, Y_2 \mid X)$ whether $\Sigma \models_S \varphi$ holds, cannot be axiomatized by a finite set of Horn rules of the form

$$I(Y_1, Z_1 \mid X_1) \wedge \cdots \wedge I(Y_k, Z_k \mid X_k) \rightarrow I(Y, Z \mid X) .$$

3   Axiomatizing GSCI Statements and Complete R.V.

In this section we show that the finite set $\mathfrak{S}$ of Horn rules from Table 2 forms a finite axiomatization for the implication problem of GSCI statements and complete random variables. Our completeness argument applies special probability models which consist of two assignments with probability one half. Special probability models will be further exploited in subsequent sections.

3.1   Sound Inference Rules

Note the following global condition that we enforce on applications of inference rules that infer GSCI statements. It ensures that sets of random variables that occur in GSCI statements are non-empty.

Remark 5.   Whenever we apply an inference rule, then we remove all empty sets Y_i from the exact position in which they occur in the sequence of independent sets of random variables. For instance, we can infer $I(\cdot \mid S)$ by an application of the weak union rule $\mathscr{W}$ to the GSCI statement $I(S \mid \emptyset)$.

The rules in $\mathfrak{S}$ are rather intuitive. The saturated trivial independence rule $\mathscr{T}$ is just $\mathscr{T}'$ when we apply the global condition above. The permutation rule $\mathscr{P}$ replaces the symmetry rule $\mathscr{S}'$ to reflect that a GSCI statement holds for a probability distribution, independently of the order in which the sets Y_i of random variables appear. For the case where $k = 2$, the only non-trivial permutation is easily captured by the symmetry rule $\mathscr{S}'$. The weak union rule $\mathscr{W}$ remains unchanged over $\mathscr{W}'$, except for the number of sets of random variables required. The restricted weak contraction rule $\mathscr{C}$ accommodates the arbitrary number of mutually independent sets of random variables. In addition, $\mathscr{C}$ can only be applied when Y-complete assignments are guaranteed. The next example shows that the condition $Y \subseteq C$ is necessary for the soundness of the restricted weak contraction rule $\mathscr{C}$. As a consequence, the implication problem of GSCI statements and complete random variables is different from the implication problem of GSCI statement where all variables are assumed to be complete.

Example 3. Recall Example 2 where $S = \{m, a, r, c, f, l, s\}$, $C = \{m, a, r, c, s\}$, $\Sigma = \{I(sar, c, fl \mid m), I(sc, ar \mid mfl)\}$, and $\varphi = I(s, ar, cfl \mid m)$. Indeed, Σ S-implies φ, but Σ does not C-imply φ.

Finally, the merging rule $\mathscr{M}$ is required to state that also the union of independent sets of random variables can be independent of other sets of random variables. In fact, the presence of $\mathscr{M}$ in $\mathfrak{S}$ is necessary since the conclusion of any other rule features at least as many independent sets as the maximum number of independent sets amongst all its premises.

Table 2 Axiomatization $\mathfrak{S} = \{\mathscr{T}, \mathscr{P}, \mathscr{M}, \mathscr{W}, \mathscr{C}\}$ of GSCI statements and complete r.v.

$\overline{I(S \mid \emptyset)}$	$\dfrac{I(Y_1, \ldots, Y_k \mid X)}{I(Y_{\pi(1)}, \ldots, Y_{\pi(k)} \mid X)}$
(saturated trivial independence, $\mathscr{T}$)	(permutation, $\mathscr{P}$)
$\dfrac{I(Y_1, \ldots, Y_{k-1}, Y_k, Z \mid X)}{I(Y_1, \ldots, Y_k Z \mid X)}$	$\dfrac{I(Y_1, \ldots, Y_{k-1}, Y_k Z \mid X)}{I(Y_1, \ldots, Y_k \mid XZ)}$
(merging, $\mathscr{M}$)	(weak union, $\mathscr{W}$)
$\dfrac{I(Y_1 \cdots Y_k, YZ_1 \cdots Z_k \mid X) \; I(Y_1 Z_1, \ldots, Y_k Z_k \mid XY)}{I(Y_1, \ldots, Y_k, YZ_1 \cdots Z_k \mid X)} \; Y \subseteq C$	
(restricted weak contraction, $\mathscr{C}$)	

The soundness of the rules in $\mathfrak{S}$ follows from the following proposition and the soundness of the rules in $\mathfrak{G}$. In particular, for the restricted weak contraction rule $\mathscr{C}$ soundness follows under the restriction that assignments must be Y-complete.

Proposition 1. *Let S denote a finite set of random variables and $C \subseteq S$. A probability distribution $\pi = (dom, P)$ over (S, C) satisfies the GSCI statement $I(Y_1, \ldots, Y_k \mid X)$ if and only if for every $i = 1, \ldots, k$, π satisfies the SCI statement $I(Y_i, S - XY_i \mid X)$.*

Proof. Assume that for every $i = 1, \ldots, k$, π satisfies the SCI statement $I(Y_i, S - XY_i \mid X)$. Let $\mathbf{x}$ be a complete assignment over X, and $\mathbf{y}_1, \ldots, \mathbf{y}_k$ be assignments for $Y_1, \ldots, Y_k$, respectively. Then we have

$$
\begin{aligned}
P(\mathbf{x}\mathbf{y}_1 \cdots \mathbf{y}_k) \cdot P(\mathbf{x})^{k-1} &= P(\mathbf{x}\mathbf{y}_1) \cdot P(\mathbf{x}\mathbf{y}_2 \cdots \mathbf{y}_k) \cdot P(\mathbf{x})^{k-2} \\
&= P(\mathbf{x}\mathbf{y}_1) \cdot P(\mathbf{x}\mathbf{y}_2) \cdot P(\mathbf{x}\mathbf{y}_3 \cdots \mathbf{y}_k) \cdot P(\mathbf{x})^{k-3} \\
&= \ldots \\
&= P(\mathbf{x}\mathbf{y}_1) \cdot \ldots \cdot P(\mathbf{x}\mathbf{y}_k) \, ,
\end{aligned}
$$

that is, $\pi = (dom, P)$ satisfies $I(Y_1, \ldots, Y_k \mid X)$.

Vice versa, assume that $\pi = (dom, P)$ over (S, C) satisfies $I(Y_1, \ldots, Y_k \mid X)$. Let $\mathbf{x}$ be a complete assignment over X and $\mathbf{y}_i, \mathbf{y}_1 \cdots \mathbf{y}_{i-1}\mathbf{y}_{i+1} \cdots \mathbf{y}_k^*$ be assignments for Y_i and $S - XY_i$, respectively. Then

$$P(\mathbf{xy}_i\mathbf{y}_1 \cdots \mathbf{y}_{i-1}\mathbf{y}_{i+1} \cdots \mathbf{y}_k) \cdot P(\mathbf{x}) = P(\mathbf{xy}_i) \cdot P(\mathbf{xy}_1 \cdots \mathbf{y}_{i-1}\mathbf{y}_{i+1} \cdots \mathbf{y}_k) \, ,$$

that is, for every $i = 1, \ldots, k$, π satisfies the SCI statement $I(Y_i, S - XY_i \mid X)$.   □

Example 4. For every probability model π over $S = \{m, a, r, c, f, l, s\}$ and every set $C \subseteq S$ of complete random variables, the GSCI statement $I(sar, c, fl \mid m)$ is satisfied by π if and only if all of the SCI statements $I(sar, cfl \mid m)$, $I(c, sarfl \mid m)$, and $I(sarc, fl \mid m)$ are satisfied by π.

Example 5. We can now prove that for $S = \{m, a, r, c, f, l, s\}$, $C = \{f, l\}$, $\Sigma = \{I(sar, c, fl \mid m), I(sc, ar \mid mfl)\}$ does indeed C-imply $\varphi = I(s, ar, cfl \mid m)$, thereby validating our statements from Example 2. In fact, the inference

$$\frac{\dfrac{I(sar, c, fl \mid m)}{\mathscr{M} : \ I(sar, cfl \mid m)} \qquad I(sc, ar \mid mfl)}{\mathscr{C} : \qquad\qquad I(s, ar, cfl \mid m)} {\scriptstyle \{f,l\} \subseteq C}$$

shows that $\Sigma \vdash_{\mathfrak{S}} \varphi$ which means that $\Sigma \models_C \varphi$ by soundness of $\mathfrak{S}$.

The following remark shows that Proposition 1 can be used to establish directly that the set $\mathfrak{S}$ of inference rules from Table 2 forms a finite axiomatization for the implication problem of GSCI statements and complete random variables.

Remark 6. The following set $\mathfrak{D}$ of inference rules

$\overline{I(S, \emptyset \mid \emptyset)}$	$\dfrac{I(Y_1, Y_2 \mid X)}{I(Y_2, Y_1 \mid X)}$
$\dfrac{I(Y_1Y_2, YZ_1Z_2 \mid X) \quad I(Y_1Z_1, Y_2Z_2 \mid XY)}{I(Y_1, Y_2YZ_1Z_2 \mid X)} \ Y \subseteq C$	$\dfrac{I(Y_1, Y_2Z \mid X)}{I(Y_1, Y_2 \mid XZ)}$

forms a finite axiomatization for the implication problem of SCI statements and complete random variables [44]. Suppose that $S, C \subseteq S$ and $\Sigma \cup \{I(Y_1, \ldots, Y_k \mid X)\}$ are given such that $\Sigma \models_C I(Y_1, \ldots, Y_k \mid X)$ holds. For

$$\Sigma_2 = \{I(V_j, S - UV_j \mid U) \mid I(V_1, \ldots, V_m \mid U) \in \Sigma\},$$

and all $i = 1, \ldots, k$ it follows from Proposition 1 that $\Sigma_2 \models_C I(Y_i, S - XY_i \mid X)$ holds, too. The completeness of $\mathfrak{D}$ for the implication of SCI statements and complete random variables means that for all $i = 1, \ldots, k$, $\Sigma_2 \vdash_{\mathfrak{D}} I(Y_i, S - XY_i \mid X)$ holds. Since $\mathfrak{D}$ is subsumed by $\mathfrak{S}$ as the special case where $k = 2$, we also have for all $i = 1, \ldots, k$ that $\Sigma_2 \vdash_{\mathfrak{S}} I(Y_i, S - XY_i \mid X)$ holds. However, the merging rule $\mathcal{M}$ shows that $\Sigma \vdash_{\mathfrak{S}} \sigma$ holds for all $\sigma \in \Sigma_2$. Consequently, for all $i = 1, \ldots, k$, $\Sigma \vdash_{\mathfrak{S}} I(Y_i, S - XY_i \mid X)$ holds. Finally, repeated applications of the restricted weak contraction rule $\mathcal{C}$ and the permutation rule $\mathcal{P}$ show that $\Sigma \vdash_{\mathfrak{S}} I(Y_1, \ldots, Y_k \mid X)$. This establishes the completeness of $\mathfrak{S}$.

Even though the last remark has already established the completeness of $\mathfrak{S}$, we want to illustrate recent techniques for proving completeness without the use of Proposition 1. This will be done in the following subsections.

3.2   The Independence Basis

For some S and $C \subseteq S$, some set Σ of GSCI statements over S, and some $X \subseteq S$ let $IDep_{\Sigma,C}(X) := \{Y \subseteq S - X \mid \Sigma \vdash_{\mathfrak{S}} I(Y, S - XY \mid X)\}$ denote the set of all $Y \subseteq S - X$ such that $I(Y, S - XY \mid X)$ can be inferred from Σ by $\mathfrak{S}$. Note that the empty set $\emptyset$ is an element of $IDep_{\Sigma,C}(X)$.

Lemma 1. *The structure* $(IDep_{\Sigma,C}(X), \subseteq, \cup, \cap, (\cdot)^{\mathcal{C}}, \emptyset, S - X)$ *forms a finite Boolean algebra, where* $(\cdot)^{\mathcal{C}}$ *maps a set W to its complement $S - (XW)$.*

Proof. It suffices to show that $IDep_{\Sigma,C}(X)$ is closed under union, intersection, and difference. The soundness of the merging rule $\mathcal{M}$ shows the closure under union. The soundness of the weak contraction rule $\mathcal{C}$ for the special case where $k = 2$ and $Y = \emptyset$ shows the closure under intersection and difference.   □

Recall that an element $a \in P$ of a poset $(P, \sqsubseteq, 0)$ with least element 0 is called an *atom* of $(P, \sqsubseteq, 0)$ precisely when $a \neq 0$ and every element $b \in P$ with $b \sqsubseteq a$ satisfies $b = 0$ or $b = a$ [23]. Further, $(P, \sqsubseteq, 0)$ is said to be *atomic* if for every element $b \in P - \{0\}$ there is an atom $a \in P$ with $a \sqsubseteq b$. In particular, every finite Boolean algebra is atomic [23]. Let $IDepB_{\Sigma,C}(X)$ denote the set of all atoms of $(IDep_{\Sigma,C}(X), \subseteq, \emptyset)$. We call $IDepB_{\Sigma,C}(X)$ the *independence basis* of X with respect to Σ. Its importance is manifested in the following result.

Theorem 2. *Let Σ be a set of GSCI statements over S and $C \subseteq S$. Then $\Sigma \vdash_{\mathfrak{S}} I(Y_1, \ldots, Y_k \mid X)$ if and only if for every $i = 1, \ldots, k$, $Y_i = \bigcup \mathcal{Y}$ for some $\mathcal{Y} \subseteq IDepB_{\Sigma,C}(X)$.*

Proof. Let $\Sigma \vdash_{\mathfrak{S}} I(Y_1, \ldots, Y_k \mid X)$. Then for all $i = 1, \ldots, k$, $\Sigma \vdash_{\mathfrak{S}} I(Y_i, S - XY_i \mid X)$ by the merging rule $\mathcal{M}$. Hence, for all $i = 1, \ldots, k$, $Y_i \in IDep_{\Sigma,C}(X)$. Since every element b of a Boolean algebra is the union over those atoms a with $a \subseteq b$ it follows that for all $i = 1, \ldots, k$, $Y_i = \bigcup \mathcal{Y}$ for $\mathcal{Y} = \{W \in IDepB_{\Sigma,C}(X) \mid W \subseteq Y_i\}$.

Vice versa, let $IDepB_{\Sigma,C}(X) = \{W_1, \ldots, W_n\}$ and for all $i = 1, \ldots, k$, let $Y_i = \bigcup \mathcal{Y}$ for some $\mathcal{Y} \subseteq IDepB_{\Sigma,C}(X)$. Since $I(W_1, \ldots, W_n \mid X) \in \Sigma_{\mathfrak{S}}^+$ holds, successive applications of the permutation rule $\mathcal{P}$ and merging rule $\mathcal{M}$ result in $I(Y_1, \ldots, Y_k \mid X) \in \Sigma_{\mathfrak{S}}^+$.   □

Example 6. Recall our example where $S = \{m, a, r, c, f, l, s\}$, $C = \{m, a, r, c, s\}$, $C' = \{f, l\}$, $\Sigma = \{I(sar, c, fl \mid m), I(sc, ar \mid mfl)\}$, and $\varphi = I(s, ar, flc \mid m)$. It follows that $IDepB_{\Sigma,C}(m) = \{sar, c, fl\}$, which we can suitably represent in the form of the single GSCI statement $I(sar, c, fl \mid m)$. According to Theorem 2, $\Sigma \not\models_C \varphi$. Moreover, $\Sigma \models_{C'} \varphi$ since $IDepB_{\Sigma,C'}(m) = \{s, ar, c, fl\}$.

3.3 Completeness

The original completeness proof for multivalued dependencies constructs a counterexample relation with 2^k tuples [5], where k denotes the elements in the (in)dependence basis $Dep_\Sigma(X)$ for the multivalued dependency $X \twoheadrightarrow Y \mid Z \notin \Sigma^+$. The original completeness proof for SCI statements constructs a probability model with $2^{|X|+1}$ values, where $I(Y, Z \mid X) \notin \Sigma_\mathfrak{S}^+$ [21]. Here, a recent technique [28] defines special probability models with two assignments of probability one half each. The technique therefore extends the existence of special probability models from the case of marginal SCI statements $I(Y, Z \mid \emptyset)$ [21] to GSCI statements and complete random variables.

Theorem 3. *The set $\mathfrak{S}$ is complete for the implication problem of GSCI statements and complete random variables.*

Proof. Let $\Sigma \cup \{I(Y_1, \ldots, Y_k \mid X)\}$ be a set of GSCI statements over S and $C \subseteq S$, and suppose that $I(Y_1, \ldots, Y_k \mid X)$ cannot be inferred from Σ using $\mathfrak{S}$. We will show that $I(Y_1, \ldots, Y_k \mid X)$ is not C-implied by Σ. For this purpose, we will construct a probability model over (S, C) that satisfies all GSCI statements of Σ, but violates $I(Y_1, \ldots, Y_k \mid X)$.

Let $IDepB_{\Sigma,C}(X) = \{W_1, \ldots, W_n\}$, in particular $S = XW_1 \cdots W_n$. Since $I(Y_1, \ldots, Y_k \mid X) \notin \Sigma_\mathfrak{S}^+$ we conclude by Theorem 2 that there is some $j \in \{1, \ldots, k\}$ such that Y_j is not the union of some elements of $IDepB_{\Sigma,C}(X)$. Consequently, there is some $i \in \{1, \ldots, n\}$ such that $Y_j \cap W_i \neq \emptyset$ and $W_i - Y_j \neq \emptyset$ hold. Let

$$T := \bigcup_{l \in \{1, \ldots, i-1, i+1, \ldots, k\}} W_l \cap C,$$

and

$$T' := \bigcup_{l \in \{1, \ldots, i-1, i+1, \ldots, k\}} W_l - C.$$

In particular, S is the disjoint union of X, T, T', and W_i. For every $v \in S - C$ we define $dom(v) = \{\mathbf{0}, \mathbf{1}, \mu\}$; and for every $v \in C$ we define $dom(v) = \{\mathbf{0}, \mathbf{1}\}$. We define the following two assignments $\mathbf{a}_1$ and $\mathbf{a}_2$ of S. We define $\mathbf{a}_1(v) = \mathbf{0}$ for all $v \in XW_iT$, $\mathbf{a}_1(v) = \mu$ for all $v \in T'$. We further define $\mathbf{a}_2(v) = \mathbf{a}_1(v)$ for all $v \in XTT'$, and

$\mathbf{a}_2(v) = \mathbf{1}$ for all $v \in W_i$. As probability measure we define $P(\mathbf{a}_1) = P(\mathbf{a}_2) = 0.5$. It follows from the construction that (dom, P) does not satisfy $I(Y_1, \ldots, Y_k \mid X)$.

It remains to show that (dom, P) satisfies every GSCI statement $I(V_1, \ldots, V_m \mid U)$ in Σ. Suppose that for some complete assignment $\mathbf{u}$ of U, $P(\mathbf{u}) = 0$. Then equation (1) will always be satisfied.

If $P(\mathbf{u}, \mathbf{v}_o) = 0$ for some complete assignment $\mathbf{u}$ of U, and for some assignment $\mathbf{v}_o$ of V_o, then $P(\mathbf{u}, \mathbf{v}_1, \ldots, \mathbf{v}_m) = 0$. Then equation (1) is also satisfied. Suppose that for some complete assignment $\mathbf{u}$ of U, $P(\mathbf{u}) = 0.5$. If for some assignments $\mathbf{v}_l$ of V_l for $l = 1, \ldots, m$, $P(\mathbf{u}, \mathbf{v}_1) = \cdots = P(\mathbf{u}, \mathbf{v}_m) = 0.5$, then $P(\mathbf{u}, \mathbf{v}_1, \ldots, \mathbf{v}_m) = 0.5$, too. Again, equation (1) is satisfied.

It remains to consider the case where $\mathbf{u}$ is some complete assignment of U such that $P(\mathbf{u}) = 1$. In this case, the construction of the probability model tells us that $U \subseteq XT$. Consequently, we can apply the weak union rule $\mathscr{W}$ and permutation rule $\mathscr{P}$ to $I(V_1, \ldots, V_m \mid U) \in \Sigma$ to infer $I(V_1 - XT, \ldots, V_m - XT \mid XT) \in \Sigma_{\mathfrak{G}}^+$. Theorem 2 also shows that $I(W_i, TT' \mid X) \in \Sigma_{\mathfrak{G}}^+$. Now we define $V_l' := V_l - XTT'$ and $Z_l := (V_l - XT) \cap T'$ for $l = 1, \ldots, m$. Consequently, $W_i = V_1' \cdots V_m'$, $T' = Z_1' \cdots Z_m'$, and $V_l - XT = V_l' Z_l'$ for $l = 1, \ldots, m$. An application of the restricted weak contraction rule $\mathscr{C}$ to $I(V_1' Z_1', \ldots, V_m' Z_m' \mid XT)$ and $I(V_1' \cdots V_m', TZ_1' \cdots Z_m' \mid X)$ results in $I(V_1', \ldots, V_m', TZ_1' \cdots Z_m' \mid X) = I(V_1 - XTT', \ldots, V_m - XTT', TT' \mid X)$. It follows from Theorem 2 that $V_l - XTT'$, for every $l = 1, \ldots, m$, is the union of elements from $IDepB_{\Sigma, C}(X)$. Consequently, $V_o - XTT' = W_i$ for some $o \in \{1, \ldots, m\}$ and $V_p - XTT' = \emptyset$ for all $p \in \{1, \ldots, m\} - \{o\}$. Therefore, $W_i \subseteq V_o$ and $W_i \cap V_p = \emptyset$ for all $p \in \{1, \ldots, m\} - \{o\}$. Therefore, we are either in the previous case where $P(\mathbf{u}, \mathbf{v}_l) = 0$ for some $l \in \{1, \ldots, m\}$; or, $P(\mathbf{u}, \mathbf{v}_o) = 0.5$, $P(\mathbf{u}, \mathbf{v}_p) = 1$ for every $p \in \{1, \ldots, m\} - \{o\}$, and $P(\mathbf{u}, \mathbf{v}_1, \ldots, \mathbf{v}_m) = 0.5$. Again, equation (1) is satisfied. This concludes the proof. $\qquad\square$

The next example illustrates the construction of the counterexample on our running example.

Example 7. Let $S = \{m, a, r, c, f, l, s\}$ denote the set of random variables from Example 1 and $C = \{m, a, r, c, s\}$, let $\Sigma = \{I(sar, c, fl \mid m), I(sc, ar \mid mfl)\}$, and $\varphi = I(s, ar, cfl \mid m)$. The assignments

$$\mathbf{a}_1 = (\text{Rashomon, T. Mifune, Tajomaru, Kurosawa}, \mu, \mu, \text{Suomi})$$

and

$$\mathbf{a}_2 = (\text{Rashomon, M. Kyo, Masako, Kurosawa}, \mu, \mu, \text{Deutsch})$$

taken together with the probability distribution $P(\mathbf{a}_1) = 0.5 = P(\mathbf{a}_2)$ define a probability model that satisfies Σ and violates φ. Indeed, this probability model is an instance of the special probability model used in the completeness proof of Theorem 3, see Table 3. In fact, $W_i = \{a, r, s\}$ and $Y_j = \{s\}$.

Table 3 Special Probability
Model from the
Completeness Proof for $\mathfrak{S}$

XT	W_i	T'	P
$0\cdots0$	$0\cdots0$	$\mu\cdots\mu$	0.5
$0\cdots0$	$1\cdots1$	$\mu\cdots\mu$	0.5

3.4 Special Probability Models

We call a probability model (dom, P) over (S, C) *special*, if for every $v \in C$, $dom(v)$ consists of two elements, for every $v \in S - C$, $dom(v)$ consists of two elements and the marker μ, and there are two assignments $\mathbf{a}_1, \mathbf{a}_2$ over (S, C) such that $P(\mathbf{a}_1) = 0.5 = P(\mathbf{a}_2)$. We say that Σ *C-implies* φ in the world of special probability models, denoted by $\Sigma \models_{2,C} \varphi$, if every special probability model over (S, C) that satisfies every GSCI statement in Σ also satisfies the GSCI statement φ. The following variant of the implication problem for GSCI statements and complete random variables emerges.

PROBLEM:	Implication problem for GSCI statements and complete r.v. in the world of special probability models
INPUT:	(S, C) with set S of random variables and subset $C \subseteq S$ of complete random variables Set $\Sigma \cup \{\varphi\}$ of GSCI statements over (S, C)
OUTPUT:	Yes, if $\Sigma \models_{2,C} \varphi$; No, otherwise

The proof of Theorem 3 implies the following result.

Corollary 1. *The implication problem for GSCI statements and complete random variables coincides with the implication problem for GSCI statements and complete random variables in the world of special probability models.*

Proof. Let $\Sigma \cup \{\varphi\}$ be a set of GSCI statements over S and $C \subseteq S$. We need to show that $\Sigma \models_C \varphi$ if and only if $\Sigma \models_{2,C} \varphi$. If it does not hold that $\Sigma \models_{2,C} \varphi$, then it also does not hold that $\Sigma \models_C \varphi$ since every special probability model is a probability model. Vice versa, if it does not hold that $\Sigma \models_C \varphi$, then it does not hold that $\Sigma \vdash_{\mathfrak{S}} \varphi$ since $\mathfrak{S}$ is sound for the implication of GSCI statements and complete random variables. However, the proof of Theorem 3 shows how to construct a special probability model over (S, C) that satisfies every GSCI statement in Σ but does not satisfy φ. Hence, it does not hold that $\Sigma \models_{2,C} \varphi$. $\qquad\square$

Corollary 1 shows that to decide the implication problem for GSCI statements and complete random variables it suffices to check special probability models.

Example 8. For the set $S = \{m, a, r, c, f, l, s\}$ of random variables, the subset $C = \{m, a, r, c, s\}$ of complete random variables, and the statements in $\Sigma = \{I(sar, c, fl \mid m), I(sc, ar \mid mfl)\}$, and $\varphi = I(s, ar, cfl \mid m)$, the probability model in Example 7 defines a special probability model that satisfies Σ and violates φ. Hence, Σ does not C-imply φ in the world of special probability models.

4   Characterization by an 𝒮-3 Fragment

In this section we establish the equivalence between the C-implication of GSCI statements and the implication of formulae in a propositional fragment 𝔉 within Cadoli and Schaerf's well-known approximation logic 𝒮-3 [54]. After repeating the syntax and semantics of 𝒮-3 logic, we define a mapping of GSCI statements to formulae in 𝔉. The core proof argument establishes an equivalence between special probability models, introduced in the previous section, and special 𝒮-3 truth assignments.

4.1   Syntax and Semantics of 𝒮-3 logic

Schaerf and Cadoli [54] introduced 𝒮-3 logics as "a semantically well-founded logical framework for sound approximate reasoning, which is justifiable from the intuitive point of view, and to provide fast algorithms for dealing with it even when using expressive languages." For a finite set L of propositional variables, let L^* denote the *propositional language* over L, generated from the unary connective ¬ (negation), and the binary connectives ∧ (conjunction) and ∨ (disjunction). Elements of L^* are also called formulae of L, and usually denoted by φ', ψ' or their subscripted versions. Sets of formulae are denoted by Σ'. We omit parentheses if this does not cause ambiguity.

Let L^ℓ denote the set of all literals over L, i.e., $L^\ell = L \cup \{\neg v' \mid v' \in L\}$. Let $\mathscr{S} \subseteq L$. An 𝒮-3 truth assignment of L is a total function $\omega : L^\ell \to \{\mathbb{F}, \mathbb{T}\}$ that maps every propositional variable $v' \in \mathscr{S}$ and its negation $\neg v'$ into opposite truth values ($\omega(v') = \mathbb{T}$ if and only if $\omega(\neg v') = \mathbb{F}$), and that does not map both a propositional variable $v' \in L - \mathscr{S}$ and its negation $\neg v'$ into *false* (we must not have $\omega(v') = \mathbb{F} = \omega(\neg v')$ for any $v' \in L - \mathscr{S}$). Accordingly, for each propositional variable $v' \in L$ and each 𝒮-3 truth assignment ω of L there are the following possibilities:

- $\omega(v') = \mathbb{T}$ and $\omega(\neg v') = \mathbb{F}$,
- $\omega(v') = \mathbb{F}$ and $\omega(\neg v') = \mathbb{T}$,
- $\omega(v') = \mathbb{T}$ and $\omega(\neg v') = \mathbb{T}$ (only if $v' \in L - \mathscr{S}$).

𝒮-3 truth assignments generalize both, standard 2-valued truth assignments as well as the 3-valued truth assignments of Levesque [36]. That is, a 2-valued truth assignment is an 𝒮-3 truth assignment where $\mathscr{S} = L$, while a 3-valued truth assignment is an 𝒮-3 truth assignment with $\mathscr{S} = \emptyset$.

An 𝒮-3 truth assignment $\omega : L^\ell \to \{\mathbb{F}, \mathbb{T}\}$ of L can be lifted to a total function $\Omega : L^* \to \{\mathbb{F}, \mathbb{T}\}$. This lifting has been defined as follows [54]. An arbitrary formula φ' in L^* is firstly converted (in linear time in the size of the formula) into its corresponding formula φ'_N in *Negation Normal Form* (NNF) using the following rewriting rules: $\neg(\varphi' \wedge \psi') \mapsto (\neg\varphi' \vee \neg\psi')$, $\neg(\varphi' \vee \psi') \mapsto (\neg\varphi' \wedge \neg\psi')$, and

$\neg(\neg\varphi') \mapsto \varphi'$. Therefore, negation in a formula in NNF occurs only at the literal level. The rules for assigning truth values to NNF formulae are as follows:

- $\Omega(\varphi') = \omega(\varphi')$, if $\varphi' \in L^{\ell}$,
- $\Omega(\varphi' \vee \psi') = \mathbb{T}$ if and only if $\Omega(\varphi') = \mathbb{T}$ or $\Omega(\psi') = \mathbb{T}$,
- $\Omega(\varphi' \wedge \psi') = \mathbb{T}$ if and only if $\Omega(\varphi') = \mathbb{T}$ and $\Omega(\psi') = \mathbb{T}$.

Thus, $\mathscr{S}$-3 logic is non-compositional. An $\mathscr{S}$-3 truth assignment ω is a *model* of a set Σ' of L-formulae if and only if $\Omega(\sigma'_N) = \mathbb{T}$ holds for every $\sigma' \in \Sigma'$. We say that Σ' $\mathscr{S}$-3 *implies* an L-formula φ', denoted by $\Sigma' \models^3_{\mathscr{S}} \varphi'$, if and only if every $\mathscr{S}$-3 truth assignment that is a model of Σ' is also a model of φ'.

4.2 The Propositional Fragment $\mathfrak{F}$

As a first step towards the anticipated duality we define the propositional fragment that corresponds to GSCI statements. Let $\phi : S \to L$ denote a bijection between a set S of random variables and the set $L = \{v' \mid v \in S\}$ of propositional variables. In particular, for $C \subseteq S$ let $\mathscr{S} = \phi(C)$. Thus, complete random variables correspond to propositional variables interpreted classically.

We extend ϕ to a mapping Φ from the set of GSCI statements over S to the fragment $\mathfrak{F}$, that is, $\mathfrak{F}$ is the range of Φ. For a GSCI statement $I(Y_1, \ldots, Y_k \mid X)$ over S, let $\Phi(I(Y_1, \ldots, Y_k \mid X))$ denote the formula

$$\bigvee_{v \in X} \neg v' \vee \bigvee_{i=1}^{k} \left(\bigwedge_{v \in \bigcup_{j \neq i} Y_j} v' \right).$$

Disjunctions over zero disjuncts are interpreted as *false*, denoted by $\mathbb{F}$, and conjunctions over zero conjuncts are interpreted as *true*, denoted by $\mathbb{T}$. We will simply denote $\Phi(\varphi) = \varphi'$ and $\Phi(\Sigma) = \{\sigma' \mid \sigma \in \Sigma\} = \Sigma'$. Note that for the special case of SCI statements $\varphi = I(Y, Z \mid X)$, that is, GSCI statements where $k = 2$, the formula φ' becomes

$$\bigvee_{v \in X} \neg v' \vee \left(\bigwedge_{v \in Y} v' \right) \vee \left(\bigwedge_{v \in Z} v' \right).$$

Example 9. Let $S = \{m, a, r, c, f, l, s\}$ denote the set of random variables from Example 1 and $C = \{m, a, r, c, s\}$, let $\Sigma = \{I(sar, c, fl \mid m), I(sc, ar \mid mfl)\}$, and $\varphi = I(s, ar, cfl \mid m)$. Then $\mathscr{L} = \{m', a', r', c', f', l', s'\}$, $\mathscr{S} = \{m', a', r', c', s'\}$, Σ' consists of

$$\neg m' \vee (c' \wedge f' \wedge l') \vee (s' \wedge a' \wedge r' \wedge f' \wedge l') \vee (s' \wedge a' \wedge r' \wedge c')$$

and

$$\neg m' \vee \neg f' \vee \neg l' \vee (a' \wedge r') \vee (s' \wedge c'),$$

and $\varphi' = \neg m' \vee (a' \wedge r' \wedge c' \wedge f' \wedge l') \vee (s' \wedge c' \wedge f' \wedge l') \vee (s' \wedge a' \wedge r')$.

4.3 Special Truth Assignments

We will now show that for every set $\Sigma \cup \{\varphi\}$ of GSCI statements over S and every $C \subseteq S$, there is a probability model $\pi = (dom, P)$ over (S, C) that satisfies Σ and violates φ if and only if there is a truth assignment ω'_π that is an $\mathscr{S}$-3 model of Σ' but not an $\mathscr{S}$-3 model of φ'. For arbitrary probability models π it is not obvious how to define the interpretation ω'_π. However, the key to showing the correspondence between counterexample probability models and counterexample truth assignments is Corollary 1. Corollary 1 tells us that for deciding $\Sigma \models_C \varphi$ it suffices to examine special probability models (instead of arbitrary probability models). For a special probability model $\pi = (dom, \{\mathbf{a}_1, \mathbf{a}_2\})$, however, we can define its corresponding special 3-valued truth assignment ω'_π of L as follows:

$$\omega_\pi(v') = \begin{cases} \mathbb{T} \text{, if } \mathbf{a}_1(v) = \mathbf{a}_2(v) \\ \mathbb{F} \text{, otherwise} \end{cases} \text{, and}$$

$$\omega_\pi(\neg v') = \begin{cases} \mathbb{T} \text{, if } \mathbf{a}_1(v) = \mu = \mathbf{a}_2(v) \text{ or } \mathbf{a}_1(v) \neq \mathbf{a}_2(v) \\ \mathbb{F} \text{, otherwise} \end{cases}.$$

Note that the 3-valued truth assignment is an $\mathscr{S}$-3 truth assignment since it is impossible to have $\mathbf{a}_1(v) = \mu = \mathbf{a}_2(v)$ for any complete random variable $v \in C$. For every $\mathscr{S}$-3 truth assignment ω of L there is some special probability model $\pi = (dom, P)$ over (S, C) such that $\omega_\pi = \omega$. In fact, if $\omega(v') = \mathbb{T} = \omega(\neg v')$ for some $v' \in \mathscr{S}$, then define $dom(v) := \{\mathbf{0}, \mathbf{1}\}$ such that the assignments of π are C-complete.

Example 10. Let $S = \{m, a, r, c, f, l, s\}$ denote the set of random variables from Example 1 and $C = \{m, a, r, c, s\}$. The special probability model π defined by

$$\mathbf{a}_1 = (\text{Rashomon, T. Mifune, Tajomaru, Kurosawa}, \mu, \mu, \text{Suomi})$$

and

$$\mathbf{a}_2 = (\text{Rashomon, M. Kyo, Masako, Kurosawa}, \mu, \mu, \text{Deutsch})$$

and the probability distribution $P(\mathbf{a}_1) = 0.5 = P(\mathbf{a}_2)$ translates into the following $\mathscr{S}$-3 interpretation of $\mathscr{L} = \{m', a', r', c', f', l', s'\}$ with $\mathscr{S} = \{m', a', r', c', s'\}$:

- $\omega_\pi(m') = \mathbb{T}$ and $\omega_\pi(\neg m') = \mathbb{F}$
- $\omega_\pi(a') = \mathbb{F}$ and $\omega_\pi(\neg a') = \mathbb{T}$
- $\omega_\pi(r') = \mathbb{F}$ and $\omega_\pi(\neg r') = \mathbb{T}$
- $\omega_\pi(c') = \mathbb{T}$ and $\omega_\pi(\neg c') = \mathbb{F}$
- $\omega_\pi(f') = \mathbb{T}$ and $\omega_\pi(\neg f') = \mathbb{T}$
- $\omega_\pi(l') = \mathbb{T}$ and $\omega_\pi(\neg l') = \mathbb{F}$
- $\omega_\pi(s') = \mathbb{F}$ and $\omega_\pi(\neg s') = \mathbb{T}$

4.4   Semantic Justification of Special Truth Assignments

Next we justify the definition of the special truth assignment and that of the propositional fragment $\mathfrak{F}$ in terms of the special probability models.

Lemma 2. *Let* $\pi = (dom, \{\mathbf{a}_1, \mathbf{a}_2\})$ *be a special probability model over* (S, C)*, and let* φ *denote a GSCI statement over* (S, C)*. Then* π *satisfies* φ *if and only if* ω_π' *is a 3-valued model of* φ'*.*

Proof. Let $\varphi = I(Y_1, \ldots, Y_k \mid X)$ and

$$\varphi' = \bigvee_{v \in X} \neg v' \vee \bigvee_{i=1}^{k} \left(\bigwedge_{v \in \cup_{j \neq i} Y_j} v' \right).$$

Suppose first that π satisfies φ. We need to show that ω_π' is a 3-valued model of φ'. Assume that $\omega_\pi'(\neg v') = \mathbb{F}$ for all $a \in X$. According to the special truth assignment we must have $\mu \neq \mathbf{a}_1(v) = \mathbf{a}_2(v) \neq \mu$ for all $v \in X$. That means $P(\mathbf{a}_1(X)) = 1$. Suppose that for all $i = 2, \ldots, k$ there is some $v \in \bigcup_{j \neq i} Y_j$ such that $\omega_\pi'(v') = \mathbb{F}$. Consequently, there is some $v \in Y_1$ such that $\omega_\pi'(v') = \mathbb{F}$. Hence, $\mathbf{a}_1(v) \neq \mathbf{a}_2(v)$ according to the special truth assignment. Then $P(\mathbf{a}_1(XY_1)) = P(\mathbf{a}_1) = 0.5$. However, since $\mathbf{a}_1(X)$ is complete on X and π satisfies φ we must have $P(\mathbf{a}_1(XY_i)) = 1$ for all $i = 2, \ldots, k$. Hence, for every $v \in Y_2 \cdots Y_k$, we have $\mathbf{a}_1(v) = \mathbf{a}_2(v)$. This means that for all $v \in Y_2 \cdots Y_k$ we have $\omega_\pi'(v') = \mathbb{T}$. This shows that ω_π' is a 3-valued model of φ'.

Suppose ω_π' is a 3-valued model of φ'. We need to show that π satisfies φ. That is, for every complete assignment $\mathbf{x}$ of X, and every assignment $\mathbf{y}_i$ of Y_i for $i = 1, \ldots, k$, we must show that $P(\mathbf{x}, \mathbf{y}_1, \ldots, \mathbf{y}_k) \cdot P(\mathbf{x})^{k-1} = P(\mathbf{x}, \mathbf{y}_1) \cdot P(\mathbf{x}, \mathbf{y}_k)$ holds. We distinguish between a few cases.

Case 1. If $P(\mathbf{x}, \mathbf{y}_i) = 0$ holds for some $i \in \{1, \ldots, k\}$, then $P(\mathbf{x}, \mathbf{y}_1, \ldots, \mathbf{y}_k) = 0$ holds, too. For the remaining cases we can therefore assume that for all $i = 1, \ldots, k$, $P(\mathbf{x}, \mathbf{y}_i) > 0$. In particular, $P(\mathbf{x}) > 0$.

Case 2. Suppose that $P(\mathbf{x}) = 0.5$. Then $P(\mathbf{x}, \mathbf{y}_i) = 0.5$ for all $i = 1, \ldots, k$. Consequently, $(\mathbf{x}, \mathbf{y}_1, \ldots, \mathbf{y}_k)$ equals $\mathbf{a}_1$ or $\mathbf{a}_2$, as $P(\mathbf{x})$ would have to be 1 otherwise. Hence, $P(\mathbf{x}, \mathbf{y}_1, \ldots, \mathbf{y}_k) = 0.5$. Therefore, we have

$$P(\mathbf{x}, \mathbf{y}_1, \ldots, \mathbf{y}_k) \cdot P(\mathbf{x})^{k-1} = (1/2)^k = P(\mathbf{x}, \mathbf{y}_1) \cdot \ldots \cdot P(\mathbf{x}, \mathbf{y}_k).$$

Case 3. Suppose $P(\mathbf{x}) = 1$. It follows that $\mathbf{a}_1(X) = \mathbf{x} = \mathbf{a}_2(X)$. Since $\mathbf{x}$ is a complete assignment of X, the special truth assignment entails that $\omega_\pi(\neg v') = \mathbb{F}$ for all $v \in X$. Since ω'_π is a 3-valued model of φ' we conclude that $\omega'_\pi(v') = \mathbb{T}$ for all $v \in S - XY_i$ for some $i \in \{1, \ldots, k\}$. This, however, would mean that $P(\mathbf{x}, \mathbf{y}_1, \ldots, \mathbf{y}_{i-1}, \mathbf{y}_{i+1}, \ldots, \mathbf{y}_k) = 1$. Since φ is saturated, it follows that $P(\mathbf{x}, \mathbf{y}_i) = 0.5$. Consequently, $(\mathbf{x}, \mathbf{y}_1, \ldots, \mathbf{y}_k)$ equals $\mathbf{a}_1$ or $\mathbf{a}_2$. That is, $P(\mathbf{x}, \mathbf{y}_1, \ldots, \mathbf{y}_k) = 0.5$. Therefore,

$$P(\mathbf{x}, \mathbf{y}_1, \ldots, \mathbf{y}_k) \cdot P(\mathbf{x})^{k-1} = 1/2 = P(\mathbf{x}, \mathbf{y}_1) \cdot \ldots \cdot P(\mathbf{x}, \mathbf{y}_k).$$

It follows that π satisfies φ. □

4.5 The Equivalence

Corollary 1 and Lemma 2 allow us to establish the anticipated equivalence between the implication problem of GSCI statements and complete random variables and the implication problem of fragment $\mathfrak{F}$ in $\mathscr{S}$-3 logic.

Theorem 4. *Let $\Sigma \cup \{\varphi\}$ be a set of GSCI statements over S and $C \subseteq S$, and let $\Sigma' \cup \{\varphi'\}$ denote the set of its corresponding propositional formulae over L. Then $\Sigma \models_C \varphi$ if and only if $\Sigma' \models^3_{\mathscr{S}} \varphi'$.*

Proof. Based on Corollary 1 it suffices to establish an equivalence between $\Sigma \models_{2,C} \varphi$ and $\Sigma' \models^3_{\mathscr{S}} \varphi'$.

Suppose first that $\Sigma \models_{2,C} \varphi$ does not hold. Then there is some special probability model π over (S, C) that satisfies every GSCI statement σ in Σ but violates φ. Let ω_π denote the special truth assignment associated with π. By Lemma 2 it follows that ω_π is a 3-valued model of every formula σ' in Σ' but not a 3-valued model of φ'. As ω_π is an $\mathscr{S}$-3 truth assignment it follows that $\Sigma' \models^3_{\mathscr{S}} \varphi'$ does not hold.

Suppose now that $\Sigma' \models^3_{\mathscr{S}} \varphi'$ does not hold. Then there is some truth assignment ω over L that is an $\mathscr{S}$-3 model of every formula σ' in Σ', but not an $\mathscr{S}$-3 model of the formula φ'. Define the following special probability model $\pi = (dom, \{\mathbf{a}_1, \mathbf{a}_2\})$ over (S, C). For $v \in C$, let $dom(v) = \{\mathbf{0}, \mathbf{1}\}$; and for $v \in S - C$, let $dom(v) = \{\mathbf{0}, \mathbf{1}, \mu\}$. We now define $\mathbf{a}_1$ and $\mathbf{a}_2$ as follows. If $\omega(v') = \mathbb{T}$ and $\omega(\neg v') = \mathbb{F}$, then $\mu \neq \mathbf{a}_1(v) = \mathbf{a}_2(v) \neq \mu$. If $\omega(v') = \mathbb{T}$ and $\omega(\neg v') = \mathbb{T}$, then $\mathbf{a}_1(v) = \mu = \mathbf{a}_2(v)$. Finally, if $\omega(v') = \mathbb{F}$ and $\omega(\neg v') = \mathbb{T}$, then $\mu \neq \mathbf{a}_1(v) \neq \mathbf{a}_2(v) \neq \mu$. Since ω

is not an $\mathscr{S}$-3 model of φ', it follows that $\mathbf{a}_1 \neq \mathbf{a}_2$. It follows now that $\omega_\pi = \omega$. By Lemma 2 it follows that π satisfies every GSCI statement σ in Σ but violates φ. Hence, $\Sigma \models_{2,C} \varphi$ does not hold.   □

Example 11. Let $S = \{m, a, r, c, f, l, s\}$ denote the set of random variables from Example 1 and $C = \{m, a, r, c, s\}$, let $\Sigma = \{I(sar, c, fl \mid m), I(sc, ar \mid mfl)\}$, and $\varphi = I(s, ar, cfl \mid m)$. The special probability model π defined by

$$\mathbf{a}_1 = (\text{Rashomon, T. Mifune, Tajomaru, Kurosawa}, \mu, \mu, \text{Suomi})$$

and

$$\mathbf{a}_2 = (\text{Rashomon, M. Kyo, Masako, Kurosawa}, \mu, \mu, \text{Deutsch})$$

shows that Σ does not C-imply φ. From a logical point of view, the special $\mathscr{S}$-3 interpretation ω_π of $\mathscr{L} = \{m', a', r', c', f', l', s'\}$ with $\mathscr{S} = \{m', a', r', c', s'\}$:

- $\omega_\pi(m') = \mathbb{T}$ and $\omega_\pi(\neg m') = \mathbb{F}$
- $\omega_\pi(a') = \mathbb{F}$ and $\omega_\pi(\neg a') = \mathbb{T}$
- $\omega_\pi(r') = \mathbb{F}$ and $\omega_\pi(\neg r') = \mathbb{T}$
- $\omega_\pi(c') = \mathbb{T}$ and $\omega_\pi(\neg c') = \mathbb{F}$
- $\omega_\pi(f') = \mathbb{T}$ and $\omega_\pi(\neg f') = \mathbb{T}$
- $\omega_\pi(l') = \mathbb{T}$ and $\omega_\pi(\neg l') = \mathbb{F}$
- $\omega_\pi(s') = \mathbb{F}$ and $\omega_\pi(\neg s') = \mathbb{T}$

shows that Σ', consisting of

$$\neg m' \vee (c' \wedge f' \wedge l') \vee (s' \wedge a' \wedge r' \wedge f' \wedge l') \vee (s' \wedge a' \wedge r' \wedge c')$$

and

$$\neg m' \vee \neg f' \vee \neg l' \vee (a' \wedge r') \vee (s' \wedge c'),$$

does not $\mathscr{S}$-3 imply $\varphi' = \neg m' \vee (a' \wedge r' \wedge c' \wedge f' \wedge l') \vee (s' \wedge c' \wedge f' \wedge l') \vee (s' \wedge a' \wedge r')$.

5 Full Hierarchical Dependencies and NOT NULL constraints

In this section we extend the duality between the implication problem of GSCI statements and complete random variables and the implication problem of the fragment $\mathfrak{F}$ under $\mathscr{S}$-3 interpretations to a trinity including the implication problem of Delobel's class of full first-order hierarchical dependencies (FOHDs) [12] and NOT NULL constraints. We adapt the technique of special probability models to establish an axiomatization $\mathfrak{H}$ for the implication problem for FOHDs and NOT

NULL constraints. The completeness proof exploits two-tuple relations. In the database context, two-tuple relations form the counterpart of special probability models, enabling us to establish the anticipated trinity of implication problems. The proof techniques in this section are different from the ones in the previous section in that they explore decomposition arguments rather than probabilities.

5.1 Defining Hierarchical Dependencies under Incomplete Data

Let $\mathfrak{A} = \{\hat{v}_1, \hat{v}_2, \ldots\}$ be a (countably) infinite set of symbols, called *attributes*. A *relation schema* is a finite set $R = \{\hat{v}_1, \ldots, \hat{v}_n\}$ of attributes from $\mathfrak{A}$. Each attribute $\hat{v}$ of a relation schema is associated with a domain $dom(\hat{v})$ which represents the set of possible values that can occur in the column named $\hat{v}$. Note that the validity of our results only depends on having at least two element values in each domain. This is a consequence of our proof techniques. In order to encompass incomplete information the domain of each attribute contains the null marker, denoted by ni $\in$ $dom(\hat{v})$. The intention of ni is to mean "no information." This is the most primitive interpretation, and it can model non-existing as well as unknown information [3, 63]. We stress that the null marker is not a domain value. In fact, it is a purely syntactic convenience that we include the null marker in the domain of each attribute as a distinguished element.

A *tuple* over R is a function $t : R \rightarrow \bigcup_{\hat{v} \in R} dom(\hat{v})$ with $t(\hat{v}) \in dom(\hat{v})$ for all $\hat{v} \in R$. The null marker occurrence $t(\hat{v}) = $ ni associated with an attribute $\hat{v}$ in a tuple t means that "no information" is available about the value $t(\hat{v})$ of t on attribute $\hat{v}$. For $X \subseteq R$ let $t(X)$ denote the restriction of the tuple t over R to X, and $dom(X) = \prod_{\hat{v} \in X} dom(\hat{v})$ the Cartesian product of the domains of attributes in X. A (partial) *relation* r over R is a finite set of tuples over R. Let t_1 and t_2 be two tuples over R. It is said that t_1 *subsumes* t_2 if for every attribute $\hat{v} \in R$, $t_1(\hat{v}) = t_2(\hat{v})$ or $t_2(\hat{v}) = $ ni holds. In consistency with previous work [3, 37, 63], the following restriction will be imposed, unless stated otherwise: No relation shall contain two tuples t_1 and t_2 such that t_1 subsumes t_2. With no null markers present this means that no duplicate tuples occur. For a tuple t over R and a set $X \subseteq R$, t is said to be X-total, if for all $\hat{v} \in X$, $t(\hat{v}) \neq $ ni. Similarly, a relation r over R is said to be X-total, if every tuple t of r is X-total. A relation r over R is said to be a *total relation*, if it is R-total.

We recall the definition of projection and join operations on partial relations [3, 37]. Let r be some relation over R. Let X be some subset of R. The *projection* $r[X]$ of r on X is the set of tuples t for which (i) there is some $t_1 \in r$ such that $t = t_1(X)$ and (ii) there is no $t_2 \in r$ such that $t_2(X)$ subsumes t and $t_2(X) \neq t$. For $Y \subseteq X$, the Y-*total projection* $r_Y[X]$ of r on X is $r_Y[X] = \{t \in r[X] \mid t$ is Y-total$\}$. Given an X-total relation r_1 over R_1 and an X-total relation r_2 over R_2 such that $X = R_1 \cap R_2$ the *natural join* $r_1 \bowtie r_2$ of r_1 and r_2 is the relation over $R_1 \cup R_2$ which contains those tuples t such that there are some $t_1 \in r_1$ and $t_2 \in r_2$ with $t_1 = t(R_1)$ and $t_2 = t(R_2)$ [3, 37]. For example, the relation

movie	actor	role	crew	feature	language	subtitle
Rashomon	T. Mifune	Tajomaru	Kurosawa	ni	ni	Suomi
Rashomon	M. Kyo	Masako	Kurosawa	ni	ni	Deutsch

is the natural join of the following three relations:

movie	actor	role	subtitle
Rashomon	T. Mifune	Tajomaru	Suomi
Rashomon	M. Kyo	Masako	Deutsch

movie	crew
Rashomon	Kurosawa

movie	feature	subtitle
Rashomon	ni	ni

Following Atzeni and Morfuni [3], a *null-free subschema* (NFS) over the relation schema R is an expression R_s where $R_s \subseteq R$. The NFS R_s over R is satisfied by a relation r over R, denoted by $\models_r R_s$, if and only if r is R_s-total. SQL, the industry standard for data management, allows attributes to be specified as NOT NULL [10].

Definition 2. A *full first-order hierarchical dependency* (FOHD) over the relation schema R is an expression $X : [Y_1 \mid \ldots \mid Y_k]$ with a non-negative integer k, $X, Y_1, \ldots, Y_k \subseteq R$ such that $Y_1, \ldots, Y_k$ form a partition of $R - X$. A relation r over R is said to *satisfy* (or said to be a *model* of) the full first-order hierarchical dependency $X : [Y_1 \mid \cdots \mid Y_k]$ over R, denoted by $\models_r X : [Y_1 \mid \cdots \mid Y_k]$, if and only if $r_X[R] = (\cdots (r_X[XY_k] \bowtie r_X[XY_{k-1}]) \bowtie \cdots) \bowtie r_X[XY_1]$ holds.

The FOHD $\emptyset : [Y_1 \mid \cdots \mid Y_k]$ expresses the fact that any relation over R is the Cartesian product over its projections to attribute sets in $\{Y_i\}_{i=1}^k$. For $k = 0$, the FOHD $X : [\,]$ is satisfied trivially, where $[\,]$ denotes the empty list.

Remark 7. In consistency with Remark 1 on GSCI statements, we assume w.l.o.g. that the sets Y_i in FOHDs are non-empty. Indeed, for all positive k we have the property that for all relations r the FOHD $X : [\emptyset, Y_2, \ldots, Y_k]$ is satisfied by r if and only if r satisfies the FOHD $X : [Y_2, \ldots, Y_k]$. In particular, if $k = 1$, then $X : [\emptyset]$ is equivalent to $X : [\,]$; more specifically, they are both satisfied by all relations.

Example 12. We use now

$$R = \{\hat{m}(\text{ovie}), \hat{v}(\text{ctor}), \hat{r}(\text{ole}), \hat{c}(\text{rew}), \hat{f}(\text{eature}), \hat{l}(\text{anguage}), \hat{s}(\text{ubtitle})\}$$

to denote a relation schema that models information about blu-rays of movies. As the NFS there are at least the two options $R_s = \{\hat{m}, \hat{v}, \hat{r}, \hat{c}, \hat{s}\}$ and $R'_s = \{\hat{f}, \hat{l}\}$. For ease of presentation in this and the following examples we denote attributes by lower-case Latin letters without the $\hat{\ }$ above them. The following full first-order hierarchical dependencies are specified to enforce consistency in database relations: $\Sigma = \{m : [sar \mid c \mid fl], mfl : [sc \mid ar]\}$. The database design team has identified an additional meaningful FOHD $\varphi = m : [s \mid ar \mid cfl]$, and is wondering whether φ must be enforced in addition to Σ, or whether it is already implicitly enforced by enforcing Σ, i.e., whether φ is R_s-implied or R'_s-implied by Σ, respectively.

5.2 Axiomatization

For the design of a relational database schema semantic constraints are defined on the relations which are intended to be instances of the schema [38]. During the design process one usually needs to determine further constraints which are logically implied by the given ones. As was the case with GSCI statements and propositional formulae before, we can speak of R_s-implication for sets of full first-order hierarchical dependencies. Similarly, we can introduce the notions of soundness and completeness for sets of inference rules. Finite sets of full first-order hierarchical dependencies are denoted by $\hat{\Sigma}$ and single FOHDs by $\hat{\varphi}$.

PROBLEM:	Implication Problem for FOHDs and NFSs
INPUT:	Relation schema R, null-free subschema R_s over R, Set $\hat{\Sigma} \cup \{\hat{\varphi}\}$ of FOHDs over R
OUTPUT:	Yes, if $\hat{\Sigma} \models_{R_s} \hat{\varphi}$; No, otherwise

Table 4 Axiomatization $\mathfrak{F} = \{\hat{\mathscr{U}}, \hat{\mathscr{P}}, \hat{\mathscr{M}}, \hat{\mathscr{A}}, \hat{\mathscr{T}}\}$ of FOHDs and NFS R_s

$\dfrac{}{\emptyset : [R]}$ (universal, $\hat{\mathscr{U}}$)	$\dfrac{X : [Y_1 \mid \cdots \mid Y_k]}{X : [Y_{\pi(1)} \mid \cdots \mid Y_{\pi(k)}]}$ (permutation, $\hat{\mathscr{P}}$)
$\dfrac{X : [Y_1 \mid \cdots \mid Y_{k-1} \mid Y_k \mid Z]}{X : [Y_1 \mid \cdots \mid Y_{k-1} \mid Y_k Z]}$ (merging, $\hat{\mathscr{M}}$)	$\dfrac{X : [Y_1 \mid \cdots \mid Y_k Z]}{XZ : [Y_1 \mid \cdots \mid Y_k]}$ (augmentation, $\hat{\mathscr{A}}$)

$$\dfrac{X : [Y_1 \cdots Y_k \mid YZ_1 \cdots Z_k] \quad XY : [Y_1 Z_1 \mid \cdots \mid Y_k Z_k]}{X : [Y_1 \mid \cdots \mid Y_k \mid YZ_1 \cdots Z_k]} Y \subseteq R_s$$
(restricted transitivity, $\hat{\mathscr{T}}$)

Remark 8. In consistency with Remark 5 on the application of inference rules to GSCI statements, note the following global condition that we enforce on all applications of inference rules that infer FOHDs. Whenever we apply such an inference rule, we remove all empty sets from the exact position in which they occur as elements in the sequence in the conclusion. For instance, we can infer $R : [\ ]$ by an application of the *augmentation rule* $\hat{\mathscr{A}}$ to the FOHD $\emptyset : [R]$.

As in the context of GSCI statements, we can define $Dep_{\hat{\Sigma},R_s}(X) := \{Y \subseteq R-X \mid \Sigma \vdash_{\mathfrak{F}} X : [Y \mid R-XY]\}$ as the set of all $Y \subseteq R-X$ such that $X : [Y \mid R-XY]$ can be inferred from $\hat{\Sigma}$ by $\mathfrak{F}$. The special case $Y = \emptyset$ of the restricted transitivity rule $\hat{\mathscr{T}}$ as well the merging rule $\hat{\mathscr{M}}$ shows that

$$(Dep_{\hat{\Sigma},R_s}(X), \subseteq, \cup, \cap, (\cdot)^{\mathscr{C}}, \emptyset, R-X)$$

forms a finite Boolean algebra where $(\cdot)^{\mathscr{C}}$ maps a set $Y \subseteq R - X$ to its complement $R - (XY)$. Let $DepB_{\hat{\Sigma},R_s}(X)$ denote the set of all atoms of $(Dep_{\hat{\Sigma},R_s}(X), \subseteq, \emptyset)$. We call $DepB_{\hat{\Sigma},R_s}(X)$ the *dependence basis* of X with respect to $\hat{\Sigma}$ and R_s [4]. The proof of the following result follows the proof of Theorem 2.

Theorem 5. *Let $\hat{\Sigma}$ be a set of FOHDs over R. Then $\hat{\Sigma} \vdash_{\mathfrak{F}} X : [Y_1 \mid \cdots \mid Y_k]$ if and only if for every $i = 1, \ldots, k$, $Y_i = \bigcup \mathscr{Y}$ for some $\mathscr{Y} \subseteq DepB_{\hat{\Sigma},R_s}(X)$.* $\qquad\square$

The completeness proof shows that an FOHD $\hat{\varphi}$ is not R_s-implied by a set of FOHDs $\hat{\Sigma}$ whenever $\hat{\varphi}$ cannot be inferred from $\hat{\Sigma}$ by $\mathfrak{F}$. We will now apply the techniques from the completeness proof for GSCI statements to construct a two-tuple relation that satisfies $\hat{\Sigma}$ but violates $\hat{\varphi}$.

Theorem 6. *The set $\mathfrak{F}$ of inference rules from Table 4 forms an axiomatization for the implication problem of full first-order hierarchical dependencies and null-free subschemata.*

Proof. It remains to show the completeness of $\mathfrak{F}$. Let R be an arbitrary relation schema, let R_s be an NFS over R, and let $\hat{\Sigma}$ be an arbitrary set of FOHDs over R. We need to show that $\hat{\Sigma}^*_{R_s} \subseteq \hat{\Sigma}^+_{\mathfrak{F}}$ holds.

Let $X : [Y_1 \mid \cdots \mid Y_k] \notin \hat{\Sigma}^+_{\mathfrak{F}}$. Let $DepB_{\hat{\Sigma},R_s}(X) = \{W_1, \ldots, W_n\}$, in particular $R = XW_1 \cdots W_n$. Since $X : [Y_1 \mid \cdots \mid Y_k] \notin \hat{\Sigma}^+_{\mathfrak{F}}$ we conclude by Theorem 5 that there is some $j \in \{1, \ldots, k\}$ such that Y_j is not the union of some elements of $DepB_{\hat{\Sigma},R_s}(X)$. Consequently, there is some $i \in \{1, \ldots, n\}$ such that $Y_j \cap W_i \neq \emptyset$ and $W_i - Y_j \neq \emptyset$ hold. Let $T := \bigcup_{l \in \{1,\ldots,i-1,i+1,\ldots,k\}} W_l \cap R_s$, and $T' := \bigcup_{l \in \{1,\ldots,i-1,i+1,\ldots,k\}} W_l - R_s$. In particular, R is the disjoint union of X, T, T', and W_i. We define the following two tuples t_1 and t_2 over R. We define $t_1(\hat{v}) = \mathbf{0}$ for all $\hat{v} \in XW_iT$, $t_1(\hat{v}) = \text{ni}$ for all $\hat{v} \in T'$. We further define $t_2(\hat{v}) = t_1(\hat{v})$ for all $\hat{v} \in XTT'$, and $t_2(\hat{v}) = \mathbf{1}$ for all $\hat{v} \in W_i$. The two-tuple relation $r = \{t_1, t_2\}$ is illustrated in Table 5. It is simple to observe that the relation r enjoys the following property: an FOHD $U : [V_1 \mid \cdots \mid V_m]$ is satisfied by r if and only if i) $U \cap T' \neq \emptyset$, or ii) $U \cap W_i \neq \emptyset$, or iii) $W_i \subseteq V_o$ for some $o \in \{V_1, \ldots, V_m\}$. Indeed, if $U \cap T' \neq \emptyset$, then $r_U[Z] = \emptyset$ for all $Z \subseteq R$. If $U \cap T' = \emptyset$ and $U \cap W_i \neq \emptyset$, then the projections $r_U[UV_l]$ contain two tuples for all $l = 1, \ldots, m$ and only the original tuples match on common attributes. If $U \subseteq XT$ and $W_i \subseteq V_o$, then the projection $r_U[UV_l]$ contains only one tuple for all $V_l \in \{V_1, \ldots, V_n\} - \{V_o\}$, and the projection $r_U[UV_o]$ contains two tuples. The join of those projections is the original relation r.

Table 5 Two-tuple Relation
from Completeness Proof
of $\mathfrak{F}$

XT	T'	W_i
$0\cdots 0$	$\texttt{ni}\cdots\texttt{ni}$	$0\cdots 0$
$0\cdots 0$	$\texttt{ni}\cdots\texttt{ni}$	$1\cdots 1$

Vice versa, if $U \subseteq XT$ and $W_i \not\subseteq V_o$ for all $V_o \in \{V_1, \ldots, V_m\}$, then the projections $r_U[UV_l]$ contain tuples whose join does not occur in the original relation r (in fact, a projection of some tuple in the joined relation to W_i contains some 0s and some 1s).

The construction ensures that r violates $X : [Y_1 \mid \cdots \mid Y_k]$ since $X \cap T' = \emptyset$, $X \cap W_i = \emptyset$, and $W_i \not\subseteq Y_s$ for $s = 1, \ldots, k$. Furthermore, r is R_s-total by construction.

It remains to show that r satisfies $\hat{\Sigma}$, that is, every FOHD $U : [V_1 \mid \cdots \mid V_m]$ in $\hat{\Sigma}$. If $U \cap T' \neq \emptyset$ or $U \cap W_i \neq \emptyset$, then r satisfies $U : [V_1 \mid \cdots \mid V_m]$. Otherwise, $U \subseteq XT$. Consequently, we can apply the augmentation rule $\hat{\mathscr{A}}$ and permutation rule $\hat{\mathscr{P}}$ to $U : [V_1 \mid \cdots \mid V_m] \in \hat{\Sigma}$ to infer $XT : [V_1 - XT \mid \cdots \mid V_l - XT] \in \hat{\Sigma}_{\mathfrak{F}}^{+}$. Theorem 5 also shows that $X : [W_i \mid TT'] \in \hat{\Sigma}_{\mathfrak{F}}^{+}$. Now we define $V_l' := V_l - XTT'$ and $Z_l := (V_l - XT) \cap T'$ for $l = 1, \ldots, m$. Consequently, $W_i = V_1' \cdots V_m'$, $T' = Z_1' \cdots Z_m'$, and $V_l - XT = V_l' Z_l'$ for $l = 1, \ldots, m$. An application of the restricted transitivity rule $\hat{\mathscr{T}}$ to $XT : [V_1' Z_1' \mid \cdots \mid V_m' Z_m']$ and $X : [V_1' \cdots V_m' \mid TZ_1' \cdots Z_m']$ results in $X : [V_1' \mid \cdots \mid V_m' \mid TZ_1' \cdots Z_m'] = X : [V_1 - XTT' \mid \cdots \mid V_m - XTT' \mid TT']$. It follows from Theorem 2 that for every $l = 1, \ldots, m$, $V_l - XTT'$ is the union of elements from $DepB_{\hat{\Sigma}, R_s}(X)$. Consequently, $V_o - XTT' = W_i$ for some $o \in \{1, \ldots, m\}$ and, therefore, $W_i \subseteq V_o$. As we have seen above, this means that r indeed satisfies $U : [V_1 \mid \cdots \mid V_m]$. This concludes the proof. $\square$

Example 13. Recall our running example: $R = \{m, a, r, c, f, l, s\}$, $R_s = \{m, a, r, c, s\}$, $\Sigma = \{m : [sar \mid c \mid fl], mfl : [sc \mid ar]\}$, and $\varphi = m : [s \mid ar \mid cfl]$. The construction from Theorem 6 may result in the following relation r

movie	actor	role	crew	feature	language	subtitle
Rashomon	T. Mifune	Tajomaru	Kurosawa	ni	ni	Suomi
Rashomon	M. Kyo	Masako	Kurosawa	ni	ni	Deutsch

that satisfies Σ and R_s, but violates φ. For example, the *movie*-total part of the join of the following projections

movie	subtitle
Rashomon	Suomi
Rashomon	Deutsch

movie	actor	role
Rashomon	T. Mifune	Tajomaru
Rashomon	M. Kyo	Masako

movie	crew	feature	language
Rashomon	Kurosawa	ni	ni

is

movie	actor	role	crew	feature	language	subtitle
Rashomon	T. Mifune	Tajomaru	Kurosawa	ni	ni	Suomi
Rashomon	M. Kyo	Masako	Kurosawa	ni	ni	Deutsch
Rashomon	T. Mifune	Tajomaru	Kurosawa	ni	ni	Deutsch
Rashomon	M. Kyo	Masako	Kurosawa	ni	ni	Suomi

which is different from r.

5.3 Implication of FOHDs and NOT NULL Constraints in the World of Two-tuple Relations

A relation r that consists of two tuples is said to be a *two-tuple relation*. We say that Σ R_s-*implies* φ in the world of two-tuple relations, denoted by $\hat{\Sigma} \models_{2,R_s} \hat{\varphi}$, if every R_s-total two-tuple relation over R that satisfies every FOHD in $\hat{\Sigma}$ also satisfies the FOHD $\hat{\varphi}$. The following variant of the implication problem for FOHDs and NFSs emerges.

PROBLEM:	Implication problem for FOHDs and NFSs in the world of two-tuple relations
INPUT:	Relation schema R, NFS R_s over R, Set $\hat{\Sigma} \cup \{\hat{\varphi}\}$ of FOHDs over R
OUTPUT:	Yes, if $\hat{\Sigma} \models_{2,R_s} \hat{\varphi}$; No, otherwise

The proof of Theorem 6 implies the following result.

Corollary 2. *The implication problem for FOHDs and NFSs coincides with the implication problem for FOHDs and NFSs in the world of two-tuple relations.*

Proof. Let $\hat{\Sigma} \cup \{\hat{\varphi}\}$ be a set of FOHDs over R. We need to show that $\hat{\Sigma} \models_{R_s} \hat{\varphi}$ if and only if $\hat{\Sigma} \models_{2,R_s} \hat{\varphi}$. If it does not hold that $\hat{\Sigma} \models_{2,R_s} \hat{\varphi}$, then it also does not hold that $\hat{\Sigma} \models_{R_s} \hat{\varphi}$ since every two-tuple relation is a relation. Vice versa, if it does not hold that $\hat{\Sigma} \models_{R_s} \hat{\varphi}$, then it does not hold that $\hat{\Sigma} \vdash_{\mathfrak{F}} \hat{\varphi}$ since $\mathfrak{F}$ is sound for the implication of FOHDs. However, the proof of Theorem 6 shows how to construct an R_s-total two-tuple relation that satisfies every FOHD in $\hat{\Sigma}$ but does not satisfy $\hat{\varphi}$. Hence, it does not hold that $\hat{\Sigma} \models_{2,R_s} \hat{\varphi}$.   □

Corollary 2 shows that to decide the implication problem for FOHDs and NFSs over R it suffices to check two-tuple relations over R.

Example 14. The two-tuple relation r from Example 13 shows that $\hat{\Sigma}$ does not R_s-imply $\hat{\varphi}_2$ in the world of two-tuple relations.

5.4 Functional and Hierarchical Dependencies

In this subsection we establish a result on the interaction of FOHDs and functional dependencies over two-tuple relations. The finding subsumes a known result on the interaction of multivalued dependencies (MVDs) and functional dependencies over two-tuple relations [1]. Recall that a functional dependency (FD) over relation schema R is an expression $X \to Y$ with $X, Y \subseteq R$. A relation r over R satisfies the FD $X \to Y$ if and only if all tuples $t, t' \in r$ with matching non-null values on all the attributes in X also have matching values on all the attributes in Y, that is, if $t(X) = t'(X)$ and t, t' are X-total, then $t(Y) = t'(Y)$ [3, 37].

Theorem 7. *Let $r = \{t_1, t_2\}$ be a two-tuple relation over relation schema R. Then r satisfies the FOHD $X : [Y_1 \mid \cdots \mid Y_k]$ if and only if there is some $i \in \{1, \ldots, k\}$ such that r satisfies the FD $X \to R - XY_i$.*

Proof. If $t_1(X) \neq t_2(X)$, or t_1 and t_2 are not both X-total, then r satisfies both the FOHD $X : [Y_1 \mid \cdots \mid Y_k]$ and the FDs $X \to R - XY_i$ for all $i = 1, \ldots, k$. For the remainder of the proof we therefore assume that $t_1(X) = t_2(X)$ holds and t_1, t_2 are both X-total, i.e., the tuples in the projections $r_X[XY_i]$ all have matching non-null values on their common attributes, i.e., the attributes in X.

Assume first that r satisfies the FD $X \to R - XY_i$ for some $i \in \{1, \ldots, k\}$. Consequently, the projections $r_X[XY_j]$ contain only one tuple for all $j \in \{1, \ldots, k\} - \{i\}$, and $r_X[XY_i]$ contains at most two tuples. The join $r_X[XY_1] \bowtie \cdots \bowtie r_X[XY_k]$ contains only tuples from r, i.e., $r_X[R] = r_X[XY_1] \bowtie \cdots \bowtie r_X[XY_k]$.

Assume now that r violates the FDs $X \to Y_i$ and $X \to Y_j$ for some $i \neq j$. Then $r_X[XY_i]$ and $r_X[XY_j]$ contain two tuples each. The join $r_X[XY_1] \bowtie \cdots \bowtie r_X[XY_k]$ thus contains tuples that are not originally in r. This concludes the proof. $\square$

Example 15. Recall the following two-tuple relation r from Example 13

movie	actor	role	crew	feature	language	subtitle
Rashomon	T. Mifune	Tajomaru	Kurosawa	ni	ni	Suomi
Rashomon	M. Kyo	Masako	Kurosawa	ni	ni	Deutsch

Indeed, r does not satisfy any of the FDs $m \to arcfl$, $m \to rcfls$, nor $m \to as$. According to Theorem 7, r does not satisfy the FOHD $\varphi = m : [s \mid a \mid crfl]$.

5.5   Equivalence to the Propositional Fragment $\mathfrak{F}$

Let $\hat{\phi} : R \to L$ denote a bijection between a relation schema R of attributes $\hat{v}$ and the set $L = \{v' \mid \hat{v} \in R\}$ of propositional variables, where $\hat{\phi}(R_s) = \mathscr{S} \subseteq L$ for an NFS R_s over R. We extend $\hat{\phi}$ to a mapping $\hat{\Phi}$ from the set of FOHDs over R to the fragment $\mathfrak{F}$, that is, $\mathfrak{F}$ is the range of $\hat{\Phi}$. For an FOHD $X : [Y_1 \mid \cdots \mid Y_k]$ over R, let $\hat{\Phi}(X : [Y_1 \mid \cdots \mid Y_k])$ denote the formula

$$\bigvee_{\hat{v} \in X} \neg v' \vee \bigvee_{i=1}^{k} \left(\bigwedge_{\hat{v} \in \bigcup_{j \neq i} Y_j} v' \right).$$

Recall from before that disjunctions over zero disjuncts are interpreted as $\mathbb{F}$ and conjunctions over zero conjuncts are interpreted as $\mathbb{T}$. We will simply denote $\Phi(\hat{\varphi}) = \varphi'$ and $\hat{\Phi}(\hat{\Sigma}) = \{\sigma' \mid \hat{\sigma} \in \hat{\Sigma}\} = \Sigma'$. Example 9 shows the $\mathfrak{F}$-formulae that correspond to the FOHDs from Example 12.

Note that for the special case of MVDs $\hat{\varphi} = X : [Y \mid Z]$, i.e., FOHDs where $k = 2$, the formula φ' becomes again

$$\bigvee_{\hat{v} \in X} \neg v' \vee \left(\bigwedge_{\hat{v} \in Y} v' \right) \vee \left(\bigwedge_{\hat{v} \in Z} v' \right).$$

We will now show that for any set $\hat{\Sigma} \cup \{\hat{\varphi}\}$ of FOHDs over R there is an R_s-total relation r over R that satisfies $\hat{\Sigma}$ and violates $\hat{\varphi}$ if and only if there is an $\mathscr{S}$-3 truth assignment ω_r' that is an $\mathscr{S}$-3 model of Σ' but not an $\mathscr{S}$-3 model of φ'. For arbitrary relations r it is not obvious how to define the truth assignment ω_r'. However, the key to showing the correspondence between counterexample relations and counterexample truth assignments is Corollary 2. Corollary 2 tells us that for deciding the implication problem of FOHDs and NFSs it suffices to examine two-tuple relations (instead of arbitrary relations). For a two-tuple relation $r = \{t_1, t_2\}$, however, we can define its corresponding special 3-valued truth assignment ω_r' of L as follows:

$$\omega_r'(v') = \begin{cases} \mathbb{T} \text{ , if } t_1(\hat{v}) = t_2(\hat{v}) \\ \mathbb{F} \text{ , otherwise} \end{cases},$$

and

$$\omega_r'(\neg v') = \begin{cases} \mathbb{T} \text{ , if } t_1(\hat{v}) \neq t_2(\hat{v}) \text{ or } t_1(\hat{v}) = \texttt{ni} = t_2(\hat{v}) \\ \mathbb{F} \text{ , otherwise} \end{cases}.$$

Next we justify the definition of the special truth assignment and that of the propositional fragment $\mathfrak{F}$ in terms of two-tuple relations.

Lemma 3. *Let* $r = \{t_1, t_2\}$ *be a two-tuple relation over R, and let* $\hat{\varphi}$ *denote an FOHD over R. Then r satisfies* $\hat{\varphi}$ *if and only if* ω'_r *is a 3-valued model of* φ'.

Proof. Let $\hat{\varphi} = X : [Y_1 \mid \cdots \mid Y_k]$ and

$$
\varphi' = \bigvee_{\hat{v} \in X} \neg a' \vee \bigvee_{i=1}^{k} \left(\bigwedge_{\hat{v} \in \bigcup_{j \neq i} Y_j} a' \right).
$$

Suppose first that r satisfies $\hat{\varphi}$. We need to show that ω'_r is a 3-valued model of φ'. Assume that $\omega'_r(\neg a') = \mathbb{F}$ for all $\hat{v} \in X$. According to the special 3-valued truth assignment we must have $\mathtt{ni} \neq t_1(\hat{v}) = t_2(\hat{v}) \neq \mathtt{ni}$ for all $\hat{v} \in X$. Suppose that for all $i = 2, \ldots, k$ there is some $\hat{v} \in \bigcup_{j \neq i} Y_j$ such that $\omega'_r(a') = \mathbb{F}$. Consequently, there is some $\hat{v} \in Y_1$ such that $\omega'_r(a') = \mathbb{F}$. Hence, $t_1(\hat{v}) \neq t_2(\hat{v})$ according to the special 3-valued truth assignment. However, since r satisfies φ, r must satisfy the FD $X \to Y_2 \cdots Y_k$ by Theorem 7. Consequently, for every $\hat{v} \in Y_2 \cdots Y_k$ we have $t_1(\hat{v}) = t_2(\hat{v})$. This means that for all $\hat{v} \in Y_2 \cdots Y_k$ we have $\omega'_r(a') = \mathbb{T}$. This shows that ω'_r is a 3-valued model of φ'.

Suppose ω'_r is a 3-valued model of φ'. We need to show that r satisfies $\hat{\varphi}$. That is, $r = r[XY_1] \bowtie \cdots \bowtie r[XY_k]$ holds. According to Theorem 7 this is equivalent to showing that r satisfies the FD $X \to R - XY_i$ for some $i \in \{1, \ldots, k\}$. Suppose that $t_1(X) = t_2(X)$ and t_1, t_2 are both X-total, otherwise there is nothing to show. This implies that $\omega'_r(a') = \mathbb{T}$ for all $\hat{v} \in X$. Assume that for $j = 2, \ldots, k$, r violates $X \to R - XY_j$, otherwise there is nothing to show. Consequently, for all $j = 2, \ldots, k$ there is some $\hat{v} \in R - XY_j$ such that $\omega'_r(a') = \mathbb{F}$. Since ω'_r satisfies φ' we must have $\omega'_r(a') = \mathbb{T}$ for all $\hat{v} \in Y_2 \cdots Y_k$. Hence, $t_1(Y_2 \cdots Y_k) = t_2(Y_2 \cdots Y_k)$, and r satisfies $X \to R - XY_1$. It follows that r satisfies $\hat{\varphi}$. □

The equivalence between two-tuple relations for FOHDs and special truth assignments extends the existing equivalence between two-tuple relations for multivalued dependencies and special truth assignments [14, 28, 52].

Corollary 2 and Lemma 3 allow us to establish the anticipated equivalence between two-tuple relations and propositional truth assignments.

Theorem 8. *Let* $\hat{\Sigma} \cup \{\hat{\varphi}\}$ *be a set of FOHDs over relation schema R with NFS R_s, and let* $\Sigma' \cup \{\varphi'\}$ *denote the set of its corresponding formulae over L with the set $\mathscr{S}$. Then* $\hat{\Sigma} \models_{R_s} \hat{\varphi}$ *if and only if* $\Sigma' \models^3_{\mathscr{S}} \varphi'$.

Proof. Based on Corollary 2 it remains to establish the equivalence between $\hat{\Sigma} \models_{2,R_s} \hat{\varphi}$ and $\Sigma' \models^3_{\mathscr{S}} \varphi'$. Suppose first that $\hat{\Sigma} \models_{2,R_s} \hat{\varphi}$ does not hold. Then there is some R_s-total relation r over R that satisfies every FOHD $\hat{\sigma}$ in $\hat{\Sigma}$ but violates $\hat{\varphi}$. Let ω'_r denote the special 3-valued truth assignment associated with r. By definition ω'_r is an $\mathscr{S}$-3 interpretation. By Lemma 3 it follows that ω'_r is an $\mathscr{S}$-3 model of every formula σ' in Σ' but not an $\mathscr{S}$-3 model of φ'. Consequently, $\Sigma' \models^3_{\mathscr{S}} \varphi'$ does not hold. Suppose now that $\Sigma' \models^3_{\mathscr{S}} \varphi'$ does not hold. Then there is some $\mathscr{S}$-3 truth assignment ω' over L that is an $\mathscr{S}$-3 model for every formula σ' in Σ', but not an

$\mathscr{S}$-3 model for the formula φ'. Define the following two-tuple relation $r = \{t_1, t_2\}$ over R: for all $\hat{v} \in R$, let $\mathtt{ni} \neq t_1(\hat{v}) = t_2(\hat{v}) \neq \mathtt{ni}$, if $\omega(a') = \mathbb{T}$ and $\omega(\neg a') = \mathbb{F}$; let $t_1(\hat{v}) = \mathtt{ni} = t_2(\hat{v})$, if $\omega(a') = \mathbb{T} = \omega(\neg a')$; and let $\mathtt{ni} \neq t_1(\hat{v}) \neq t_2(\hat{v}) \neq \mathtt{ni}$, if $\omega(a') = \mathbb{F}$ and $\omega(\neg a') = \mathbb{T}$. In particular, it follows that $\omega'_r = \omega'$. By Lemma 3 it follows that r satisfies every FOHD $\hat{\sigma}$ in $\hat{\Sigma}$ but violates $\hat{\varphi}$. In addition, r is R_s-total since the construction ensures that the null marker $\mathtt{ni}$ can only occur on attributes outside of R_s. Hence, $\hat{\Sigma} \models_{2,R_s} \hat{\varphi}$ does not hold. □

Examples 9 and 13 illustrate the equivalences between the implication problem of FOHDs and NFSs and the implication problem of formulae in $\mathfrak{F}$ under $\mathscr{S}$-3 interpretations.

6 Related Work

Dawid [11] has started to investigate fundamental properties of conditional independence, leading to a claim that "rather than just being another useful tool in the statistician's kitbag, conditional independence offers a new language for the expression of statistical concepts and a framework for their study." Geiger and Pearl [20, 21, 50] have systematically investigated the implication problem for fragments of conditional independence statements over different probability models. In particular, they have established an axiomatization of saturated conditional independence (SCI) statements by a finite set of Horn rules [20]. Studený [56] showed that no axiomatization by a finite set of Horn rules exists for general conditional independence statements. Recently, Naumov and Nicholls [46] established a complete infinite recursively enumerable axiomatization of the propositional theory for conditional independence statements. Niepert et al. [47, 48] established an axiomatization for stable conditional independence statements, which subsume saturated statements, and showed that their associated implication problem is coNP-complete. Recently, this line of work has been extended to incomplete data, in which the implication problem changes [7, 32–34, 41–44]. Figure 2 shows a classification of this work by distinguishing between implication problems in fixed and undetermined sets of attributes, random variables, or propositional variables, respectively (referred to collectively as features), and by distinguishing between the sets of features that can be declared complete (either only the empty set $\emptyset$, or the entire set S, or an arbitrary subset C of S). The present article is a summary of the results and techniques applied to fixed sets of features. Similar results hold when the set of features remains undetermined [7, 32, 33]. In particular, the results establish strong bonds with database semantics and approximation logics.

In fact, database theory has studied more than 100 different classes of database dependencies [57] over strictly relational data, where incomplete data must not occur. These dependencies enforce the semantics of application domains within a database system [38]. Here, multivalued dependencies [15] are an expressive class whose implication problem can be decided in almost linear time [4, 18, 51].

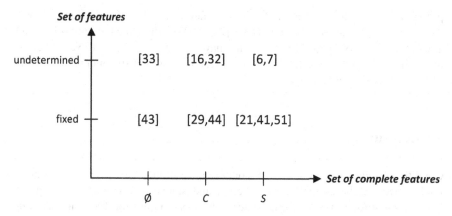

Fig. 2 Classification of Related Work on Conditional Independence and Hierarchical Dependence

In particular, they form the basis for the Fourth Normal Form in database design which characterizes database schemata whose instances are free from data redundancy [15, 60, 62]. The implication problem of multivalued dependencies is equivalent to that of a Boolean propositional fragment [53], and to that of SCI statements [61]. Furthermore, it is known that the equivalence between MVD implication and that of their corresponding propositional counterpart cannot be extended to an equivalence between the implication problem of embedded MVDs and that of any Boolean propositional fragment [53]. We also note that the implication problem of embedded multivalued dependencies is undecidable [29, 30] and not axiomatizable by a finite set of Horn rules [49]. Studený also showed that the implication problem of embedded MVDs and that of CI statements does not coincide [56]. Again, this line of work has been extended to incomplete data [16, 27, 28, 31, 35] and the present article can be understood as a summary of these findings.

It is important to point out that the results in this article can be proven more directly in different ways. Firstly, for a set $\Sigma \cup \{I(Y_1, \ldots, Y_k \mid X)\}$ of GSCI statements over S with $C \subseteq S$ it holds that $\Sigma \models_C I(Y_1, \ldots, Y_k \mid X)$ if and only if $\Sigma[XC] \models_S I(Y_1, \ldots, Y_k \mid X)$, where $\Sigma[U] = \{I(W_1, \ldots, W_m \mid V) \in \Sigma \mid V \subseteq U\}$. This embedding translates every instance of the implication problem for GSCI statements and complete random variables into an instance of the implication problem for GSCI statements. This illustrates the significance of the special case where $C = S$. Secondly, every instance of the implication problem for GSCI statements and complete random variables can be translated into an instance of an implication problem for SCI statements and complete random variables, see Remark 6. Finally, the results for full first-order hierarchical dependencies from Section 5 can be obtained by exploiting a strong correspondence between relations that satisfy an FOHD $X : [Y_1 | \cdots | Y_k]$ and probability models that satisfy $I(Y_1, \ldots, Y_k \mid X)$. For instance, a two-tuple relation $r = \{t_1, t_2\}$ satisfies $X : [Y_1 | \cdots | Y_k]$ if and only if the special probability model $\tau(r)$ satisfies the GSCI statement $I(Y_1, \ldots, Y_k \mid X)$, where $\tau(r)$ is obtained by stipulating $P(t_1) = 0.5 = P(t_2)$. Vice versa, the special probability model π satisfies the GSCI statement $I(Y_1, \ldots, Y_k \mid X)$ if and only if the relation $\tau'(\pi)$ satisfies the FOHD $X : [Y_1 \mid \cdots \mid Y_k]$, where the two tuples in $\tau'(\pi)$ are simply the two assignments of π that have probability one half. Nevertheless, the main focus of this article is not on the results for GSCI statements, but on the techniques used to obtain them.

7 Conclusion

Conditional independence is a core concept in disciplines as diverse as artificial intelligence, databases, probability theory, and statistics. The implication problem for conditional independence statements is paramount for many applications including Bayesian networks and database design. It is known that the implication problem for general conditional independence statements cannot be axiomatized by a finite set of Horn rules, and is coNP-complete to decide for their stable fragment, already in the idealized case where all data is complete. This article showcases the equivalences between three different implication problems: i) generalized saturated conditional independence statements in the presence of a set of complete random variables, ii) a fragment of propositional logic under $\mathscr{S}$-3 interpretations, and iii) Delobel's class of full first-order hierarchical database dependencies in the presence of a set of attributes declared NOT NULL. Axiomatizations in the form of finite sets of Horn rules were established, and algorithms to decide the associated implication problems in almost linear time are also available [44]. The key to these equivalences are special probability models and two-tuple relations. It is further known that none of these equivalences holds between the frameworks of conditional independence statements, any fragment of Boolean propositional logic, and general first-order hierarchical dependencies, already in the case of complete data [44].

This body of work is a strong advocate for investigating notions of dependence and independence as first-class citizens within standard frameworks for reasoning, as successfully started in dependence and independence logics.

Acknowledgements I would like to thank the anonymous reviewer for feedback that largely improved the presentation of the paper. This research is supported by the Marsden Fund Council from Government funding, administered by the Royal Society of New Zealand.

References

1. Abiteboul, S., Hull, R., Vianu, V.: Foundations of Databases. Addison-Wesley, Boston, MA (1995)
2. Abramsky, S., Kontinen, J., Väänänen, J.A., Vollmer, H.: Dependence logic: theory and applications (Dagstuhl seminar 13071). Dagstuhl Reports **3**(2), 45–54 (2013)
3. Atzeni, P., Morfuni, N.: Functional dependencies and constraints on null values in database relations. Inf. Control. **70**(1), 1–31 (1986)

4. Beeri, C.: On the membership problem for functional and multivalued dependencies in relational databases. ACM Trans. Database Syst. **5**(3), 241–259 (1980)

5. Beeri, C., Fagin, R., Howard, J.H.: A complete axiomatization for functional and multivalued dependencies in database relations. In: Proceedings of the SIGMOD International Conference on Management of Data, pp. 47–61. ACM, Toronto (1977)

6. Biskup, J., Link, S.: Appropriate inferences of data dependencies in relational databases. Ann. Math. Artif. Intell. **63**(3–4), 213–255 (2012)

7. Biskup, J., Hartmann, S., Link, S.: Probabilistic conditional independence under schema certainty and uncertainty. In: Proceedings of the 6th International Conference on Scalable Uncertainty Management (SUM). Lecture Notes in Computer Science, vol. 7520, pp. 365–378. Springer, Berlin (2012)

8. Codd, E.F.: Extending the database relational model to capture more meaning. ACM Trans. Database Syst. **4**(4), 397–434 (1979)

9. Codd, E.F.: Missing information (applicable and inapplicable) in relational databases. SIG-MOD Rec. **15**(4), 53–78 (1986)

10. Date, C., Darwen, H.: A guide to the SQL standard. Addison-Wesley, Reading, MA (1997)

11. Dawid, A.P.: Conditional independence in statistical theory. J. R. Stat. Soc. Ser. B Methodol. **41**(1), 1–31 (1979)

12. Delobel, C.: Normalization and hierarchical dependencies in the relational data model. ACM Trans. Database Syst. **3**(3), 201–222 (1978)

13. Durand, A., Kontinen, J.: Hierarchies in dependence logic. ACM Trans. Comput. Log. **13**(4), 31 (2012)

14. Fagin, R.: Functional dependencies in a relational data base and propositional logic. IBM J. Res. Dev. **21**(6), 543–544 (1977)

15. Fagin, R.: Multivalued dependencies and a new normal form for relational databases. ACM Trans. Database Syst. **2**(3), 262–278 (1977)

16. Ferrarotti, F., Hartmann, S., Link, S.: Reasoning about functional and full hierarchical dependencies over partial relations. Inf. Sci. **235**, 150–173 (2013)

17. Fienberg, S.: The analysis of incomplete multi-way contingency tables. Biometrics **28**(1), 177–202 (1972)

18. Galil, Z.: An almost linear-time algorithm for computing a dependency basis in a relational database. J. ACM **29**(1), 96–102 (1982)

19. Galliani, P.: Inclusion and exclusion dependencies in team semantics - on some logics of imperfect information. Ann. Pure Appl. Logic **163**(1), 68–84 (2012)

20. Geiger, D., Pearl, J.: Logical and algorithmic properties of independence and their application to Bayesian networks. Ann. Math. Artif. Intell. **2**, 165–178 (1990)

21. Geiger, D., Pearl, J.: Logical and algorithmic properties of conditional independence and graphical models. Ann. Stat. **21**(4), 2001–2021 (1993)

22. Grädel, E., Väänänen, J.A.: Dependence and independence. Stud. Logica **101**(2), 399–410 (2013)

23. Graetzer, G.: General Lattice Theory. Birkhäuser, Boston, MA (1998)

24. Halpern, J.: Reasoning About Uncertainty. MIT, Cambridge, MA (2005)

25. Hartmann, S., Link, S.: On a problem of Fagin concerning multivalued dependencies in relational databases. Theor. Comput. Sci. **353**(1–3), 53–62 (2006)

26. Hartmann, S., Link, S.: Characterising nested database dependencies by fragments of propositional logic. Ann. Pure Appl. Log. **152**(1–3), 84–106 (2008)

27. Hartmann, S., Link, S.: When data dependencies over SQL tables meet the Logics of Paradox and S-3. In: Proceedings of the 29th ACM SIGMOD-SIGART-SIGACT Symposium on Principles of Database Systems (PoDS), pp. 317–326. ACM, Indianapolis, IN (2010)

28. Hartmann, S., Link, S.: The implication problem of data dependencies over SQL table definitions: axiomatic, algorithmic and logical characterizations. ACM Trans. Database Syst. **37**(2), pp. 13:1–13:40 (2012)

29. Herrmann, C.: On the undecidability of implications between embedded multivalued database dependencies. Inf. Comput. **122**(2), 221–235 (1995)

30. Herrmann, C.: Corrigendum to "on the undecidability of implications between embedded multivalued database dependencies". Inf. Comput. **204**(12), 1847–1851 (2006)

31. Hartmann, S., Kirchberg, M., Link, S.: Design by example for SQL table definitions with functional dependencies. VLDB J. **21**(1), 121–144 (2012)

32. Köhler, H., Link, S.: Logics for approximating implication problems of saturated conditional independence. In: Fermé, E., Leite, J. (eds.) 14th European Conference on Logics in Artificial Intelligence JELIA 2014, Funchal, Madeira, September 24–26, 2014. Proceedings. Lecture Notes in Computer Science, vol. 8761, pp. 224–238. Springer, Berlin (2014)

33. Köhler, H., Link, S.: Saturated conditional independence with fixed and undetermined sets of incomplete random variables. In: Zhang, M., Tian, J. (eds.) 30th International Conference on Uncertainty in Artificial Intelligence, UAI 2014, Quebec City, Quebec, July 23–27, 2014. Proceedings, pp. 224–238. AUAI Press, Edinburgh (2014)

34. Kontinen, J., Link, S., Väänänen, J.A.: Independence in database relations. In: Proceedings of the 20th International Workshop on Logic, Language, Information, and Computation (WoLLIC). Lecture Notes in Computer Science, vol. 8071, pp. 179–193. Springer, Berlin (2013)

35. Le, V.B.T., Link, S., Ferrarotti, F.: Effective recognition and visualization of semantic requirements by perfect SQL samples. In: Proceedings of the 32th International Conference on Conceptual Modeling (ER). Lecture Notes in Computer Science, vol. 8217, pp. 227–240. Springer, Berlin (2013)

36. Levesque, H.: A knowledge-level account of abduction. In: Proceedings of the 11th International Joint Conference on Artificial Intelligence (IJCAI), pp. 1061–1067. Morgan Kaufmann, Detroit (1989)

37. Lien, E.: On the equivalence of database models. J. ACM **29**(2), 333–362 (1982)

38. Link, S.: Consistency enforcement in databases. In: Bertossi, L.E., Katona, G.O.H., Schewe, K., Thalheim, B. (eds.) Semantics in Databases, Second International Workshop, Dagstuhl Castle, Germany, January 7–12, 2001, Revised Papers, Lecture Notes in Computer Science, vol. 2582, pp. 139–159. Springer, Berlin (2003)

39. Link, S.: Charting the completeness frontier of inference systems for multivalued dependencies. Acta Inf. **45**(7–8), 565–591 (2008)

40. Link, S.: Characterizations of multivalued dependency implication over undetermined universes. J. Comput. Syst. Sci. **78**(4), 1026–1044 (2012)

41. Link, S.: Propositional reasoning about saturated conditional probabilistic independence. In: Proceedings of the 19th International Workshop on Logic, Language, Information and Computation (WoLLIC). Lecture Notes in Computer Science, vol. 7456, pp. 257–267. Springer, Buenos Aires (2012)

42. Link, S.: Approximate reasoning about generalized conditional independence with complete random variables. In: Proceedings of the 7th International Conference on Scalable Uncertainty Management (SUM). Lecture Notes in Computer Science, vol. 8078, pp. 269–282. Springer, Berlin (2013)

43. Link, S.: Reasoning about saturated conditional independence under uncertainty: axioms, algorithms, and Levesque's situations to the rescue. In: Proceedings of the Twenty-Seventh AAAI Conference on Artificial Intelligence (AAAI). AAAI Press, Menlo Park, CA (2013)

44. Link, S.: Sound approximate reasoning about saturated conditional probabilistic independence under controlled uncertainty. J. Appl. Log. **11**(3), 309–327 (2013)

45. Lohmann, P., Vollmer, H.: Complexity results for modal dependence logic. Stud. Logica **101**(2), 343–366 (2013)

46. Naumov, P., Nicholls, B.: R.E. axiomatization of conditional independence. In: Proceedings of the 14th Conference on Theoretical Aspects of Rationality and Knowledge (TARK), pp. 148–155 (2013)

47. Niepert, M., Van Gucht, D., Gyssens, M.: Logical and algorithmic properties of stable conditional independence. Int. J. Approx. Reason. **51**(5), 531–543 (2010)

48. Niepert, M., Gyssens, M., Sayrafi, B., Gucht, D.V.: On the conditional independence implication problem: a lattice-theoretic approach. Artif. Intell. **202**, 29–51 (2013)

49. Parker, D.S. Jr., Parsaye-Ghomi, K.: Inferences involving embedded multivalued dependencies and transitive dependencies. In: Proceedings of the International Conference on Management of Data (SIGMOD), pp. 52–57. ACM, New York (1980)
50. Pearl, J.: Probabilistic Reasoning in Intelligent Systems: Networks of Plausible Inference. Morgan Kaufmann, San Francisco, CA (1988)
51. Sagiv, Y.: An algorithm for inferring multivalued dependencies with an application to propositional logic. J. ACM 27(2), 250–262 (1980)
52. Sagiv, Y., Delobel, C., Parker, D.S. Jr., Fagin, R.: An equivalence between relational database dependencies and a fragment of propositional logic. J. ACM 28(3), 435–453 (1981)
53. Sagiv, Y., Delobel, C., Parker, D.S. Jr., Fagin, R.: Correction to "An equivalence between relational database dependencies and a fragment of propositional logic". J. ACM 34(4), 1016–1018 (1987)
54. Schaerf, M., Cadoli, M.: Tractable reasoning via approximation. Artif. Intell. 74, 249–310 (1995)
55. Stefanini, F.: Graphical models for eliciting structural information. In: Classification and Data Mining, pp. 139–146. Springer, Berlin (2013)
56. Studený, M.: Conditional independence relations have no finite complete characterization. In: Kubik, S., Visek, J. (eds.) Transactions of the 11th Prague Conference on Information Theory, Statistical Decision Functions and Random Processes, pp. 377–396. Kluwer, Dordrecht (1992)
57. Thalheim, B.: Dependencies in Relational Databases. Teubner, Stuttgart (1991)
58. Väänänen, J.A.: Dependence Logic - A New Approach to Independence Friendly Logic. London Mathematical Society student texts, vol. 70. Cambridge University Press, Cambridge (2007)
59. Väänänen, J.A., Hodges, W.: Dependence of variables construed as an atomic formula. Ann. Pure Appl. Log. 161(6), 817–828 (2010)
60. Vincent, M.: Semantic foundations of 4NF in relational database design. Acta Inf. 36(3), 173–213 (1999)
61. Wong, S., Butz, C., Wu, D.: On the implication problem for probabilistic conditional independency. Trans. Syst. Man Cybern. Part A Syst. Humans 30(6), 785–805 (2000)
62. Wu, M.: The practical need for fourth normal form. In: Proceedings of the Twenty-third Technical Symposium on Computer Science Education (SIGCSE), pp. 19–23. ACM, New York (1992)
63. Zaniolo, C.: Database relations with null values. J. Comput. Syst. Sci. 28(1), 142–166 (1984)

Context-Specific and Local Independence in Markovian Dependence Structures

Henrik Nyman, Johan Pensar, and Jukka Corander

Abstract Directed acyclic graphs (DAGs) have been established as one of the primary tools for characterizing dependencies and causality among variables in multivariate systems. However, it has also been recognized that DAGs may hide more nuanced forms of independence that are important for interpretation and operational efficiency of the dependence models. Such independencies are typically context-specific, meaning that a variable may lose its connection to another variable in a particular context determined by some other set of variables. Here we review context-specific independence in different classes of Markovian probability models both for static and spatially or temporally organized variables, including Bayesian networks, Markov networks, and higher-order Markov chains. The generality of the context-specific independence as a concept may spawn new ways to characterize dependence systems also beyond these traditional models, for example, in dependence logic.

1 Introduction

Markovian assumptions about conditional independence between stochastic variables are ubiquitous throughout science and technology, in principle they are found wherever statistical models are being used. Such assumptions generally make models more stable and tractable for statistical inference and the use of conditional independence as a concept is often well-founded by the existing knowledge about the phenomenon for which a model is being built. Here we consider three widely used classes of statistical models where Markovian properties are a central feature of the model. The first two are directed and undirected graphical models for finite sets of variables, known as Bayesian networks and Markov networks, respectively. The third class is Markov chains, which are used to model stochastic variables organized according to a time or spatial indexing. In particular, our interest is targeted towards

H. Nyman (✉) • J. Pensar
Department of Mathematics and Statistics, Åbo Akademi University, Turku, Finland
e-mail: hennyman@abo.fi; jopensar@abo.fi

J. Corander
Department of Mathematics and Statistics, University of Helsinki, Helsinki, Finland
e-mail: jukka.corander@helsinki.fi

© Springer International Publishing Switzerland 2016
S. Abramsky et al. (eds.), *Dependence Logic*, DOI 10.1007/978-3-319-31803-5_10

existence of independencies in any such systems which are hidden by the basic Markov assumptions, but which can be characterized using a common notion of context-specific independence (CSI).

Directed acyclic graphs (DAGs) quickly established their position as the basic workhorse for characterizing dependence in multivariate systems after pioneering works in the 1980s [23]. Such models are often cited as Bayesian networks, albeit the statistical inference related to structural learning and parameter estimation for such models need not be strictly Bayesian, i.e., be based on formal use of probabilistic prior statements about the unknowns present in the models. For a comprehensive discussion about Bayesian networks see the books by Koller and Friedman, as well as Koski and Noble [17, 19]. CSI in Bayesian networks has been introduced at various levels of generality, each of which has its strengths and limitations [6, 12, 24, 26]. We will review and illustrate several aspects related to these different types of CSI assumptions.

Markov networks, also known as Markov random field models, represent a class of undirected graphical models [20] for multivariate systems where directionality of dependence among variables is not encoded. Such models are widely used, for example, in applications related to spatial data and image analysis. The notion of CSI is much less widely recognized for Markov networks compared with Bayesian networks [9, 21]. Here we review undirected graphical models incorporating CSI and consider how they differ from the directed models embedding similar local independence structures.

Markov chains represent the backbone of models for time-ordered stochastic variables, originally even giving rise to the very notion of a Markov property. Higher-order Markov chains are versatile models for many applications, examples include modeling natural languages and DNA sequences. However, their rigid and parametric rich structure is challenging for statistical inference as they do not capture specific instances of independence. This led to the recognition of the importance of CSI and the development of variable-order and variable-length Markov chains (VLMC) [1, 4, 16, 28, 32]. Recently, a generalization of the VLMC model class was introduced to allow for sparsity that can extend beyond the basic form of CSI. We review the basic concepts related to CSI in Markov chain models and how they may be generalized.

This article is structured as follows. In the next section we introduce Bayesian networks and the concept of CSI that allows generalization of such directed models. Sections 3 and 4 present the use of CSI for Markov networks and Markov chains, respectively. The last section summarizes some of our observations and discusses possibilities for further research in this area.

2   Bayesian networks

For a set of stochastic variables, $X_\Delta = (X_1, X_2, \ldots, X_n)$, a Bayesian network is defined by a pair (G, P_Δ) where G is a directed acyclic graph (DAG) and P_Δ is a probability distribution over the variables X_Δ satisfying a set of marginal and conditional (in)dependencies induced by G. In turn, the graph G consists of a pair (Δ, E), where Δ denotes a set of nodes and E a set of edges such that $E \subseteq (\Delta \times \Delta)$. Each node $j \in \Delta$ corresponds to a variable $X_j \in X_\Delta$ and each edge represents a direct dependence between two variables. For an in-depth review of the theory surrounding Bayesian networks see, for instance, [17] and [19].

The parents Π_j of a node j in a graph are defined as $\Pi_j = \{i : (i,j) \in E\}$. We use $\mathscr{X}_A$, where $A \subseteq \Delta$, to denote the outcome space of the set of variables X_A and x_A to denote a specific outcome, i.e., $x_A \in \mathscr{X}_A$. Following the local directed Markov property [17, 19, 23], each variable X_j is conditionally independent of variables corresponding to non-descendants of node j given X_{Π_j}. Consequently, the joint probability distribution of a Bayesian network (G, P_Δ) can be factorized according to

$$P(X_\Delta = x_\Delta) = P(X_1 = x_1, X_1 = x_2, \ldots, X_n = x_n) = \prod_{j=1}^{n} P(X_j = x_j | X_{\Pi_j} = x_{\Pi_j}),$$

(1)

for any $x_\Delta \in \mathscr{X}_\Delta$. Due to this factorization it is natural to consider P_Δ as being constructed from a series of conditional probability distributions. A set of conditional probability distributions for a variable X_j can readily be presented using a conditional probability table (CPT).

To illustrate the use of CPTs we introduce a classic example that we will return to later in this section. Consider a system containing the four binary variables X_A, X_B, X_E, and X_S corresponding to the events "alarm armed", "burglary", "earthquake", and "loud alarm sound", respectively. The value 1 for these variables indicates an affirmative response while 0 indicates a negative response. The graph used to illustrate the dependence between these variables is depicted in Figure 1. The nodes A, B, and E are all parents of S, i.e., $\Pi_S = \{A, B, E\}$. Each row of the CPT corresponding to X_S, displayed in Table 1, determines the probability $P(X_S = 1 | X_A = x_A, X_B = x_B, X_E = x_E)$ for a specific set of values x_A, x_B, and x_E. As seen

Fig. 1 Graph associated with burglar alarm example. The nodes signify different events, A - "alarm armed", B - "burglary", E - "earthquake", and S - "loud alarm sound"

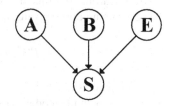

Table 1 CPT corresponding to X_S in burglar alarm example

X_A	X_B	X_E	$P(X_S = 1 | X_A, X_B, X_E)$	X_A	X_B	X_E	$P(X_S = 1 | X_A, X_B, X_E)$
0	0	0	0.00	1	0	0	0.05
0	0	1	0.00	1	0	1	0.20
0	1	0	0.00	1	1	0	0.95
0	1	1	0.00	1	1	1	0.95

from the CPT, some of the listed conditional probability distributions are identical, meaning that some configurations of the outcomes of the parent variables affect X_S identically. These types of "regularities", which cannot be explicitly modeled using ordinary Bayesian networks, serve as the motivation for formalizing the notion of CSI in Bayesian networks [2].

Definition 1 (CSI). Consider four disjoint sets of variables (X_A, X_B, X_C, X_D) in X_Δ. The variables X_A and X_B are defined as contextually independent given $X_C = x_C$ and X_D, denoted by $X_A \perp X_B | X_C = x_C, X_D$, if

$$P(X_A = x_A | X_B = x_B, X_C = x_C, X_D = x_D) = P(X_A = x_A | X_C = x_C, X_D = x_D),$$

for all possible values x_A, x_B, and x_D for which $P(X_B = x_B, X_C = x_C, X_D = x_D) > 0$.

When studying CPTs we only consider a variable X_j and its corresponding set of parents X_{Π_j}. In this situation a CSI between X_j and $X_B \subset X_{\Pi_j}$ occurs given the context $X_{\Pi_j \setminus B} = x_{\Pi_j \setminus B}$, if

$$P(X_j = x_j | X_B = x_B, X_{\Pi_j \setminus B} = x_{\Pi_j \setminus B}) = P(X_j = x_j | X_{\Pi_j \setminus B} = x_{\Pi_j \setminus B}),$$

for all possible values x_j and x_B for which $P(X_B = x_B, X_{\Pi_j \setminus B} = x_{\Pi_j \setminus B}) > 0$. This type of CSI which can be verified by a direct examination of the CPT is often referred to as a local CSI. From the CPT in Table 1 it is possible to deduce the following (local) CSIs:

$$X_S \perp \{X_B, X_E\} | X_A = 0 \quad \text{and} \quad X_S \perp X_E | X_A = 1, X_B = 1.$$

Before we look at different models capable of capturing CSIs, we consider a formula first introduced in [15] which is used to calculate the marginal likelihood of an observed dataset given a DAG. Using a Bayesian paradigm to ascertain the optimal dependence structure, in the form of a DAG G, for a set of stochastic variables given a dataset $\mathbf{X}$, it is necessary to consider the posterior distribution $P(G|\mathbf{X})$. Throughout this article we assume that the dataset is complete, i.e., the data contain no missing values. The posterior distribution can be determined as

$$P(G|\mathbf{X}) = \frac{P(\mathbf{X}|G)P(G)}{\sum_{G' \in \mathscr{G}} P(\mathbf{X}|G')P(G')},$$

where $\mathscr{G}$ denotes the model space containing all possible graphs and $P(G)$ is a prior distribution over the model space. In [15] a formula based on earlier work in [7] and [5] is introduced for calculating $P(\mathbf{X}|G)$. This formula utilizes the properties of the Dirichlet distribution, which is a conjugate prior for the multinomial distribution, and can be written in closed form as

$$P(\mathbf{X}|G) = \prod_{j=1}^{n} \prod_{l=1}^{k_j} \frac{\Gamma(\sum_{i=1}^{r_j} \alpha_{jil})}{\Gamma(n_{jl} + \sum_{i=1}^{r_j} \alpha_{jil})} \prod_{i=1}^{r_j} \frac{\Gamma(n_{jil} + \alpha_{jil})}{\Gamma(\alpha_{jil})}. \tag{2}$$

The function $\Gamma(\cdot)$ is the Gamma function which is defined as $\Gamma(x) = \int_0^\infty t^{x-1}e^{-t}dt$ and satisfies the properties $\Gamma(1) = 1$ and $\Gamma(x+1) = x\Gamma(x)$. As before, n denotes the number of variables in the system, k_j is the number of possible outcomes of the variables X_{Π_j}, and r_j is the number of possible outcomes of X_j. The number of times that the variables X_{Π_j} assume the value l in the data is denoted by n_{jl}. Similarly, n_{jil} denotes the number of times that X_{Π_j} assume the value l and X_j the value i. The values α_{jil} are the hyperparameters used in a prior Dirichlet distribution. For a discussion concerning the choice of these hyperparameters, see [6, 12, 15], and [24].

Next we will consider how CSIs and other local independencies, corresponding to a merging of outcomes in a CPT, can be implemented in Bayesian networks. The sources of the considered methods provide a revised version of (2) allowing for the calculation of the posterior model probabilities enabling learning of the model structure from data.

2.1 Bayesian networks with structured CPTs

2.1.1 Default tables and decision trees

In [12] two methods for learning local structures in Bayesian networks are considered, this paper also pioneered the introduction of a revised version of (2) for use with CSI models. The first method considered uses so-called default tables to merge rows in a CPT. The structure of a default table is very similar to a CPT, with the difference that in a default table not all of the possible outcomes for the parents of a variable are listed. Instead, the table provides a default value for the outcomes not explicitly listed. For instance, in the previously considered burglar alarm example all instances where $X_A = 0$, corresponding to a state where the burglar alarm is not armed, result in the probability 0.00 that the alarm will sound. Thus, it is possible to merge all such outcomes into a default outcome, as shown in Table 2. The resulting amended CPT for X_S requires only five free parameters, compared to eight free parameters for the original CPT.

Using a default table it is, however, not possible to also merge the outcomes $(X_A = 1, X_B = 1, X_E = 0)$ and $(X_A = 1, X_B = 1, X_E = 1)$, which induce the same probability $P(X_S = 1 | X_A, X_B, X_E) = 0.95$. For this reason [12], as well as [2], considers so-called decision trees, see [3] and [27]. A tree is a DAG containing exactly one root node. Each node in a tree has exactly one parent, except for the

Table 2 Default table corresponding to X_S in burglar alarm example

X_A	X_B	X_E	$P(X_S = 1 | X_A, X_B, X_E)$
1	0	0	0.05
1	0	1	0.20
1	1	0	0.95
1	1	1	0.95
	*		0.00

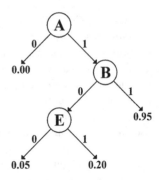

Fig. 2 Decision tree used in burglar alarm example. The decision tree only contains four leaf nodes, corresponding to four unique conditional distributions for X_S

root node which has no parents. In a decision tree, a leaf node is defined as a node that is not the parent of any other node. In terms of CPT representation, each leaf node is associated with a conditional probability distribution for the variable under consideration. The internal nodes and edges of the decision tree encode the information on how to choose among the leaf nodes. The tree in Figure 2 is the decision tree used for X_S in the burglar alarm example. The root node corresponds to X_A and the outgoing edges from node A corresponds to the different outcomes of X_A. In the context $X_A = 0$, X_S is independent of X_B and X_E, resulting in the outcome $X_A = 0$ leading directly to a leaf node. In the context $X_A = 1$, $X_B = 1$, X_S is independent of X_E, resulting in another leaf node. Finally, the two different outcomes of X_E also result in two different leaf nodes. For this example the tree-based CPT only requires four free parameters.

In [12] equation (2) is slightly modified to enable the calculation of the marginal likelihood of a dataset when the local structure in a DAG is defined through a default table or a decision tree. The modification consists of replacing k_j with q_j in the formula, where q_j equals the number of conditional distributions for variable X_j defined via the default table or decision tree. The term n_{jl} then no longer indicates the number of times in the data that the variables X_{Π_j} assume the value l, but rather the number of times these variables assume an outcome included in a group of outcomes that have been merged together. The same adjustment is made to the definition of n_{jil}.

2.1.2 Decision graphs

We start this section by considering a Bayesian network over four binary variables (X_1, X_2, X_3, X_4) where the parents of node 4 consist of the other three nodes, i.e., $\Pi_4 = \{1, 2, 3\}$. The CPT for the variable X_4 is given in Table 3. From the CPT it is possible to deduce the following CSIs:

Table 3 CPT corresponding to X_4 in decision graph example

X_1	X_2	X_3	$P(X_4 = 1\|X_1, X_2, X_3)$	X_1	X_2	X_3	$P(X_4 = 1\|X_1, X_2, X_3)$
0	0	0	p_1	1	0	0	p_3
0	0	1	p_1	1	0	1	p_4
0	1	0	p_2	1	1	0	p_2
0	1	1	p_2	1	1	1	p_5

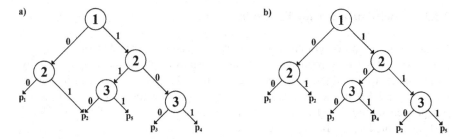

Fig. 3 Different models used to represent the CPT in Table 3. In a) a decision graph and in b) a decision tree

$$X_1 \perp X_4 | X_2 = 1, X_3 = 0,$$
$$X_3 \perp X_4 | X_1 = 0, X_2 = 0, \qquad (3)$$
$$X_3 \perp X_4 | X_1 = 0, X_2 = 1.$$

The regularities in this CPT cannot be compactly represented using either default tables or decision trees. In [6] a generalization of decision trees, termed decision graphs, is used to expand the set of CPTs that can be represented. Decision graphs have largely the same structure as decision trees with the difference that a node in a decision graph may have multiple parents. Using the decision graph shown in Figure 3a, containing only five leaf nodes, the CPT is compactly represented. If we instead were to use a decision tree (Figure 3b) the minimum amount of required leaf nodes would be six, and p_2 would feature in two leaf nodes.

Decision graphs can be used to represent a wider range of CPTs than decision trees, as any combination of CSIs between the variable under consideration and one of its parents can be captured using a decision graph. However, decision graphs are even more general than that, as also merging of rows in a CPT that do not correspond to a CSI may be represented. Consider a situation where node 3 has two parents, corresponding to the binary variables X_1 and X_2, and the outcomes $(X_1 = 0, X_2 = 0)$ and $(X_1 = 1, X_2 = 1)$ induce identical conditional distributions for X_3. While this case can readily be presented using a decision graph, or a default table, it does not correspond to a CSI.

Fig. 4 Labeled directed
acyclic graph corresponding
to the CPT in Table 3

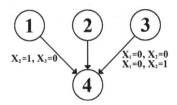

2.1.3 Labeled directed acyclic graphs

In [24] a new type of graphical representation, termed labeled directed acyclic
graphs (LDAGs), is introduced. An LDAG displays CSI in the graph itself by
adding labels to the edges. More specifically, a CSI between X_i and X_j is displayed
by adding a label to the edge between i and j detailing for which outcomes the
independence holds. An LDAG-based model can represent any collection of local
CSIs associated with a Bayesian network.

As an example of an LDAG consider again the CPT in Table 3 inducing the CSIs
listed in (3). The LDAG corresponding to this CPT is shown in Figure 4. LDAGs can
be considered a middle ground between decision trees and decision graphs. LDAGs
are more general than decision trees as demonstrated by the above example, but
not as general as decision graphs, since LDAGs operate within the scope of CSI
restrictions. In order to perform model learning the modified version of (2) provided
in [12] is applied to decision graphs and LDAGs in [6] and [24], respectively.

2.2 The role of CSI in model learning and probabilistic inference

The effect of including structured CPTs in the process of learning models from
data has been investigated by several authors. Here we give a short summary of
the key results from the work in [6, 12, 24]. The posterior probability is generally
the preferred score function, as the marginal likelihood $P(\mathbf{X}|G)$ can be calculated
analytically using (2) for any Bayesian network with a compact CPT representation.

In [12] default tables and decision trees are used to demonstrate the positive
effects local representation may have on the global model learning process. To
quantify the generalization error of the different procedures, Kullback-Leibler
divergence (or entropy distance) is used to measure the distance between the true
generating distribution and the approximate induced distribution. It is concluded
that the structured CPTs induce more accurate models for two reasons. Firstly, the
need for fewer parameters facilitates the parameter estimation process by making
it more stable. Secondly, the added flexibility for including larger families of
distributions without an exponential penalty results in networks that better emulate
the (in)dependencies in the real distribution.

In [6] decision graphs are used to represent the CPTs during the model learning process. In the experimental section various search operators are applied to identify both tree- and graph-based local structures. Since the main goal of the experiments is to find the model that maximizes the posterior probability, the marginal likelihood is used to assess the fit of the learned structures. It is shown that it is possible to identify graph-based models with higher marginal likelihood than tree-based models. This is a rather natural result considering that the class of decision graphs is a generalization of the class of decision trees.

In [24] compact CPTs are achieved by merging rows according to CSI-based rules. Similar as in [12], the identified models are evaluated by the Kullback-Leibler divergence. The experimental results agree with the previous research in the sense that by including CSI and structured CPTs in the learning procedure the model quality can be improved. However, it is also shown that models optimized with respect to the marginal likelihood alone may suffer from poor out-of-sample performance due to overfitting. To improve the model quality an adjustable prior is designed to counteract this phenomenon by penalizing overly specific CSI-structures. A cross-validation based scheme is used to adjust the prior appropriately.

In addition to improving the model learning process, CSI can also be exploited to make probabilistic inference more efficient. In the context of Bayesian networks, probabilistic inference refers to the process of computing the posterior probability distribution for a list of query variables given some variables with either fixed values or uncertain outcomes (hard or soft evidence, respectively).

The key to efficient probabilistic inference lies in the concept of factorization of the joint distribution. Incorporating CSIs into the models allows a further decomposition of (1) into a finer-grained factorization which can improve the inference. In [2] it is investigated how decision trees can be used to improve various inference algorithms. As a consequence of the replication problem it is concluded in [25] that rule-based versions may be more efficient than tree-based. In [34] a more general analysis of the computational advantages that CSIs can deliver is presented.

Finally, in [26] the concept of contextual belief networks is introduced as a class of Bayesian networks for which the conditional probability distributions are associated with parent contexts rather than explicit parent configurations. Based on the introduced model class, the work in [26] further improves the efficiency of the method presented in [25] by using a combination of contexts and tables.

3 Markov networks

Similar to a Bayesian network, a Markov network is a graphical model defined by a pair (G, P_Δ), the difference being that in a Markov network G is an undirected graph. In an undirected graph the set of edges are defined as $E \subseteq \{\Delta \times \Delta\}$ and $\{i, j\} \in E \Leftrightarrow \{j, i\} \in E$. In [9], and later in [21], Markov networks are generalized to encompass CSI, resulting in a novel model class termed as stratified graphical

Fig. 5 Stratified graphical
model encompassing the CSI
$X_2 \perp X_3 | X_1 = 1$

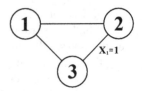

models (SGMs). In an SGM two nodes, i and j, are defined as adjacent if $\{i, j\} \in E$.
For SGMs the set of conditioning variables in a CSI statement between X_i and X_j
is constituted by the set of variables corresponding to the nodes adjacent to both i
and j, denoted by $L_{\{i,j\}}$. In order to ensure the validity of this definition of CSI, G is
required to be a chordal graph.

Just as for LDAGs, CSIs can be represented in the graph structure by adding
labels to the edges of G. As an example consider the graph in Figure 5. This
graph induces a dependence structure where all three variables are marginally and
conditionally dependent of each other. However, given the context $X_1 = 1$, X_2 and
X_3 are independent of each other, i.e., $X_2 \perp X_3 | X_1 = 1$.

In [21] a class of decomposable SGMs, which restrict the set of edges to which
labels can be added, are introduced. The primary reason for imposing the restrictions
is that the revised version of (2) can be used to calculate the marginal likelihood of a
dataset given a decomposable SGM. The restrictions also allow for a factorization of
the joint distribution according to the cliques and separators found in G, analogous
to the factorization used for undirected chordal graphs [13, 19].

Markov networks are closely related to graphical log-linear models. A log-linear
parameterization [20, 33] is defined by a parameter vector ϕ, such that the joint
distribution of the variables X_Δ can be written as

$$\log P(X_\Delta = x_\Delta) = \sum_{A \subseteq \Delta} \phi_{A(x_A)},$$

where x_A denotes the marginal outcome of variables X_A in the outcome x_Δ. For the
log-linear parameterization we have the restriction that if $x_j = 0$ for any $j \in A$,
then $\phi_{A(x_A)} = 0$ [33]. One of the reasons for using a log-linear parameterization
is that marginal and conditional independencies are expressed very succinctly in
the parameters. It holds for graphical log-linear models that if the edge $\{i, j\}$ is not
present in G, then all parameters $\phi_{A(x_A)}$, where $\{i, j\} \subseteq A$, are equal to zero [33].
The restrictions imposed to the log-linear parameters by a CSI are also clearly
defined [9]. Consider the CSI $X_i \perp X_j \mid X_{L_{\{i,j\}}} = x_{L_{\{i,j\}}}$. Let $A \subseteq L_{\{i,j\}} \cup \{i, j\}$
denote the set of variables containing the pair $\{i, j\}$ and the set of all variables with
non-zero values in $x_{L_{\{i,j\}}}$. The restrictions imposed on the log-linear parameters are
then of the form $\sum_{B \subseteq A} \phi_{B(x_B)} = 0$, where $\{i, j\} \subseteq B$.

In [22] it is demonstrated that some SGMs belong to the class of non-hierarchical
models [33], a class of models that has been given only restricted attention due to
its superficial lack of interpretability. Additionally, a cyclical projection algorithm,
based on the theory in [10] and similar to the methods used in [8, 29] for non-chordal

graphs, is developed which enables the evaluation of the maximum likelihood estimate of a distribution given a non-decomposable SGM. This, in turn, enables model learning also outside the class of decomposable SGMs when the BIC score [30] is used to determine model fit.

4 Markov chains

A Markov chain of order k is a sequence of stochastic variables $(X_0, X_1, \ldots)$, such that the outcome space of X_i is $\mathscr{X}$, for $i = 0, 1, \ldots$ and

$$P(X_n = x_n | X_{n-1} = x_{n-1}, X_{n-2} = x_{n-2}, \ldots, X_0 = x_0) =$$

$$P(X_n = x_n | X_{n-1} = x_{n-1}, X_{n-2} = x_{n-2}, \ldots, X_{n-k} = x_{n-k}),$$

for all $n \geq k$ and $x_i \in \mathscr{X}$, $i = 0, 1, \ldots, n$. The notation X_i^j will henceforth be used to denote the sequence of variables $(X_i, X_{i+1}, \ldots, X_j)$ and similarly x_i^j will be used to denote a specific outcome for these variables. A Markov chain of order k is said to be time-homogeneous, if

$$P(X_n = x_n | X_{n-k}^{n-1} = x_{n-k}^{n-1}) = P(X_t = x_n | X_{t-k}^{t-1} = x_{n-k}^{n-1}),$$

for all n, $t \geq k$ and possible values $x_i \in \mathscr{X}$, $i = n - k, \ldots, n$. There exist several approaches to compress the history of a Markov chain, corresponding to CSI or other methods for merging parent outcomes, see, for instance, [1, 4, 16, 28, 32].

In [4] a class of Markov chains called variable-length Markov chains (VLMCs) is introduced.

Definition 2 (VLMC). First, the function $l = l(x_0^{n-1})$ is defined as

$$l(x_0^{n-1}) = \min\{m : P(X_n = x_n | X_0^{n-1} = x_0^{n-1}) = P(X_n = x_n | X_{n-m}^{n-1} = x_{n-m}^{n-1})\},$$

for all $x_n \in \mathscr{X}$. Next, the context function c is defined as $c(x_0^{n-1}) = x_{n-l}^{n-1}$. For any value n, $c(x_0^{n-1})$ is called the context of variable X_n. A VLMC is defined by the pair (P, c) where c is a context function and P is a probability distribution determining the probability $P(X_n = x_n | X_{n-l}^{n-1} = c(x_0^{n-1}))$, for all $x_i \in \mathscr{X}$, $i = n - l, \ldots, n$.

If $0 < k < \infty$ is the smallest integer such that $l(x_0^\infty) \leq k$, for all possible outcomes x_0^∞, then the VLMC is said to be of order k.

Contrary to ordinary kth order Markov chains, for a VLMC, like the name suggests, the length of the relevant history may vary. Comparing an ordinary Markov chain and a VLMC of order k, the context function can be seen as a function that merges the outcomes of the parents $(X_{n-1}, X_{n-2}, \ldots, X_{n-k})$ of the variable X_n. The context function can be represented using a decision tree, including some additional constraints, meaning that it induces CSIs between X_n and its parents.

Fig. 6 Decision tree
corresponding to the context
function in (4)

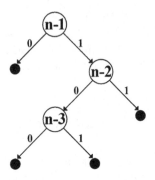

The additional constraints arise from the fact that if X_n is independent of X_{n-i} in
a context $(X_{n-1} = x_{n-1}, \ldots, X_{n-i+1} = x_{n-i+1})$ then X_n is also independent of all
variables X_{n-j}, where $j > i$, in the same context.

Consider a VLMC of order 3 where $\mathscr{X} = \{0, 1\}$, and the context function c is
defined as

$$
c(x_0^{n-1}) = \begin{cases}
0, & \text{if } x_{n-1} = 0, \\
1, 0, 0, & \text{if } x_{n-1} = 1, x_{n-2} = 0, x_{n-3} = 0, \\
1, 0, 1, & \text{if } x_{n-1} = 1, x_{n-2} = 0, x_{n-3} = 1, \\
1, 1, & \text{if } x_{n-1} = 1, x_{n-2} = 1.
\end{cases}
\tag{4}
$$

The context function induces the CSIs

$$
X_n \perp \{X_{n-2}, X_{n-3}\} | X_{n-1} = 0 \quad \text{and} \quad X_n \perp X_{n-3} | X_{n-1} = 1, X_{n-2} = 1,
$$

and can be presented using the decision tree in Figure 6.

In [16] a more general class of Markov chains, termed sparse Markov chains
(SMCs), are considered.

Definition 3 (SMC). Let X_0^{∞} be a time-homogeneous Markov chain of order k,
and $\mathscr{X}^k = \times_{i=1}^k \mathscr{X}$. Further, let $S = (s_1, \ldots, s_d)$ be a partition of $\mathscr{X}^k$, such that iff
$y_{n-k}^{n-1}, z_{n-k}^{n-1} \in s_c$ for some $s_c \in S$ it holds that

$$
P(X_n = x_n | X_{n-k}^{n-1} = y_{n-k}^{n-1}) = P(X_n = x_n | X_{n-k}^{n-1} = z_{n-k}^{n-1}),
$$

for all $x_n \in \mathscr{X}$.

This definition allows SMCs to be non-hierarchical in the sense that in some
contexts X_n may be independent of X_{n-i} given X_{n-j}, with $j > i$. The dependence
structures that can be presented using SMCs bear a strong resemblance to those that
can be presented using a decision graph. In fact, just as decision trees are used to
display the structure of a VLMC, a decision graph can be used for SMCs.

Table 4 CPT corresponding to the partition S defined in (5)

X_{n-1}	X_{n-2}	$P(X_n|X_{n-1}, X_{n-2})$	X_{n-1}	X_{n-2}	$P(X_n|X_{n-1}, X_{n-2})$	X_{n-1}	X_{n-2}	$P(X_n|X_{n-1}, X_{n-2})$
0	0	p_1	1	0	p_1	2	0	p_5
0	1	p_2	1	1	p_2	2	1	p_2
0	2	p_3	1	2	p_4	2	2	p_4

Fig. 7 Decision graph corresponding to the CPT in Table 4 and the partition S defined in (5)

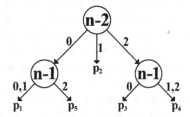

Until now we have operated under the assumption that the number of outgoing edges in an internal node in a decision tree or decision graph coincides with the number of possible outcomes of the corresponding variable. Removing this assumption, and letting an outgoing edge correspond to more than one outcome, allows for a new kind of restriction to the considered probability distribution, a restriction that falls outside the scope of CSI. Consider two variables X_1 and X_2 with the outcome space $\mathscr{X} = \{0, 1, 2\}$ and a conditional distribution of X_2 such that

$$P(X_2 = x_2|X_1 = 0) = P(X_2 = x_2|X_1 = 1),$$

for all $x_2 \in \mathscr{X}$ and

$$P(X_2 = x_2|X_1 = 0) \neq P(X_2 = x_2|X_1 = 2),$$

for some $x_2 \in \mathscr{X}$. These types of restriction can be portrayed in a CPT, and for SMCs in the partition S of $\mathscr{X}^k$. Consider an SMC where $\mathscr{X} = \{0, 1, 2\}$, and S is defined according to

$$S = \{\{00, 10\}, \{01, 11, 21\}, \{02\}, \{12, 22\}, \{20\}\}. \tag{5}$$

This non-hierarchical SMC corresponds to the CPT in Table 4 and can also be displayed using the decision graph in Figure 7.

5 Discussion

We have reviewed the concept of context-specific independence appearing in superficially different model classes all exploiting Markovian independence structures to enable a compact representation of the joint distribution in multivariate systems that can potentially be very high-dimensional. These include both static representations that are directed or undirected, and dynamic representations incorporating time-indexing of the variables. In general, the notion of context-specific independence allows one to capture local characteristics of multivariate distributions that may be hidden by the standard Markov assumptions. This is useful for many purposes and may lead to simpler model interpretation as well as more efficient, sparser parametrization of the distributions.

An unchartered territory for the theory of CSI is represented by both static and dynamic Markovian dependence models in continuous state spaces. Despite of corresponding to a considerably more challenging problem than representation of a context in discrete space, it is possible that sufficiently rigid forms of CSI do exist, such that they lead to identifiable and tractable model learning and parameter estimation. Similarly, it would be attractive to derive more general rules for the correspondence between non-hierarchical log-linear models and local restrictions to conditional probability distributions, upon the observation made in [22].

Another field where the introduction of CSI might lead to an improved understanding of the dependence structure is marked point processes. Marked point processes are generally used to model the occurrence of some specific events in a continuous time space. In [11], graphical models that can convey so-called local independencies in marked point processes are introduced. While these local independencies are not directly related to CSI, [11] shows that it can beneficial to use more refined dependence structure when considering marked point processes. CSI could possibly be used to further improve the notion of local independence.

Largely independently of the theory of graphical dependence models, dependence logic has emerged as a branch of mathematical logic [14, 18, 31]. An interesting target for future research would be to develop a notion of context-specific independence for dependence logic. This could also spawn further research directions through various generalizations into forms of independence that extend beyond context-specific independence which are not representable through ordinary factorization of conditional probabilities in accordance with basic Markov properties.

Acknowledgements H.N. and J.P. were funded by the Foundation of Åbo Akademi University, as part of the grant for the Center of Excellence in Optimization and Systems Engineering. J.P. was also funded by the Magnus Ehrnrooth foundation. J.C was funded by ERC grant 239784 and the COIN centre of excellence (grant 251170 from Academy of Finland).

References

1. Bacallado, S.: Bayesian analysis of variable-order, reversible Markov chains. Ann. Stat. **39**(2), 838–864 (2011)
2. Boutilier, C., Friedman, N., Goldszmidt, M., Koller, D.: Context-specific independence in Bayesian networks. In: Proceedings of the Twelfth Annual Conference on Uncertainty in Artificial Intelligence, pp. 115–123. Morgan Kaufmann, San Francisco, CA (1996)

3. Breiman, L., Friedman, J., Stone, C.J., Olshen, R.A.: Classification and regression trees. CRC, Wadsworth (1984)
4. Bühlmann, P., Wyner, A.J.: Variable length Markov chains. Ann. Stat. **27**(2), 480–513 (1999)
5. Buntine, W.: Theory refinement on Bayesian networks. In: Proceedings of the Seventh Conference on Uncertainty in Artificial Intelligence, pp. 52–60. Morgan Kaufmann, San Francisco, CA (1991)
6. Chickering, D.M., Heckerman, D., Meek, C.: A Bayesian approach to learning Bayesian networks with local structure. In: Proceedings of the Thirteenth Conference on Uncertainty in Artificial Intelligence, pp. 80–89. Morgan Kaufmann, San Francisco, CA (1997)
7. Cooper, G.F., Herskovits, E.: A Bayesian method for the induction of probabilistic networks from data. Mach. Learn. **9**(4), 309–347 (1992)
8. Corander, J.: Bayesian graphical model determination using decision theory. J. Multivar. Anal. **85**(2), 253–266 (2003)
9. Corander, J.: Labelled graphical models. Scand. J. Stat. **30**, 493–508 (2003)
10. Csiszár, I.: I-divergence geometry of probability distributions and minimization problems. Ann. Probab. **3**(1), 146–158 (1975)
11. Didelez, V.: Graphical models for marked point processes based on local independence. J. R. Stat. Soc. Ser. B (Stat Methodol.) **70**(1), 245–264 (2008)
12. Friedman, N., Goldszmidt, M.: Learning Bayesian networks with local structure. In: Proceedings of the Twelfth Annual Conference on Uncertainty in Artificial Intelligence, pp. 252–262. Morgan Kaufmann, San Francisco, CA (1996)
13. Golumbic, M.C.: Algorithmic Graph Theory and Perfect Graphs, 2nd edn. Elsevier, Amsterdam (2004)
14. Grädel, E., Väänänen, J.: Dependence and independence. Stud. Logica **101**(2), 300–410 (2012)
15. Heckerman, D., Geiger, D., Chickering, D.M.: Learning Bayesian networks: the combination of knowledge and statistical data. Mach. Learn. **20**(3), 197–243 (1995)
16. Jääskinen, V., Xiong, J., Corander, J., Koski, T.: Sparse Markov chains for sequence data. Scand. J. Stat. **41**, 639–655 (2013)
17. Koller, D., Friedman, N.: Probabilistic Graphical Models: Principles and Techniques. MIT, Cambridge (2009)
18. Kontinen, J., Väänänen, J.: Axiomatizing first-order consequences in dependence logic. Ann. Pure Appl. Logic **164**(11), 1101–1117 (2013)
19. Koski, T., Noble, J.: Bayesian Networks: An Introduction. Wiley, Chippenham (2009)
20. Lauritzen, S.L.: Graphical Models. Oxford University Press, Oxford (1996)
21. Nyman, H., Pensar, J., Koski, T., Corander, J.: Stratified graphical models - context-specific independence in graphical models. Bayesian Anal. **9**(4), 883–908 (2014)
22. Nyman, H., Pensar, J., Koski, T., Corander, J.: Context-specific independence in graphical log-linear models. Comput. Stat. (2015). doi:10.1007/s00180-015-0606-6
23. Pearl, J.: Probabilistic Reasoning in Intelligent Systems: Networks of Plausible Inference. Morgan Kaufmann, San Francisco, CA (1988)
24. Pensar, J., Nyman, H., Koski, T., Corander, J.: Labeled directed acyclic graphs: a generalization of context-specific independence in directed graphical models. Data Min. Knowl. Disc. **29**(2), 503–533 (2015)
25. Poole, D.: Probabilistic partial evaluation: exploiting rule structure in probabilistic inference. In: Proceedings of the Fifteenth International Joint Conference on Artificial Intelligence, pp. 1284–1291 (1997)
26. Poole, D., Zhang, N.: Exploiting contextual independence in probabilistic inference. J. Artif. Intell. Res. **18**, 263–313 (2003)
27. Quinlan, J.R., Rivest, R.L.: Inferring decision trees using the minimum description length principle. Inf. Comput. **80**, 227–248 (1989)
28. Rissanen, J.: A universal data compression system. IEEE Trans. Inf. Theory **29**(5), 656–664 (1983)
29. Rudas, T.: A new algorithm for the maximum likelihood estimation of graphical log-linear models. Comput. Stat. **13**, 529–537 (1998)

30. Schwarz, G.: Estimating the dimension of a model. Ann. Stat. **6**, 461–464 (1978)
31. Väänänen, J.: Dependence Logic. Cambridge University Press, Cambridge (2007)
32. Weinberger, M.J., Rissanen, J., Feder, M.: A universal finite memory source. IEEE Trans. Inf. Theory **41**(3), 643–652 (1995)
33. Whittaker, J.: Graphical Models in Applied Multivariate Statistics. Wiley, Chichester (1990)
34. Zhang, N., Poole, D.: On the role of context-specific independence in probabilistic inference. In: Proceedings of the Sixteenth International Joint Conference on Artificial Intelligence, pp. 1288–1293 (1999)

Dependence and Independence in Social Choice: Arrow's Theorem

Eric Pacuit and Fan Yang

Abstract One of the goals of social choice theory is to study the group decision methods that satisfy two types of desiderata. The first type ensures that the group decision depends in the right way on the voters' opinions. The second type ensures that the voters are free to express any opinion, as long as it is an admissible input to the group decision method. Impossibility theorems, such as Arrow's Theorem, point to an interesting tension between these two desiderata. In this paper, we argue that dependence and independence logic offer an interesting new perspective on this aspect of social choice theory. To that end, we develop a version of independence logic that can express Arrow's properties of preference aggregation functions. We then prove that Arrow's Theorem is derivable in a natural deduction system for the first-order consequences of our logic.

1 Introduction

The modern era in social choice theory started with Kenneth Arrow's groundbreaking *impossibility theorem* [3]. Arrow showed that there is no method that a group can use to rank a set of alternatives satisfying a minimal set of desirable properties. Much has been written about this theorem (see, for instance, [20, 30, 43]) and its implications for theories of democracy [9, 27, 36] and beyond [29, 31, 33, 44]. Social choice theory has since grown into a large and multi-faceted research area (see [26] for an overview). In this chapter, we focus on one type of theorem studied by social choice theorists: *axiomatic characterizations* of group decision methods. We will present a version of independence logic [14] that we use to formalize these theorems. This is not merely an exercise in applying a logical framework to a new area. We will argue that dependence and independence logic offers an interesting new perspective on the axiomatic characterization of group decision methods.

E. Pacuit (✉)
Department of Philosophy, University of Maryland, College Park, MD, USA
e-mail: epacuit@umd.edu

F. Yang
Department of Philosophy and Religious Studies, Utrecht University, Janskerkhof 13, 3512 BL
Utrecht, The Netherlands
e-mail: fan.yang.c@gmail.com

© Springer International Publishing Switzerland 2016
S. Abramsky et al. (eds.), *Dependence Logic*, DOI 10.1007/978-3-319-31803-5_11

One of the main goals of social choice theory is to identify principles of group decision making that ensure that group decisions depend *in the right way* on the voters' preferences.[1] That is, group decision methods should be designed in such a way that no individual voter should have any undue influence over the group decision. At the same time, it is important to devise a group decision method without placing any restrictions on the inputs. That is, group decision methods should be designed under the assumption that the voters' opinions are *independent*. From this perspective, the so-called impossibility theorems in social choice theory highlight an interesting conflict between dependence and independence.

This chapter is organized as follows. Section 2 briefly recounts the basic mathematical framework used in social choice theory. Sections 3 and 4 are extended discussions of the notions of dependence and independence found in social choice theory. In Section 5, we present a version of independence logic that we use to formalize Arrow's Theorem. We show in Section 5.3 that Arrow's Theorem can be derived in a natural deduction system for this logic [16, 25]. Section 6 contains some concluding remarks.

2   The Social Choice Framework

Let us start by recalling some notions concerning relations.

Definition 1. A *relation* R on X is a subset of $X \times X$. We write $a\,R\,b$ when $(a, b) \in R$ and write $a\,\not{R}\,b$ when $(a, b) \notin R$. We write $a\,R\,b\,R\,c$ when $a\,R\,b$, $b\,R\,c$, and $a\,R\,c$.

Definition 2. Let $R \subseteq X \times X$ be a relation. R is said to be

- *transitive* provided for all $a, b, c \in X$, if $a\,R\,b$ and $b\,R\,c$, then $a\,R\,c$;
- *complete* provided for all $a, b \in X$, either $a\,R\,b$ or $b\,R\,a$;
- *antisymmetric* provided for all $a, b \in X$, if $a\,R\,b$ and $b\,R\,a$, then $a = b$;
- *linear* provided that it is transitive, complete, and antisymmetric.

Throughout this chapter, we fix a set $V = \{x_1, x_2, x_3, \ldots, x_n\}$ of n *voters* (or *individuals*) and a (finite[2]) set X of *alternatives* (e.g., candidates, restaurants, social states, etc.). Each voter $x_i \in V$ is asked to rank the elements of X, where a **ranking** is a transitive and complete relation on X. Let $O(X)$ denote the set of all rankings of X. Each ranking $R \in O(X)$ is associated with two special subrelations: The **strict subrelation** defined as

$$P_R = \{(a, b) \in X \times X \mid a\,R\,b \text{ and } b\,\not{R}\,a\},$$

---

[1] In formal work on social choice theory, it is common to identify a voter's *preference* over a set of alternatives X with her *ranking* over the set of alternatives. In general, a ranking of the alternatives is only one way in which a voter may express her preference over the set of alternatives. Consult [17] for a discussion of the main philosophical issues here.

[2] For simplicity, we restrict attention to a finite set of alternatives. This restriction is not necessary for what follows, though it does have some implications on the design of the formal language used to describe a social choice model.

and the **indifference subrelation** defined as

$$I_R = \{(a, b) \in X \times X \mid a\,R\,b \text{ and } b\,R\,a\}.$$

Let $R \in O(X)$ represents voter x_i's ranking of X. If $a\,R\,b$, then we say that "x_i weakly prefers a to b"; if $a\,P_R\,b$, then we say that "x_i strictly prefers a to b"; if $a\,I_R\,b$, then we say that "x_i is indifferent between a and b." In case a voter is indifferent between two alternatives, we say that the two alternatives are *tied* in the voter's ranking. A **linear ranking** is a ranking that is a linear relation and let $L(X)$ denote the set of all linear rankings of X. Clearly, ties are not allowed in linear rankings.

A **profile** for the set of voters V is a sequence of (linear) rankings of X that assigns to each voter x_i a ranking R_i, denoted $\mathbf{R} = (R_1, \ldots, R_n)$. The set of all profiles of rankings for n voters is denoted $O(X)^n$ (similarly for $L(X)^n$). If $\mathbf{R} = (R_1, \ldots, R_n) \in O(X)^n$, then P_i denotes the strict subrelation P_{R_i} of R_i. Similarly, I_i denotes the indifference subrelation I_{R_i} of R_i. For a profile $\mathbf{R} = (R_1, \ldots, R_n) \in O(X)^n$, let $\mathbf{V_R}(a\,P\,b) = \{x_i \in V \mid a\,P_i\,b\}$ be the set of voters that rank a strictly above b (similarly for $\mathbf{V_R}(a\,I\,b)$ and $\mathbf{V_R}(a\,R\,b)$).

A group decision method associates an *outcome* ("the group decision") with each profile of *ballots*. Typically, the ballots are (linear) rankings of the alternatives. The social choice literature has largely focused on two types of outcomes. The first are (linear) rankings of X representing the overall group ranking of the alternatives. The second are non-empty subsets of X representing the "social choice" (or the "winning" alternatives). As we mentioned in Section 1, the starting assumption in the social choice literature is that group decisions should be *completely determined* by the voters' reported[3] rankings. This means that group decision methods should be represented by *functions* from sets of profiles to the possible outcomes. There are two types of functions corresponding to the two types of group outcomes:

- A **preference aggregation function** is a function $F : \mathscr{B} \to \mathscr{O}$, where $\mathscr{O}$ is a set of relations on X (typically, $\mathscr{O}$ is either $L(X)$ or $O(X)$) and $\mathscr{B} \subseteq L(X)^n$ or $\mathscr{B} \subseteq O(X)^n$.
- A **social choice function** is a function $F : \mathscr{B} \to \mathscr{O}$ where $\mathscr{O} = \wp(X) \setminus \{\emptyset\}$ and $\mathscr{B} \subseteq L(X)^n$ or $\mathscr{B} \subseteq O(X)^n$.

We illustrate the above definitions with the following examples. For each profile of linear rankings $\mathbf{P} = (P_1, \ldots, P_n)$, define a relation $R_{maj}^{\mathbf{P}}$ on the set X of candidates as follows: for all $a, b \in X$,

$$a\,R_{maj}^{\mathbf{P}}\,b \text{ iff } |\mathbf{V_P}(a\,P\,b)| > |\mathbf{V_P}(b\,P\,a)|.$$

The relation $R_{maj}^{\mathbf{P}}$ is called the *majority ordering*, ranking candidate a above candidate b provided more voters rank a above b than b above a. Note that

---

[3]In this article, we set aside any game-theoretic issues around whether voters have an incentive to report their *true* preferences.

$R^{\mathbf{P}}_{maj} \notin O(X)$, since $R^{\mathbf{P}}_{maj}$ is not necessarily transitive.[4] This is illustrated by the famous *Condorcet Paradox*: Consider a profile $\mathbf{P} = (P_1, P_2, P_3)$ for three voters $V = \{x_1, x_2, x_3\}$ and three candidates $X = \{a, b, c\}$. Suppose that ranking for voter x_1 is $a\ P_1\ b\ P_1\ c$; the ranking for voter x_2 is $b\ P_2\ c\ P_2\ a$; and the ranking for voter x_3 is $c\ P_3\ a\ P_3\ b$. Then, $\mathbf{V_P}(a\ P_1\ b) = \{x_1, x_3\}$, $\mathbf{V_P}(b\ P_1\ c) = \{x_1, x_2\}$, and $\mathbf{V_P}(a\ P_1\ c) = \{x_1\}$. Thus, $a\ R^{\mathbf{P}}_{maj}\ b$ and $b\ R^{\mathbf{P}}_{maj}\ c$, but it is not the case that $a\ R^{\mathbf{P}}_{maj}\ c$ (in fact, we have $c\ R^{\mathbf{P}}_{maj}\ a$ producing a cycle: $a\ R^{\mathbf{P}}_{maj}\ b\ R^{\mathbf{P}}_{maj}\ c\ R^{\mathbf{P}}_{maj}\ a$).

The **Borda ranking** is an example of a social ranking that is transitive. Let $\mathbf{P} = (P_1, \ldots, P_n)$ be an arbitrary profile of linear rankings of a k-element set X of candidates. For each voter x_i and each $m = 1, \ldots, k$, let $P_i(m)$ be the candidate ranked in the mth-position by the voter x_i. For example, if $a\ P_i\ b\ P_i\ c$, then $P_i(1) = a$, $P_i(2) = b$, and $P_i(3) = c$. For each $d \in X$, define the *Borda score* $BS(\mathbf{P}, d)$ as

$$BS(\mathbf{P}, d) = \sum_{m=1}^{k} (k - m) \cdot |\{x_i \in V \mid P_i(m) = d\}|.$$

Now, the **Borda ranking** $R^{\mathbf{P}}_b$ is defined as follows:

$$a\ R^{\mathbf{P}}_B\ b \quad \text{iff} \quad BS(\mathbf{P}, a) \geq BS(\mathbf{P}, b).$$

The function $F_B : L(X)^n \to O(X)$, defined as $F_B(\mathbf{P}) = R^{\mathbf{P}}_B$, is a preference aggregation function. An example of a social choice function is plurality rule: $F_{pl} : L(X)^n \to \wp(X) \setminus \{\emptyset\}$, defined as: for each $\mathbf{P} = (P_1, \ldots, P_n) \in L(X)^n$,

$$F_{pl}(\mathbf{P}) = \{c \in X : |\{x_i \mid P_i(1) = c\}| \geq |\{x_i \mid P_i(1) = b\}| \text{ for all } b \in X\}.$$

Thus, F_{pl} selects the candidate(s) ranked first by the most voters. Consult [34] for an overview of preference aggregation and social choice rules and their properties.

3   Dependence in Social Choice Theory

Generally speaking, axiomatic characterization results proceed in two steps. The first step is to identify an interesting class of functions, each of which is intended to represent a possible group decision method. Different classes of functions build in different assumptions about the *structure* of the group decision problem. For instance, fix a set X of at least three candidates and a set $V = \{x_1, \ldots, x_n\}$ of n

---

[4]Also, $R^{\mathbf{P}}_{maj}$ may not be complete if there is an even number of voters. There are a variety of ways to modify the definition of the majority ordering to ensure completeness when there are an even number of voters.

voters. Then, the set

$$\mathfrak{F}(n, X) = \{F \mid F : O(X)^n \to O(X)\}$$

represents group decision problems in which the n voters are asked to rank the alternatives, and, based on the voters' rankings, identify the group ranking of the alternatives. Furthermore, since the domain of each function $F \in \mathfrak{F}(n, X)$, denoted $dom(F)$, is the set $O(X)^n$ of *all* profiles of rankings (i.e., all functions are assumed to be *total*), the voters' input is not restricted in any way. In particular, a voter's choice of ranking is *independent* of the other voters' choice of rankings (this will be discussed in more detail in Section 4). The second step is to characterize the desired set of group decision methods in terms of principles expressible as properties of the given class of functions. The goal is to find a set of properties of group decision rules that makes the group decision depend *in the right way* on the voters' inputs. Many different principles of group decision making have been discussed in the social choice literature. We discuss some of these properties in this section (see [20, 26, 30, 34] for discussions of additional properties). The statement of these properties will be tailored to the class $\mathfrak{F}(n, X)$ of preference aggregation functions. We leave it to the reader to adapt the principles to different classes of preference aggregation functions.

Since group decision methods are assumed to be *functions*, the output (a group ranking) does *functionally depend* on the rankings of the voters. However, this dependence is much too weak. There are many functions in $\mathfrak{F}(n, X)$ that are defective in some way. For instance, the constant function $F_R : O(X)^n \to O(X)$ for a fixed $R \in O(X)$, defined as $F_R(\mathbf{R}) = R$, is in $\mathfrak{F}(n, X)$. An obvious problem with a constant function is that the group decision is insensitive to any unanimous agreement among the voters. Suppose that $X = \{a, b, c\}$ and $a \, R \, b \, R \, c$, and consider the constant function F_R. If $\mathbf{R}$ is a profile in which every voter ranks b strictly above a (i.e., $\mathbf{V_R}(b \, P \, a) = V$), then $F_R(\mathbf{R}) = R$ is an outcome that does not truly reflect the voters' opinions (at least with respect to alternatives a and b). This suggests the following property:

Unanimity   For all alternatives $a, b \in X$, for all profiles $\mathbf{R} = (R_1, \ldots, R_n) \in O(X)^n$, if $a \, P_i \, b$ for all $x_i \in V$ (i.e., $\mathbf{V_R}(a \, P \, b) = V$), then $a \, P_{F(\mathbf{R})} \, b$.

This principle ensures that any unanimous agreement among the voters' strict rankings is reflected in the group ranking.[5] Unanimity is a fundamental principle of group decision methods.

An important distinction that was prevalent early on in the burgeoning social choice literature is between *single-profile* and *multi-profile* properties [21, 35, 37, 38]. Unanimity is an example of a single-profile property. It rules out specious pairings of group rankings with profiles in terms of properties of

---

[5]One can also explore alternative definitions of *Unanimity* of varying strengths. For example, if all voters weakly rank candidate a above candidate b, then society does so as well.

the given profile (e.g., the property that all voters rank candidate a strictly above candidate b). Thus, as will become clear in Section 5, Unanimity can be formalized in dependence logic using just a first-order formula without *dependency atoms*.

The notion of dependency found in dependence and independence logic is best exemplified by *multi-profile* properties. The general form of a multi-profile property runs as follows: If (two or more) profiles are related in a certain way, then the outcomes associated with these profiles must be related in some way. The most prominent multi-profile property is *independence of irrelevant alternatives* (IIA).

Independence of Irrelevant Alternatives   For all profiles, $\mathbf{R}, \mathbf{R}' \in O(X)^n$, for all $a, b \in X$, if $\mathbf{V_R}(a \ R \ b) = \mathbf{V_{R'}}(a \ R \ b)$ and $\mathbf{V_R}(b \ R \ a) = \mathbf{V_{R'}}(b \ R \ a)$, then $a \ F(\mathbf{R}) \ b$ iff $a \ F(\mathbf{R}') \ b$.

IIA ensures that the group ranking of two candidates depends only on how the individual voters rank those candidates. That is, if all voters agree on the relative rankings of a and b in two profiles, then the group rankings under each profile must rank a and b in the same way. IIA plays a crucial role in Arrow's impossibility theorem and many other results in social choice theory. Intuitions vary about the reasonableness of the IIA requirement for group decision methods. Many well-known voting methods do *not* satisfy IIA (the most prominent such example is the *Borda's ranking* that we defined in Section 2, see [39, 40] for an extensive argument in favor of using the Borda score to make group decisions, and see [34] for a discussion and further examples). Nonetheless, there are persuasive arguments that IIA is a natural requirement for a group decision method (see, for instance, [51, pg. 58] and [3, Chapter III, section 3]). Furthermore, Muller and Satterthwaite [32, 41] showed that IIA is equivalent to *strategy proofness* (strategy proofness means that voters do not have an incentive to misrepresent their preferences[6]).

Various authors have explored the implications of weakening IIA. An important result along these lines is from Blau [4]. Given a profile $\mathbf{R} \in O(X)^n$ and $Y \subseteq X$, let $\mathbf{R}_Y = ((R_1)_Y, \ldots, (R_n)_Y)$, where each $(R_i)_Y = R_i \cap (Y \times Y)$, i.e., $(R_i)_Y$ is the restriction of R_i to Y. Then, IIA can be reformulated as follows:

Binary Independence   For all profiles, $\mathbf{R}, \mathbf{R}' \in O(X)^n$, for all $a, b \in X$, if $\mathbf{R}_{\{a,b\}} = \mathbf{R}'_{\{a,b\}}$, then $F(\mathbf{R})_{\{a,b\}} = F(\mathbf{R}')_{\{a,b\}}$.

Blau studied the following generalization of IIA:

m-ary Independence   For all profiles, $\mathbf{R}, \mathbf{R}' \in O(X)^n$, for all m-element sets $Y \subseteq X$, if $\mathbf{R}_Y = \mathbf{R}'_Y$, then $F(\mathbf{R})_Y = F(\mathbf{R}')_Y$.

Of course, if $m = |X|$, then m-ary independence simply amounts to the usual requirement for any function. Blau showed that if a preference aggregation function satisfies m-ary independence (where $2 \leq m < |X|$), then it must also satisfy binary independence (the converse is obvious). An alternative way to define m-ary

---

[6]A full discussion of this result is beyond the scope of this article. See [46] for a precise statement of the Müller-Satterthwaite Theorem (including the additional assumptions needed to prove the equivalence) and a discussion of the relevant literature.

independence is as follows: Let $\mathscr{S}_m = \{Y \subseteq X \mid |Y| = m\}$ be the collection of all sets of m candidates from X. Then, m-ary independence says that for any $S \in \mathscr{S}_m$, if the relative rankings of all candidates in S are the same in two profiles, then the outcomes associated with these profiles must agree on the rankings of all candidates in S. Recently, Susumo Cato [7] showed that Blau's results hold for any collection $\mathscr{S}$ of sets of candidates satisfying the following connectedness property: For all candidates $a, b \in X$ there are sets $S_1, \ldots, S_k \in \mathscr{S}$ such that $\{a, b\} = \bigcap_{i=1}^{k} S_i$.

Preference aggregation rules that do not satisfy IIA take a global perspective when determining the social ranking of the candidates. For example, the Borda ranking of candidates a and b depends on the voters' rankings of *all* the candidates. A weaker version of IIA requires that the group ranking of candidates a and b depends on the voters' rankings of *some* subset of candidates (which may contain more candidates than just a and b). Campbell and Kelly studied preference aggregation methods that satisfy this weaker version of IIA together with additional multi-profile properties [6, 20].

A second multi-profile property, Neutrality, requires that the aggregation method treats all the candidates equally. To state this formally, suppose that $\mu : X \to X$ is a permutation of the candidates (i.e., a one-to-one function from X onto X). Given a relation R on X, define the relation R^μ as follows: For all $a, b \in X$,

$$\mu(a) \; R^\mu \; \mu(b) \text{ iff } a \; R \; b$$

For any profile $\mathbf{R} = (R_1, \ldots, R_n) \in O(X)^n$, a permutation μ applied to $\mathbf{R}$ is defined as $\mathbf{R}^\mu = (R_1^\mu, \ldots, R_n^\mu)$. Now, we define Neutrality as follows:

Neutrality For all profiles $\mathbf{R} \in dom(F)$ and all permutations $\mu : X \to X$, $F(\mathbf{R}^\mu) = F(\mathbf{R})^\mu$.

Neutrality ensures that a social ranking of the candidates depends only on where the candidates fall in the rankings in a given profile. Suppose that there are two profiles $\mathbf{R}$ and $\mathbf{R}'$ and two candidates a and b such that the positions that a occupies in the rankings in $\mathbf{R}$ are the same as the positions that b occupies in $\mathbf{R}'$. That is $\mathbf{R}' = \mathbf{R}^\mu$ where μ is the permutation such that $\mu(a) = b$, $\mu(b) = a$ and for all $c \neq a, b$, $\mu(c) = c$. Then, the social ranking of a given the profile $\mathbf{R}$ must be the same as the social ranking of b given the profile $\mathbf{R}'$.

While Neutrality requires that the candidates are treated equally, another property, Anonymity, requires that the voters are treated equally. A permutation of the voters is a one-to-one function $\pi : V \to V$. Anonymity requires that the group decision does not depend on the *name* of the voters.

Anonymity For all profiles $\mathbf{R} \in dom(F)$ and all permutations of $\pi : V \to V$, $F(\mathbf{R}) = F(\pi(\mathbf{R}))$, where $\pi(\mathbf{R}) = (R_{\pi(1)}, \ldots, R_{\pi(n)})$.

Anonymity is a fundamental requirement of the democratic process and is strictly enforced in most elections. The overall tally of a ranking $R \in O(X)$ in a profile $\mathbf{R}$ is the number of voters that submitted the ranking R (i.e., the tally of R in

$\mathbf{R} = (R_1, \ldots, R_n)$ is $|\{x_i \in V \mid R_i = R\}|$). Anonymity requires that group decisions depend on the tallies of the rankings in a profile rather than the profiles themselves (which identify the voter associated with each ranking).

In many voting situations, anonymity is dropped when the group decision results in a tie. Often, one voter is chosen (perhaps at random) to be the designated "tie-breaker." In such a case, the tie-breaker imposes her strict ranking of the candidates on the rest of the group. An egregious failure of anonymity occurs when there is a voter that imposes her strict rankings of the candidates on the group *no matter what rankings the other voters submit*. Of course, in any given profile, there will often be a number of voters that completely agree with the social ranking. There is nothing wrong with this. Indeed, it may very well be that, for every profile, there is some voter that completely agrees with the group ranking associated with that profile. It is a problem only when the quantifiers are reversed: there is a voter x_d such that for all profiles, voter x_d's strict rankings of the candidates agree with the social ranking. Such a voter is called an **Arrovian dictator**.

Non-Dictator   There is no $x_d \in V$ such that for all profiles $\mathbf{R} \in dom(F)$, for all $a, b \in X$, if $a\ P_d\ b$, then $a\ P_{F(\mathbf{R})}\ b$.

Non-Dictatorship ensures that the strict social ranking does not depend on only one voter (cf. the discussion in Section 6).

****

The properties introduced in this section ensure that group decisions depend *in the right way* on the voters' reported rankings. Arrow's ground-breaking theorem identified a surprising conflict between these principles:

Theorem 1 (Arrow [3]). *There are no preference aggregation functions* $F : O(X)^n \to O(X)$, *with* $|X| > 2$, *satisfying Independence of Irrelevant Alternatives, Unanimity, and Non-Dictatorship.*

Much of the subsequent work in social choice theory has focused on finding properties[7] that characterize interesting group decision rules. Amartya Sen adeptly explains the social choice problem in his Nobel Prize lecture:

> When a set of axioms regarding social choice can all be simultaneously satisfied, there may be several possible procedures that work, among which we have to choose. In order to choose between different possibilities through the use of discriminating axioms, we have to introduce further axioms, until one and only one possible procedure remains. This is something of an exercise in brinkmanship. We have to go on and on cutting alternative possibilities, moving—implicitly—towards an impossibility, but then stop just before all possibilities are eliminated, to wit, when one and only one option remains.   [42, pg. 354]

---

[7]Properties of group decision methods are often called "axioms" in the social choice literature. However, the principles studied in the social choice literature do not have the same status as the axioms of, for example, Peano arithmetic or the axioms defining a group. As should be clear from the discussion in this section, many of the so-called axioms of social choice are certainly not "self-evident," and may require extensive justification.

There is much more to say about Arrow's Theorem (cf. [5, 20, 30]). We return to this theorem in Section 5.3, showing how it can be formalized in independence logic.

4   Independence in Social Choice Theory

Arrow's Theorem is directed at preference aggregation functions $F : O(X)^n \rightarrow O(X)$. A key assumption, which we only briefly mentioned in the previous section, is that the domain of F is $O(X)^n$. This is the **Universal Domain** (UD) assumption. Thus, F must assign a group ranking to any possible profile of rankings. Arrow argued that, without specialized knowledge about the group decision problem, preference aggregation functions must be designed to handle any possible input:

> If we do not wish to require any prior knowledge of the tastes of individuals before specifying our social welfare function, that function will have to be defined for every logically possible set of individual orderings.   [3, pg. 24]

There are two aspects of UD that can be studied separately. The first is that there are no restrictions on the rankings available to a voter. This imposes a richness condition on the domain of a preference aggregation function $F : \mathscr{B} \rightarrow \mathscr{O}$, where $\mathscr{B}$ is a set of profiles of (linear) rankings of X:

All rankings   For any voter $x_i \in V$ and any (linear) ranking R, there is a profile $\mathbf{R} = (R_1, \ldots, R_n) \in dom(F)$ such that $R_i = R$.

Consider the following set of profiles of linear rankings for three candidates $X = \{a, b, c\}$ and three voters $V = \{x_1, x_2, x_3\}$. To simplify the notation, we write $a \ b \ c$ to denote the ranking $a \ P \ b \ P \ c$. Let $\mathscr{E}$ be the profiles displayed in Table 1 (each row corresponds to a profile).

As the reader is invited to check, $\mathscr{E}$ satisfies the **all rankings** property. Of course, $\mathscr{E}$ does not contain all possible profiles of linear rankings (i.e., $\mathscr{E} \subsetneq L(X)^n$). In particular, voters x_1 and x_3 have the same ranking in each profile. This means that for each $\mathbf{P} \in \mathscr{E}$, the majority ordering $R_{maj}^{\mathbf{P}}$ is transitive. Thus, $F_{maj} : \mathscr{E} \rightarrow L(X)$ is a well-defined preference aggregation rule. The problem with this domain is that voters x_1's and x_3's rankings are not chosen *independently*. They form a winning coalition ensuring that the group decision always agrees with their rankings. Thus, both x_1 and x_3 are Arrovian dictators. This suggests an additional constraint on domains of preference aggregation functions.

Independence   For any profiles $\mathbf{R} = (R_1, \ldots, R_n), \mathbf{R}' = (R_1', \ldots, R_n') \in dom(F)$ and any voter $x_i \in V$, there is a profile $\mathbf{R}'' = (R_1'', \ldots, R_n'') \in dom(F)$ such that $R_i'' = R_i$ and $R_j'' = R_j'$ for all $j \neq i$.

This constraint ensures that the voters' choice of rankings is not correlated in any way. The domain in Table 1 does not satisfy the independence property: There are profiles $\mathbf{P} = (a \ b \ c, a \ b \ c, a \ b \ c)$ and $\mathbf{P}' = (a \ c \ b, a \ b \ c, a \ c \ b)$, but no profile $\mathbf{P}'' = (P_1'', P_2'', P_3'')$ such that $P_1'' = a \ b \ c = P_2''$ and $P_3'' = a \ c \ b$.

Table 1 A set $\mathscr{E}$ of profiles satisfying the **All rankings** property.

x_1	x_2	x_3	x_1	x_2	x_3
a b c	a b c	a b c	a b c	a c b	a b c
a c b	a b c	a c b	a c b	a c b	a c b
b a c	a b c	b a c	b a c	a c b	b a c
b c a	a b c	b c a	b c a	a c b	b c a
c a b	a b c	c a b	c a b	a c b	c a b
c b a	a b c	c b a	c b a	a c b	c b a
a b c	b a c	a b c	a b c	b c a	a b c
a c b	b a c	a c b	a c b	b c a	a c b
b a c	b a c	b a c	b a c	b c a	b a c
b c a	b a c	b c a	b c a	b c a	b c a
c a b	b a c	c a b	c a b	b c a	c a b
c b a	b a c	c b a	c b a	b c a	c b a
a b c	c a b	a b c	a b c	c b a	a b c
a c b	c a b	a c b	a c b	c b a	a c b
b a c	c a b	b a c	b a c	c b a	b a c
b c a	c a b	b c a	b c a	c b a	b c a
c a b	c a b	c a b	c a b	c b a	c a b
c b a	c a b	c b a	c b a	c b a	c b a

It is not hard to see that imposing both **All rankings** and **Independence** ensures that the domain of the preference aggregation function is the set of *all* profiles of (linear) rankings. However, the **All rankings** and **Independence** constraints are stronger than what is needed to prove Arrow's Theorem. A weaker constraint on the domain that is sufficient to prove Arrow's Theorem was identified by Kalai, Muller, and Satterthwaite [19]. Their approach is to weaken the **All rankings** property while maintaining the **Independence** property.

We say that the Independence property is satisfied for a domain $\mathscr{B}$ whenever there is a set $\Omega \subseteq O(X)$ of "admissible" rankings for each voter and $\mathscr{B} = \Omega^n$. The following example from [19] illustrates this. Suppose that $Y = A \cup B$, where $A = \{a_1, a_2, a_3\}$ and $B = \{b_1, b_2, b_3\}$, and let

$$\Omega = \{R \in O(Y) \mid a \, P_R \, b \text{ for } a \in A \text{ and } b \in B\}.$$

So, Ω is the set of all rankings that rank all candidates in A strictly above all candidates in B. Then, $\mathscr{E} = \Omega^n$ satisfies **Independence**. Note that there is a preference aggregation function $F : \mathscr{E} \to O(Y)$ satisfying Unanimity, IIA, and Non-Dictatorship: For each $\mathbf{R} = (R_1, \ldots, R_n) \in \mathscr{E}$, let $F(\mathbf{R}) = (R_1)_A \cup (R_2)_B \cup \{(a,b) \mid a \in A \text{ and } b \in B\}$. That is, F ranks all candidates in A strictly above all candidates in B, ranks the candidates in A according to voter x_1, and ranks the candidates in B according to voter x_2. Thus, voter x_1 is an Arrovian dictator over the set A and x_2 is an Arrovian dictator over the set B, but there is no dictator for the entire set of candidates. This example shows that **Independence** alone is not sufficient to prove Arrow's Theorem.

To see that **All rankings** is not necessary to prove Arrow's Theorem, consider the following variant of the above example. Define $\Omega' \subseteq O(Y)$ as follows:

$$\Omega' = \{R \in O(Y) \mid a \; P_R \; b \text{ for } a \in A \text{ and } b \in B, \text{ and } b_1 \; P_R \; b_2 \; P_R \; b_3\}.$$

Thus, Ω' contains any ranking R that ranks all candidates in A strictly above all candidates in B, ranks b_1 strictly above b_2 and b_3, and ranks b_2 above b_3. Suppose that $\mathscr{E}' = (\Omega')^n$. Then, $\mathscr{E}'$ satisfies **Independence** but not **All rankings**. By inspecting the proof of Arrow's Theorem, it is not hard to show that there is no preference aggregation function $F : \mathscr{E}' \to O(Y)$ satisfying **Unanimity**, **IIA**, and **Non-Dictatorship**.

The difference between the two domains is that $\mathscr{E}'$ is based on a set of admissible rankings that satisfies a "saturation" property. We need some notation to formally state this property. Suppose that $\Omega \subseteq O(X)$ is a set of rankings for a set X of at least three candidates. A set of three candidates $\{a, b, c\} \subseteq X$ is called a **free triple** for Ω provided a, b, and c are distinct (i.e., $a \neq b \neq c \neq a$) and for each $R \in O(\{a, b, c\})$ there is a $R' \in \Omega$ such that $R'_{\{a,b,c\}} = R$. A pair of distinct candidates $\{a, b\}$ is said to be **trivial** in Ω provided for all $R, R' \in \Omega$, $R_{\{a,b\}} = R'_{\{a,b\}}$ (i.e., all rankings in Ω agree on the ranking of a and b). Two non-trivial pairs of candidates $A = \{a, b\}$ and $B = \{c, d\}$ are **strongly connected** in Ω provided $|A \cup B| = 3$ and $A \cup B$ is a free triple for Ω. Two pairs of candidates A and B are said to be **connected** provided there is a sequence of $B_1, \ldots, B_k$ of pairs of candidates such that $A = B_1$, $B = B_n$, and for all $i = 1, \ldots k - 1$, B_i and B_{i+1} are strongly connected. Finally, say that Ω is **saturated** provided there are at least two non-trivial pairs of candidates and every two non-trivial pairs of candidates are connected. Saturated domains are sufficient to prove Arrow's Theorem:

Theorem 2 (Kalai, Muller, and Satterthwaite [19]). *There is no $F : \mathscr{E} \to O(X)$, where $\mathscr{E} = \Omega^n$ and Ω is saturated, satisfying Unanimity, Independence of Irrelevant Alternatives, and Non-Dictatorship.*

5 Dependence and Independence Logic for Social Choice Theory

In this section, we use dependence and independence logic to formalize the notions of dependence and independence discussed in Sections 3 and 4. The initial idea to use dependence and independence logic to formalize results from social choice theory, such as Arrow's Theorem, is from Jouko Väänänen [49].

We think of the set of voters $V = \{x_1, \ldots, x_n\}$ as a set of distinguished first-order variables. In addition, we include a fresh first-order variable y that is intended to represent the group decision. Suppose that $\mathbf{R} = (R_1, \ldots, R_n) \in O(X)^n$ is a profile for

V and $F : \mathcal{B} \to \mathcal{O}$ is a preference aggregation function with $\mathbf{R} \in \mathcal{B}$. The pair $(\mathbf{R}, F)$ induces an assignment on $V^+ = \{x_1, \ldots, x_n, y\}$, denoted $s_{\mathbf{R},F} : V^+ \to \mathcal{B} \cup \mathcal{O}$, defined as follows:

$$s_{\mathbf{R},F}(x_1) = R_1, \quad \ldots, \quad s_{\mathbf{R},F}(x_n) = R_n \text{ and } s_{\mathbf{R},F}(y) = F(\mathbf{R}). \tag{1}$$

Then, any group decision function F is associated with a set of assignments:

$$S_F = \{s_{\mathbf{R},F} \mid \mathbf{R} \in dom(F)\} \tag{2}$$

Such a set of assignments is called a *team*, which is the central object of study in dependence and independence logic. Thus, there is a natural link with social choice theory: The properties of preference aggregation functions discussed in Sections 3 and 4 can be viewed as properties of teams, expressible in the language of dependence and independence logic.

Teams of assignments for the variables $\{x_1, \ldots, x_n, y\}$ are intended to represent election scenarios. Each assignment in the team represents a choice of ballot (typically, a ranking of the set of candidates) for each voter and the resulting group decision. Of course, not every team corresponds to some preference aggregation function. In particular, the rankings associated with y must be a function of the rankings associated with $\{x_1, \ldots, x_n\}$. In the language of dependence logic, this means that y *depends* on $\{x_1, \ldots, x_n\}$. Consider the team of assignments displayed in Table 2 assigning to $\{x_1, x_2, y\}$ linear rankings over the set $X = \{a, b, c\}$. That is, each assignment is a map $s : \{x_1, x_2, y\} \to L(X)$. (Recall that we write $a\ b\ c$ for the ranking $a\ P\ b\ P\ c$.)

Since the rankings associated with y do functionally depend on the rankings associated with the variables $\{x_1, x_2\}$, this team does represent possible election scenarios for 2 voters and 3 candidates. In the remainder of this section, we show how to use dependence and independence logic to reason about group decision methods.

Table 2 An example of a team for 2 voters.

	x_1	x_2	y
s_1	$a\ b\ c$	$c\ b\ a$	$b\ a\ c$
s_2	$a\ c\ b$	$b\ c\ a$	$c\ b\ a$
s_3	$c\ a\ b$	$b\ a\ c$	$a\ c\ b$
s_4	$b\ c\ a$	$a\ c\ b$	$c\ a\ b$
s_5	$a\ b\ c$	$b\ c\ a$	$b\ a\ c$

5.1 The Logic

In this section, we define the syntax and semantics of the version of independence logic (IndS) that we use to formalize Arrow's Theorem.

Suppose that Var is an infinite set of variables with distinguished elements $x_1, \ldots, x_n$ and y (where n is the number of voters). We use $u, v, w, \ldots$ (with or without subscripts) as metalanguage symbols that stand for first-order variables. Suppose that X is a finite set of candidates containing at least three elements. To simplify the presentation of the logical framework, we focus on logics for reasoning about preference aggregation functions. Note that both n (the number of voters) and X (the set of candidates) are parameters in the definition of our language. For simplicity, in this section, we restrict attention to linear orders. Recall that $L(X)$ is the (finite) set of all linear relations on X. Thus, our language is intended to describe properties of functions of the form $F : L(X)^n \rightarrow L(X)$. The definitions below can be adapted to reason about other types of group decision functions, such as social choice functions or functions where the domain and/or range is $O(X)$ (the set of all rankings on X).

The signature $\mathscr{L}_X$ contains an equality symbol $=$, unary predicate symbols E_R for each ranking $R \in L(X)$ and unary predicate symbols R_{ab} for each pair (a, b) of elements from X. Since our signature does not contain function symbols or constant symbols, variables are the only $\mathscr{L}$-terms. A **first-order atomic $\mathscr{L}$-formula** is a string of the form $u = v$, $E_R(w)$ or $R_{ab}(w)$.

Definition 3 (Syntax). A well-formed $\mathscr{L}$-formula of **independence logic** for social choice theory (IndS) is a string generated by the following grammar:

$$\varphi ::= \alpha \mid \neg \alpha \mid \bot \mid =(w_1, \ldots, w_k, u) \mid w_1 \ldots w_k \perp u_1 \ldots u_m \mid w_1 \ldots w_k \subseteq u_1 \ldots u_k$$
$$\mid \varphi \wedge \varphi \mid \varphi \vee \varphi \mid \forall x \varphi \mid \exists x \varphi,$$

where α is an $\mathscr{L}_X$-atomic first-order formula.

The formulas $=(w_1, \ldots, w_k, u)$, $w_1 \ldots w_k \perp u_1 \ldots u_m$, and $w_1 \ldots w_k \subseteq u_1 \ldots u_k$ are called *dependence atom, independence atom* and *inclusion atom*, respectively. We refer to them in this chapter as *atoms of dependence and independence*. The original independence logic as introduced in [14] does not have dependence atoms or inclusion atoms in the language. Since these two atoms are definable in Ind ([11, 14]), we will include these atoms in the language of our logic IndS.

The set $Fv(\varphi)$ of free variables of a formula φ of IndS is defined in the standard way except that we have the new cases for the atoms of dependence and independence:

- $Fv(=(w_1, \ldots, w_k, u)) = \{w_1, \ldots, w_k, u\}$
- $Fv(w_1 \ldots w_k \perp u_1 \ldots u_m) = \{w_1, \ldots, w_k, u_1, \ldots, u_m\}$
- $Fv(w_1 \ldots w_k \subseteq u_1 \ldots u_k) = \{w_1, \ldots, w_k, u_1, \ldots, u_k\}$

A formula φ is called a *sentence* if $Fv(\varphi) = \emptyset$.

Formulas of IndS are interpreted in standard first-order models M. We assume that the domain of a model M, denoted $dom(M)$, always has at least two elements.

Our formalization of Arrow's Theorem requires that the domain contains all linear rankings of (at least three) candidates. An **intended $\mathscr{L}_X$-model** for IndS is an $\mathscr{L}_X$-model M where $dom(M) = L(X)$. The set of intended models is first-order definable using the unary predicates E_R (this will be shown in Section 5.2). For any $e \in dom(M)$ and any linear ranking $R \in L(X)$, the intended interpretation of $E_R^M(e)$ is that e is the linear ranking R, i.e., $E_R^M = \{e \in dom(M) \mid e = R\}$. For each $e \in dome(M)$, the intended interpretation of $R_{ab}^M(e)$ is that the ranking associated with the element e ranks a above b. More formally, for $a, b \in X$, the unary $R_{ab}^M = \{R \in L(X) \mid a \ R \ b\}$. For example, if $X = \{a, b, c\}$, then $M = (L(X), \{E_R \mid R \in L(X)\}, \{R_{de}^M \mid d, e \in X\})$ is an intended $\mathscr{L}_X$-model for IndS. Suppose that R_1 is the relation $a \ R \ b \ R \ c$; R_2 is the relation $a \ R \ c \ R \ b$; R_3 is the relation $b \ R \ a \ R \ c$; R_4 is the relation $b \ R \ c \ R \ a$; R_5 is the relation $c \ R \ a \ R \ b$; and R_6 is the relation $c \ R \ b \ R \ a$. Then:

- $E_{R_i}^M = \{R_i\}$, for $i = 1, \ldots, 6$; and
- $R_{ab}^M = \{R_1, R_2, R_5\}, R_{ac}^M = \{R_1, R_2, R_3\}, R_{bc}^M = \{R_1, R_3, R_4\}, \ldots$

Definition 4 (Assignments, Teams). An **assignment** on M is a map $s : \mathsf{Var} \longrightarrow dom(M)$. A **team** S on M is a set of assignments on M.

For any assignment s and any element $a \in dom(M)$, we write $s(a/w)$ for the assignment defined as $s(a/w)(w) = a$ and $s(a/w)(u) = s(u)$ if $u \neq w$. We now define the team semantics for our logic IndS.

Definition 5 (Semantics). Suppose that M is an $\mathscr{L}$-model for IndS and S is a team on M. For each $\mathscr{L}_X$-formula φ of IndS, we define $M \models_S \varphi$ inductively as follows:

- $M \models_S \alpha$ with α a first-order atomic formula iff for all $s \in S$, $M \models_s \alpha$ in the usual sense;
- $M \models_S \neg\alpha$ with α a first-order atomic formula iff for all $s \in S$, $M \not\models_s \alpha$ in the usual sense;
- $M \models_S \bot$ iff $S = \emptyset$;
- $M \models_S \ =(w_1, \ldots, w_k, u)$ iff for all $s, s' \in S$,

$$\text{if } \langle s(w_1), \ldots, s(w_k) \rangle = \langle s'(w_1), \ldots, s'(w_k) \rangle, \text{ then } s(u) = s'(u);$$

- $M \models_S w_1 \ldots w_k \perp u_1 \ldots u_m$ iff for all $s, s' \in S$, there is $s'' \in S$ such that

$$\langle s''(w_1), \ldots, s''(w_k) \rangle = \langle s(w_1), \ldots, s(w_k) \rangle$$

and

$$\langle s''(u_1), \ldots, s''(u_m) \rangle = \langle s'(u_1), \ldots, s'(u_m) \rangle;$$

- $M \models_S w_1 \ldots w_k \subseteq u_1 \ldots u_k$ iff for all $s \in S$, there is $s' \in S$ such that

$$\langle s'(w_1), \ldots, s'(w_k)\rangle = \langle s(u_1), \ldots, s(u_k)\rangle;$$

- $M \models_S \varphi \wedge \psi$ iff $M \models_S \varphi$ and $M \models_S \psi$;
- $M \models_S \varphi \vee \psi$ iff there exist teams $S_0, S_1 \subseteq S$ with $S = S_0 \cup S_1$ such that $M \models_{S_0} \varphi$ and $M \models_{S_1} \psi$;
- $M \models_S \forall w\varphi$ iff $M \models_{S(M/w)} \varphi$, where $S(M/w) = \{s(a/w) \mid s \in S$ and $a \in M\}$;
- $M \models_S \exists w\varphi$ iff $M \models_{S[F/w]} \varphi$ for some function $F : S \longrightarrow \wp(M) \setminus \{\emptyset\}$, where $S[F/w] = \{s(a/w) \mid s \in S$ and $a \in F(s)\}$;

A sentence φ is said to be *true* in M if the team $\{\emptyset\}$ of the empty assignment satisfies φ, i.e., $M \models_{\{\emptyset\}} \varphi$. We say that a formula φ is a **logical consequence** of a set Γ of formulas provided, for all models M and all teams S on M, if $M \models_S \psi$ for all $\psi \in \Gamma$, then $M \models_S \varphi$. We write $\psi \models \varphi$ for $\{\psi\} \models \varphi$. If $\varphi \models \psi$ and $\psi \models \varphi$, then we say that φ and ψ are *logically equivalent*, in symbols $\varphi \equiv \psi$.

For any team S on a model M and any set $V \subseteq \mathsf{Var}$ of variables, the set $S \upharpoonright V = \{s \upharpoonright V : s \in S\}$ is called a *team on V*. It is straightforward to check that our logic IndS has the Locality Property and the Empty Team Property:

(Locality Property)   If $S \upharpoonright \mathrm{Fv}(\varphi) = S' \upharpoonright \mathrm{Fv}(\varphi)$, then $M \models_S \varphi \iff M \models_{S'} \varphi$.
(Empty Team Property) $M \models_\emptyset \varphi$ for all models M.

We refer the reader to [11, 16, 24, 25, 48] for other properties of the logic. In our formalization of Arrow's Theorem, most formulas will have free variables only from the set $V^+ = \{x_1, \ldots, x_n, y\}$ of the distinguished variables that we fixed. By the Locality Property, in most cases, it is then sufficient to consider teams on the set V^+ only. These teams, as discussed, are in one-to-one correspondence to the sets of profiles together with a preference aggregation rule (which may or may not be a function).

We say that a formula of IndS is **first-order**, if it does not contain any atoms of dependence and independence. First-order formulas have the Flatness Property:

(Flatness Property) $M \models_S \varphi$ if, and only if, $M \models_{\{s\}} \varphi$ for all $s \in S$.

For any first-order formula φ, we write $\neg\varphi$ for the (first-order) formula inductively defined as follows:

$$\neg(\alpha) := \neg\alpha \qquad \neg(\psi \wedge \chi) := \neg\psi \vee \neg\chi \qquad \neg(\forall x\varphi) := \exists x\neg\varphi$$
$$\neg(\neg\alpha) := \alpha \qquad \neg(\psi \vee \chi) := \neg\psi \wedge \neg\chi \qquad \neg(\exists x\varphi) := \forall x\neg\varphi$$

where α is an atomic first-order formula. The reader is invited to check that for any first-order formulas φ and ψ, the following clause holds:

$$M \models_S \neg\varphi \vee \psi \iff \text{ for all } s \in S, \text{ if } M \models_{\{s\}} \varphi, \text{ then } M \models_{\{s\}} \psi. \qquad (3)$$

In this sense, the formula $\neg\varphi \vee \psi$, abbreviated as $\varphi \supset \psi$, expresses a type of classical material implication that will play a role in the sequel.

It is known that independence logic has the same expressive power as *existential second-order logic* (Σ_1^1) [11]. Therefore, all Σ_1^1-properties of social choice theory can be expressed in our logic. In what follows, we will demonstrate how to express the properties needed to prove Arrow's Theorem.

Our goal is to find a set of formulas Γ_{Arrow} expressing the assumptions of Arrow's Theorem and a first-order formula θ_D expressing that there is an Arrovian dictator such that $\Gamma_{Arrow} \vdash \theta_D$. That is, θ_D is **derivable** in independence logic using the assumptions in Γ_{Arrow}. Due to its strong expressive power, the full independence logic is not axiomatizable (see [16] and also [25]). However, the first-order consequences of IndS are axiomatizable. A complete natural deduction system for the first-order consequence relation over sentences of IndS was given in [16]. More recently, Kontinen [23] generalized this result to open formulas by adding an extra predicate symbol to the signature. Our main goal in this section is to demonstrate that Arrow's Theorem not only can be formalized in IndS, but also can be derived syntactically using the system of [16] and [23].

5.2   *Expressing Arrow's Conditions*

There are three types of properties that we need to express in order to formalize Arrow's Theorem. The first type consists of properties that do not involve any atoms of dependence or independence. These are expressible using first-order formulas only. The second type is intended to capture the notion of dependence from Section 3. The third type captures the notion of independence from Section 4.

First-Order Properties

The first step is to find formulas guaranteeing that the domain contains all linear rankings of the set of candidates X. Let Γ_{DM} of be the following set of sentences:

$$\{\exists w(E_R(w)) \mid R \in L(X)\} \cup \{\forall w \bigvee \{E_R(w) \mid R \in L(X)\}\}$$

(*Domain Requirement*) $\cup \ \{\forall w \forall u((E_R(w) \wedge E_R(u)) \supset (w = u)) \mid R \in L(X)\}$

$$\cup \ \{\forall w \bigwedge \{\neg E_R(w) \vee \neg E_{R'}(w) \mid R, R' \in L(X) \text{ and } R \neq R'\}\}$$

Any model M of Γ_{DM} has the property that (1) each linear ranking $R \in L(X)$ corresponds to a unique element e in the model and (2) each element e of the model corresponds to a unique ranking $R \in L(X)$.

The next step is to characterize the intended meaning of the unary predicates R_{ab}. Recall that the atomic formula $R_{ab}(x_i)$ is intended to express the property that voter x_i ranks a above b. Let Γ_{RK} be the following set of sentences:

(*Ranking*) $\bigcup_{P \in L(X)} \{\forall w((E_P(w) \supset R_{ab}(w)) \wedge (R_{ab}(w) \supset E_P(w))) \mid a \ P \ b\}$

It is not hard to see that for any model M of $\Gamma_{DM} \cup \Gamma_{RK}$ and any $a, b \in X$, the interpretation of the predicate R_{ab} is $R_{ab}^M = \{P \in L(X) \mid a \ P \ b\}$. Thus, any model M of $\Gamma_{DM} \cup \Gamma_{RK}$ is an intended $\mathscr{L}_X$-model for IndS. Note that the order-theoretic properties of the relations are logical consequences of $\Gamma_{DM} \cup \Gamma_{RK}$. For instance, if M is a model of $\Gamma_{DM} \cup \Gamma_{RK}$, then the following formula that defines transitivity is true:

$$\forall w \left(\bigwedge \{(R_{ab}(w) \wedge R_{bc}(w)) \supset R_{ac}(w) \mid a, b, c \in X\} \right)$$

We introduce the following notation to express strict preference and indifference, respectively:

(Strict preference) For each $a, b \in X$, let $P_{ab}(w) := R_{ab}(w) \wedge \neg R_{ba}(w)$
(Indifference) For each $a, b \in X$, let $I_{ab}(w) := R_{ab}(w) \wedge R_{ba}(w)$

Thus, $P_{ab}(x_i)$ means that voter x_i strictly ranks a above b and $I_{ab}(x_i)$ means that voter x_i is indifferent between a and b. Similarly, $P_{ab}(y)$ means that society strictly ranks a above b and $I_{ab}(y)$ means that the society is indifferent between a and b.

Another property that is expressible using only first-order formulas is Unanimity.

(Unanimity) $\theta_U := \bigwedge \{(P_{ab}(x_1) \wedge \cdots \wedge P_{ab}(x_n)) \supset P_{ab}(y) \mid a, b \in X\}.$

To see why the above formula expresses Unanimity, suppose that S_F is a team induced by a preference aggregation function F. If $M \models_{S_F} \theta_U$, then for each $a, b \in X$, we have $M \models_{S_F} (P_{ab}(x_1) \wedge \cdots \wedge P_{ab}(x_n)) \supset P_{ab}(y)$. According to equation (3), this means that for all $a, b \in X$ and all $s_{\mathbf{R},F} \in S_F$,

$$\text{if } M \models_{\{s_{\mathbf{R},F}\}} P_{ab}(x_1) \wedge \cdots \wedge P_{ab}(x_n), \text{ then } M \models_{\{s_{\mathbf{R},F}\}} P_{ab}(y).$$

Unpacking the above definitions gives us the definition of Unanimity for a preference aggregation function F: For all candidates $a, b \in X$, and all profiles $\mathbf{R} = (R_1, \ldots, R_n) \in dom(F)$, if $a \, P_i \, b$ for all voters x_i, then $a \, P_{F(\mathbf{R})} \, b$.

Dependence Properties

The first dependence property we will express concerns the functional dependence of the group decision on the voters' rankings. In our setting, this non-trivial property is easily expressed using a simple dependence atom:

(Functionality of Preference Aggregation Rule) $\theta_F := \, =\!(x_1, \ldots, x_n, y)$

Recall that a team S on a model M satisfies θ_F iff for any two assignments $s, s' \in S$,

$$\text{if } s(x_i) = s'(x_i) \text{ for all } 1 \leq i \leq n, \text{ then } s(y) = s'(y).$$

To see that any team S_G induced by a preference aggregation function G satisfies θ_F, the key observation is that if $s, s' \in S_G$, then $s = s_{\mathbf{R},G}$ and $s' = s_{\mathbf{R}',G}$ for some profiles $\mathbf{R}, \mathbf{R}' \in dom(G)$. If $s_{\mathbf{R},G}(x_i) = s_{\mathbf{R}',G}(x_i)$ for all $1 \leq i \leq n$, then $\mathbf{R}$ and $\mathbf{R}'$ are the same profile, and, since G is a function, $s_{\mathbf{R},G}(y) = G(\mathbf{R}) = G(\mathbf{R}') = s_{\mathbf{R}',G}(y)$. We leave it for the reader to check that, conversely, any team satisfying θ_F is associated with a preference aggregation function.

As we argued in Section 3, the notion of dependence found in social choice theory goes beyond simple functional dependence of the group decision on the voters' inputs. The principles from Section 3 are defined to ensure that *properties* of the group ranking depend only on *properties* of the voters' rankings. To express this stronger form of dependence, for any first-order formulas $\varphi_1, \ldots, \varphi_k, \psi$, we introduce a new formula $=(\varphi_1, \ldots, \varphi_k, \psi)$. To define the semantics of this formula we need some notations. Suppose that Γ is a set of first-order formulas and s, s' are two assignments for a model M. We write $s \sim_\Gamma s'$ when

$$\text{for all } \gamma \in \Gamma, M \models_{\{s\}} \gamma \text{ if, and only if, } M \models_{\{s'\}} \gamma.$$

The semantics for $=(\varphi_1, \ldots, \varphi_k, \psi)$ is given by the clause:

- $M \models_s =(\varphi_1, \ldots, \varphi_k, \psi)$ iff for all $s, s' \in S$, if $s \sim_{\{\varphi_1, \ldots, \varphi_k\}} s'$, then $s \sim_{\{\psi\}} s'$.

Without going into any detail, we remark that this new formula $=(\varphi_1, \ldots, \varphi_k, \psi)$ is definable in our logic IndS, as

$$=(\varphi_1, \ldots, \varphi_k, \psi) \equiv \exists w_1 \ldots \exists w_k \exists u \exists v_0 \exists v_1 \Big(=(w_1, \ldots, w_k, u) \wedge =(v_0) \wedge =(v_1)$$

$$\wedge (v_0 \neq v_1) \wedge \bigwedge_{i=1}^{k} \big(\theta(w_i, v_0, v_1) \wedge \delta(w_i, \varphi_i, v_0, v_1) \big) \wedge \theta(u, v_0, v_1) \wedge \delta(u, \psi, v_0, v_1) \Big),$$

where $\theta(v, v_0, v_1) := (v = v_0) \vee (v = v_1)$ and

$$\delta(v, \chi, v_0, v_1) := \big(\chi \supset (v = v_1) \big) \wedge \big(\neg \chi \supset (v = v_0) \big).$$

Now, using this generalized dependence formula, we can state IIA in our logic.

(*Independence of Irrelevant Alternatives*)

$$\theta_{IIA} := \bigwedge \{ =(R_{ab}(x_1), R_{ba}(x_1) \ldots, R_{ab}(x_n), R_{ba}(x_n), R_{ab}(y)) \mid a, b \in X \}.$$

To see that this corresponds to binary independence, note that

$$\text{if } s_{\mathbf{R}, F}(x_i) = R_i \in R_{ab}^M \Leftrightarrow s_{\mathbf{R}', F}(x_i) = R_i' \in R_{ab}^M \text{ and}$$

$$s_{\mathbf{R}, F}(x_i) = R_i \in R_{ba}^M \Leftrightarrow s_{\mathbf{R}', F}(x_i) = R_i' \in R_{ba}^M, \text{ then } (R_i)_{\{a,b\}} = (R_i')_{\{a,b\}},$$

where $\mathbf{R} = (R_1, \ldots, R_n)$ and $\mathbf{R}' = (R_1', \ldots, R_n')$.

Remark 1 (Alternative Definitions of IIA). One may be tempted to simplify the definition of IIA as follows:

$$\theta_{IIA}' := \bigwedge \{ =(R_{ab}(x_1), \ldots, R_{ab}(x_n), R_{ab}(y)) \mid a, b \in X \}$$

This formulas says that, for each pair of alternatives $a, b \in X$, the truth of $R_{ab}(y)$ depends only on the truth of $R_{ab}(x_1), \ldots R_{ab}(x_n)$.

Suppose that S_F is a team on a model M induced by a preference aggregation function F that satisfies the formula $=(R_{ab}(x_1), \ldots, R_{ab}(x_n), R_{ab}(y))$. Then, for any $s_{\mathbf{R},F}, s_{\mathbf{R}',F} \in S_F$,

$$\text{if } s_{\mathbf{R},F}(x_i) = R_i \in R_{ab}^M \Leftrightarrow s_{\mathbf{R}',F}(x_i) = R_i' \in R_{ab}^M \text{ for all } 1 \leq i \leq n,$$

$$\text{then } F(\mathbf{R}) \in R_{ab}^M \Leftrightarrow F(\mathbf{R}') \in R_{ab}^M,$$

where $\mathbf{R} = (R_1, \ldots, R_n)$ and $\mathbf{R}' = (R_1', \ldots, R_n')$. While this does express a sense in which the social ranking of a and b depends on the individual rankings of a and b, it does *not* express Arrow's IIA property from Section 3. In particular, the above property is more demanding than binary independence. To see why, suppose that a team S contains two assignments $s_{\mathbf{R},F}$ and $s_{\mathbf{R}',F}$, where

- $\mathbf{R} = (R_1, \ldots, R_n)$ with for all i, $a \, I_i \, b$; and
- $\mathbf{R}' = (R_1', \ldots, R_n')$ with for all i, $a \, P_i' \, b$.

Since $a \, I_i \, b$ is defined as $a \, R_i \, b$ and $b \, R_i \, a$, $a \, P_i' \, b$ is defined as $a \, R_i' \, b$ and $b \, \not{R}_i' \, a$, it is true that for all i, $R_i \in R_{ab}^M$ iff $R_i' \in R_{ab}^M$. If θ_{IIA} is satisfied in the team, this would require that $F(\mathbf{R}) \in R_{ab}^M$ iff $F(\mathbf{R}') \in R_{ab}^M$. However, since for all $1 \leq i \leq n$,

$$(R_i)_{\{a,b\}} = \{(a,b), (b,a)\} \neq \{(a,b)\} = (R_i')_{\{a,b\}},$$

binary independence does not impose any constraints on the social ranking of a and b.

We leave a full discussion of different versions of IIA, including a formalization of **m-ary independence** and a derivation in our logic of Blau's Theorem mentioned in Section 3, for an extended version of this paper.

We conclude this subsection by finding a formula that expresses the existence of an Arrovian dictator. The existence of an Arrovian dictator means that there is a strong form of dependence of the social outcome on a single voter. In particular, if x_d is an Arrovian dictator, then all of x_d's strict rankings are reflected in the social ranking. This is characterized by the following first-order formula:

- $\theta_{D_0}(x_d) := \bigwedge_{a,b \in X} (P_{ab}(x_d) \supset P_{ab}(y))$.

To express that there exists a dictator among the n voters, we need a new connective: The intuitionistic disjunction, denoted by $\lor\!\!\!\lor$, whose semantics is given by the clause:

- $M \models_S \varphi \lor\!\!\!\lor \psi$ iff $M \models_S \varphi$ or $M \models_S \psi$.

Without going into any detail, we remark that the intuitionistic disjunction is definable in our logic IndS:

$$\varphi \lor\!\!\!\lor \psi \equiv \exists w \exists u \Big(=(w) \land =(u) \land \big((w = u) \lor \varphi\big) \land \big((w \neq u) \lor \psi\big) \Big),$$

where $w, u \notin \mathrm{Fv}(\varphi) \cup \mathrm{Fv}(\psi)$. The following formula expresses the existence of an Arrovian dictator among the n voters:

$$(Dictator) \quad \theta_D := \bigvee_{i=1}^{n} \theta_{D_0}(x_i).$$

Independence Properties

As explained in Section 4, a key assumption in Arrow's Theorem is the **Universal Domain** condition. This is characterized by the All Rankings condition and the Independence condition. Our logic IndS can express these two properties:

$$(All\ Rankings) \quad \theta_{AR} := \bigwedge \{\forall u(u \subseteq x_i) : 1 \leq i \leq n\}$$

$$(Independence) \quad \theta_I := \bigwedge \{\langle x_j \rangle_{j \neq i} \perp x_i : 1 \leq i \leq n\}$$

To see that θ_{AR} corresponds to the All Rankings condition, let M be a model and S_F be a team on M induced by a preference aggregation function F. Suppose that $M \models_{S_F} \theta_{AR}$. Then, for each voter x_i, we have $M \models_{S_F(M/u)} u \subseteq x_i$. The value of u ranges over all possible elements of the domain of M. Since the domain of (an intended model) M is the set of all (linear) rankings, the values of u range over all (linear) rankings. The inclusion atom ensures that each such (linear) ranking must occur in the team S_F as a value for the voter x_i. That is, for each ranking $R \in dom(M)$, there is an $s \in S_F$ such that $s(x_i) = R$. This is exactly the **All ranking** property. The correspondence between θ_I and the **Independence** condition is more straightforward, so we leave it for the reader to verify. Note that in our formalization of the Universal Domain assumption, we make essential use of the dependence and independence atoms.

5.3 Arrow's Theorem

There are two additional necessary assumptions for the proof of Arrow's Theorem. The first is that there are at least three candidates (i.e., $|X| \geq 3$). Indeed, if there are only two candidates, then majority rule satisfies Unanimity, Independence of Irrelevant Alternatives, Non-Dictatorship, and Universal Domain (cf. May's Theorem [28] for a characterization of majority rule). This is not a property that can be expressed in our logic. Rather, it is an implicit assumption built into the definition of our logic, as we have fixed a set X containing at least three alternatives and assumed that our signature $\mathscr{L}_X$ has predicate symbols R_{ab} for each pair $a, b \in X$.

The second assumption is that there are only finitely many voters. It can be shown that Arrow's Theorem does not hold (if the Axiom of Choice is assumed) when there are infinitely many voters (see [10, 22]). However, there are analogues of Arrow's Theorem for countably many voters ([46, Chapter 6] and [18]). Again, this is an assumption that is built into the definition of our logic. In the above presentation of our logic IndS, we started by distinguishing a finite set $V^+ = \{x_1, \ldots, x_n, y\}$ of variables. The fact that V^+ is a finite set of variables was implicitly used when we showed that the Arrow conditions are expressible in our logic. In particular, our logic IndS is finitary and we do not see a way to define the propositional dependence formula $=(\bar{\varphi}, \psi)$ when $\bar{\varphi}$ is an infinite sequence of formulas.

A complete discussion of the different ways in which these last two assumptions can be formally represented in independence logic will be left for an extended version of this paper. For the remainder of this paper, we assume that our logic IndS satisfies the above two assumptions which are needed to prove Arrow's Theorem.

Theorem 3 (Arrow's Theorem, semantic version). $\Gamma_{Arrow} \models \theta_D$, where $\Gamma_{Arrow} = \Gamma_{DM} \cup \Gamma_{RK} \cup \{\theta_U, \theta_F, \theta_{IIA}, \theta_{AR}, \theta_I\}$.

The proof of this theorem follows by adapting the standard proofs of Arrow's Theorem (see, for instance, [12] or [1] for a category-theoretic perspective). In the remainder of this section, we will demonstrate that Arrow's Theorem can also be derived syntactically in the natural deduction system of [16] and [23].

We write $\Gamma \vdash \varphi$ if the formula φ can be derived from the set Γ of formulas in the natural deduction system given in [16] and [23]. [8]

Theorem 4 ([16, 23]). *If φ is a first-order formula and Γ a set of formulas of independence logic, then we have $\Gamma \models \varphi$ iff $\Gamma \vdash \varphi$.*

Unfortunately, the above completeness theorem cannot be directly applied to Theorem 3 to show that Arrow's Theorem is derivable in our logic IndS. The problem is that the formula θ_D, which expresses the existence of an Arrovian dictator, is not a first-order formula. Nonetheless, it is possible to transform the formalization of Arrow's Theorem so that we can apply Theorem 4.

Consider a unary connective $\sim$, called **weak classical negation**, whose semantics is given by the clause:

- $M \models_S \sim \varphi$ iff $M \not\models_S \varphi$ whenever $S \neq \emptyset$.[9]

We invite the reader to check the following crucial fact. Note that since our logic has the empty team property, the additional condition "whenever $S \neq \emptyset$" in the semantics of $\sim$ is essential to establish this fact.

Fact 5. $\Gamma_{Arrow} \models \theta_D \iff \Gamma_{Arrow}, \sim \theta_D \models \bot$.

Since the atom $\bot$ (**falsum**) is a first-order formula, we are almost ready to apply Theorem 4. The remaining issue is that we need to make sure that the formula $\sim \theta_D$ is definable in our original independence logic Â IndS, or, equivalently, that it is Σ_1^1.

---

[8] The interested reader can consult [16] and [23] for the details of the natural deduction system. We do not include the system here since we are only proving the existence of a derivation of Arrow's Theorem rather than providing a derivation. We will take up this challenge in the extended version of this paper.

[9] Note that a slightly different connective $\sim'$ with the semantics $M \models_S \sim' \varphi$ iff $M \not\models_S \varphi$ is known as **classical negation** in the dependence logic literature.

To establish this, let us take a closer look at the formula $\sim \theta_D$, which expresses the fact that there is no dictator. Unpacking the definitions, we obtain

$$\sim \theta_D = \sim \bigvee_{i=1}^{n} \theta_{D_0}(x_i) \equiv \bigwedge_{i=1}^{n} \sim \theta_{D_0}(x_i) \equiv \bigwedge_{i=1}^{n} \bigvee_{a,b \in X} \sim (P_{ab}(x_i) \supset P_{ab}(y)).$$

This means that the problem reduces to defining the formula $\sim (P_{ab}(x_i) \supset P_{ab}(y))$ in our logic IndS. We present this translation in the following proposition, whose proof is left to the reader.

Proposition 6. $\sim (P_{ab}(x_i) \supset P_{ab}(y)) \equiv \exists w \exists u ((wu \subseteq x_i y) \wedge P_{ab}(x_i) \wedge \neg P_{ab}(y)).$

Now, since $\Gamma_A, \sim \theta_D \models \bot$ and all the formulas in the set $\Gamma_A \cup \{\sim \theta_D\}$ are expressible in IndS, by Theorem 4 we conclude that $\Gamma_A, \sim \theta_D \vdash \bot$. In order to derive that $\Gamma_A \vdash \theta_D$, we need a weak classical negation elimination rule ($\sim$ E) defined as follows:

$$[\sim \psi]$$
$$\vdots$$
$$\frac{\varphi \qquad \bot}{\psi} \sim E$$

We do not see how to derive this rule from the natural deduction system given in [16, 23]. Our solution is to add this rule (which is sound) to the natural deduction system of [16, 23]. We write $\Gamma \vdash^{\sim E} \varphi$ if φ can be derived from Γ in this extended system. This gives us a syntactic version of Arrow's Theorem:

Theorem 7 (Arrow's Theorem, syntactic version). $\Gamma_{Arrow} \vdash^{\sim E} \theta_D$.

6   Concluding Remarks

One of the goals of social choice theory is to develop group decision methods that satisfy two main desiderata. The first is that the group decision should depend in the right way on the voters' opinions. The second is that the voters should be free to express any opinion, as long as it is an admissible input to the group decision method. Impossibility theorems, such as Arrow's Theorem, point to an interesting tension between these two desiderata. Properties of group decision methods that ensure that group decisions depend on voters' opinions and that the voters' opinions are independent cannot be simultaneously satisfied. We argued that dependence and independence logic offers an interesting new perspective on this aspect of social choice theory.

Our main focus in the chapter was Arrow's ground-breaking theorem. We developed a version of independence logic that can express Arrow's properties of

Table 3 An example of the reasoning of Arrow's Theorem.

	x_1	x_2	y
s_1	$a\, P_1\, b\, P_1\, c$	$c\, P_2\, b\, P_2\, a$	$b\, P\, a\, I\, c$
s_2	$a\, P_1'\, c\, P_1'\, b$	$c\, P_2'\, b\, P_2'\, a$	??

preference aggregation functions. We then proved that Arrow's Theorem is derivable in a natural deduction system for the first-order consequences of our logic. Our work highlights a number of topics that deserve further study.

The most pressing topic is to find a derivation of Arrow's Theorem in the natural deduction system for our logic. This would not only lead to a potentially new proof of Arrow's Theorem, but it could also identify interesting patterns of reasoning used throughout the social choice literature. To illustrate, consider the following example. Suppose that $S = \{s_1, s_2\}$ is a team on a model M for the set $V = \{x_1, x_2\}$ of two voters and the set $X = \{a, b, c\}$ of three candidates. The assignments are given in Table 3.

Assuming that S satisfies Unanimity and IIA, the question is: What are the possible social ranking for s_2 (i.e., what are the possible values for $s_2(y)$?). Since Unanimity holds for S, i.e., $M \models_S \theta_U$, we have $M \models_{\{s_2\}} P_{cb}(x_1) \wedge P_{cb}(x_2) \supset P_{cb}(y)$. From Table 3 we know that $M \models_{\{s_2\}} P_{cb}(x_1) \wedge P_{cb}(x_2)$, thus we must conclude that $M \models_{\{s_2\}} P_{cb}(y)$, i.e., $s_2(y) \in P_{cb}^M$.

Now, since IIA holds for S, i.e., $M \models_S \theta_{IIA}$, we have

$$M \models_S =(R_{ac}(x_1), R_{ca}(x_1), R_{ac}(x_2), R_{ca}(x_2), R_{ac}(y)) \tag{4}$$

and

$$M \models_S =(R_{ba}(x_1), R_{ab}(x_1), R_{ba}(x_2), R_{ab}(x_2), R_{ba}(y)). \tag{5}$$

Let us examine (4). By examining Table 3, we have

$$M \models_{\{s_1\}} R_{ac}(x_1) \wedge \neg R_{ca}(x_1), \quad M \models_{\{s_2\}} R_{ac}(x_1) \wedge \neg R_{ca}(x_1),$$

$$M \models_{\{s_1\}} \neg R_{ac}(x_2) \wedge R_{ca}(x_2) \text{ and } M \models_{\{s_2\}} \neg R_{ac}(x_2) \wedge R_{ca}(x_2).$$

Thus, $s_1 \sim_{\{R_{ac}(x_1), R_{ca}(x_1), R_{ac}(x_2), R_{ca}(x_2)\}} s_2$. Hence, we conclude that $s_1 \sim_{\{R_{ac}(y)\}} s_2$. Now, since $M \models_{\{s_1\}} R_{ac}(y)$, we obtain $M \models_{\{s_2\}} R_{ac}(y)$, meaning $s_2(y) \in R_{ac}^M$. By a similar reasoning, since, $s_1 \sim_{\{R_{ba}(x_1), R_{ab}(x_1), R_{ba}(x_2), R_{ab}(x_2)\}} s_2$, we conclude from (5) that $s_1 \sim_{\{R_{ba}(y)\}} s_2$. Thus, $s_2(y) \in R_{ba}^M$.

Putting everything together, we have $s_2(y) \in P_{cb}^M$ and $s_2(y) \in R_{ac}^M \cap R_{ba}^M$. If, in addition, the team satisfies the transitivity axiom, then $s_2(y)$ cannot be assigned any element of the domain of M. The general approach is to use the dependence and independence properties to generate constraints on the group decision. These constraints may or may not be jointly satisfiable, depending on the form of the group decision (e.g., whether the group decision is a ranking).

A second topic for further investigation is to explore to what extent our logic can be a unifying framework to reason about principles of group decision making. Our analysis has identified three types of dependence found in the social choice literature. Suppose that $w_1, \ldots, w_k$ and v are variables, and φ is a first-order formula.

1. $=(w_1, \ldots, w_k, v)$: The value assigned to v is completely determined by the values assigned to the w_i.
2. $=(\varphi(w_1), \ldots, \varphi(w_k), \varphi(v))$: The truth value of $\varphi(v)$ is completely determined by the truth values of the $\varphi(w_i)$.
3. $(\bigwedge\limits_{i=1}^{k} \varphi(w_i)) \supset \varphi(v)$: If each of the w_i satisfy φ, then v must also satisfy φ.

The logic from Section 5 is ideally suited to explore the relationship between these different levels of dependence, especially in conjunction with the independence properties discussed in Section 4. We further conjecture that our logic can capture the reasoning underlying many results related to Arrow's Theorem (e.g., the Muller-Satterthwaite Theorem [32], Wilson's Theorem [50], the Gibbard-Sattherthwaite Theorem [13, 41], and versions of Arrow's Theorem for an infinite population [10, 22]).

Finally, it is important to compare our formalization of Arrow's Theorem with other approaches using modal logic [2, 8, 47], first-order logic [15], and computer-aided proofs [45]. A complete comparison with these different logics for social choice will be left for future work.

References

1. Abramsky, S.: Arrow's theorem by arrow theory. In: Villaveces, A., Kossak, R., Kontinen, J. Hirvonen, Å. (eds.) Logic Without Borders: Essays on Set Theory, Model Theory, Philosophical Logic and Philosophy of Mathematics, pp. 15–30. De Gruyter, Boston, MA (2015)
2. Agotnes, T., van der Hoek, W., Wooldridge, M.: On the logic of preference and judgment aggregation. Auton. Agent. Multi-Agent Syst. **22**(1), 4–30 (2009)
3. Arrow, K.J.: Social Choice and Individual Values. Yale University Press, New Haven, CT (1951)
4. Blau, J.H.: Arrow's theorem with weak independence. Economica **38**(152), 413–420 (1971)
5. Campbell, D.E., Kelly, J.S.: Impossibility theorems in the Arrovian framework. In: Arrow, K.J., Suzumura, A.S.K. (eds.) Handbook of Social Choice and Welfare, Elsevier, vol. 1, pp. 35–94 (2002)
6. Campbell, D.E., Kelly, J.S.: Social welfare functions that satisfy Pareto, anonymity, and neutrality, but not independence of irrelevant alternatives. Soc. Choice Welf. **29**(1), 69–82 (2007)
7. Cato, S.: Independence of irrelevant alternatives revisited. Theor. Decis. **76**(4), 511–527 (2014)
8. Ciná, G., Endriss, U.: A syntactic proof of Arrow's theorem in a modal logic of social choice functions. In: Proceedings of the 14th International Conference on Autonomous Agents and Multiagent Systems (AAMAS-2015) (2015)
9. Coleman, J.L., Ferejohn, J.: Democracy and social choice. Ethics **97**(1), 6–25 (1986)

10. Fishburn, P.: Arrow's impossibility theorem: concise proof and infinitely many voters. J. Econ. Theory **2**, 103–106 (1970)
11. Galliani, P.: Inclusion and exclusion in team semantics: on some logics of imperfect information. Ann. Pure Appl. Log. **163**(1), 68–84 (2012)
12. Geanakoplos, J.: Three brief proofs of Arrow's theorem. Econ. Theory **26**(1), 211–215 (2005)
13. Gibbard, A.: Manipulation of voting schemes: a general result. Econometrica **41**(3), 587–601 (1973)
14. Grädel, E., Väänänen, J.: Dependence and independence. Stud. Logica **101**(2), 399–410 (2013)
15. Grandi, U., Endriss, U.: First-order logic formalisation of impossibility theorems in preference aggregation. J. Philos. Log. **42**(4), 595–618 (2013)
16. Hannula, M.: Axiomatizing first-order consequences in independence logic. Ann. Pure Appl. Log. **166**(1), 61–91 (2015)
17. Hausman, D.M.: Preference, Value, Choice and Welfare. Cambridge University Press, Cambridge (2011)
18. Herzberg, F., Eckert, D.: The model-theoretic approach to aggregation: impossibility results for finite and infinite electorates. Math. Soc. Sci. **1**(2012), 41–47 (64)
19. Kalai, E., Muller, E., Satterthwaite, M.A.: Social welfare functions when preferences are convex and continuous: Impossibility results. Public Choice **34**, 87–97 (1979)
20. Kelly, J.S.: Arrow Impossibility Theorems. Academic, New York (1978)
21. Kemp, M.C., Ng, Y.K.: On the existence of social welfare functions, social orderings and social decision functions. Economica **43**, 59–66 (1976)
22. Kirman, A.P., Sondermann, D.: Arrow's theorem, many agents and invisible dictators. J. Econ. Theory **5**, 267–277 (1972)
23. Kontinen, J.: On natural deduction in dependence logic. In: Villaveces, A., Kossak, R., Kontinen, J., Hirvonen, A. (eds.) Logic Without Borders: Essays on Set Theory, Model Theory, Philosophical Logic and Philosophy of Mathematics, pp. 297–304. De Gruyter, Boston, MA (2015)
24. Kontinen, J., Väänänen, J.: On definability in dependence logic. J. Log. Lang. Inf. **18**(3), 317–332 (2009)
25. Kontinen, J., Väänänen, J.: Axiomatizing first-order consequences in dependence logic. Ann. Pure Appl. Log. **164**(11), 1101–1117 (2013)
26. List, C.: Social choice theory. In: Zalta, E.N. (ed.) The Stanford Encyclopedia of Philosophy, Winter 2013 edn. (2013)
27. Mackie, G.: Democracy Defended. Cambridge University Press, Cambridge (2003)
28. May, K.: A set of independent necessary and sufficient conditions for simply majority decision. Econometrica **20**(4), 680–684 (1952)
29. Morreau, M.: It simply does not add up: trouble with overall similarity. J. Philos. **107**, 469–490 (2010)
30. Morreau, M.: Arrow's theorem. In: Zalta, E.N. (ed.) The Stanford Encyclopedia of Philosophy, Winter 2014 edn. (2014)
31. Morreau, M.: Mr. Fit, Mr. Simplicity and Mr. Scope: from social choice to theory choice. Erkenntnis **79**, 1253–1268 (2014)
32. Muller, E., Satterthwaite, M.A.: The equivalence of strong positive association and strategy-proofness. J. Econ. Theory **14**, 412–418 (1977)
33. Okasha, S.: Theory choice and social choice: Kuhn versus Arrow. Mind **120**, 83–115 (2011)
34. Pacuit, E.: Voting methods. In: Zalta, E.N. (ed.) The Stanford Encyclopedia of Philosophy, Winter 2012 edn. (2012)
35. Parks, R.P.: An impossibility theorem for fixed preferences: a dictatorial Bergson-Samuelson welfare function. Rev. Econ. Stud. **43**(3), 447–450 (1976)
36. Riker, W.H.: Liberalism Against Populism. W. H. Freeman, San Francisco, CA (1982)
37. Roberts, K.W.S.: Social choice theory: the single-profile and multi-profile approaches. Rev. Econ. Stud. **47**(2), 441–450 (1980)
38. Rubinstein, A.: The single profile analogues to multi profile theorems: mathematical logic's approach. Int. Econ. Rev. **25**(3), 719–730 (1984)

39. Saari, D.: Mathematical structure of voting paradoxes: II. Positional voting. J. Econ. Theory **15**(1), 55–102 (2000)
40. Saari, D.: Chaotic Elections! A Mathematician Looks at Voting. American Mathematical Society, Providence, RI (2001)
41. Satterthwaite, M.A.: Strategy-proofness and Arrow's conditions: existence and correspondence theorems for voting procedures and social welfare functions. J. Econ. Theory **10**(2), 187–217 (1975)
42. Sen, A.: The possibility of social choice. Am. Econ. Rev. **89**(3), 349–378 (1999)
43. Sen, A., Maskin, E.: The Arrow Impossibility Theorem. Columbia University Press, New York (2014)
44. Stegenga, J.: An impossibility theorem for amalgamating evidence. Synthese **190**(2), 2391–2411 (2013)
45. Tang, P., Lin, F.: Computer-aided proofs of Arrow's and other impossibility theorems. Artif. Intell. **173**(11), 1041–1053 (2009)
46. Taylor, A.D.: Social Choice and the Mathematics of Manipulation. Cambridge University Press, Cambridge (2005)
47. Troquard, N., van der Hoek, W., Wooldridge, M.: Reasoning about social choice functions. J. Philos. Log. **40**(4), 473–498 (2011)
48. Väänänen, J.: Dependence Logic: A New Approach to Independence Friendly Logic. Cambridge University Press, Cambridge (2007)
49. Väänänen, J.: Dependence logic. Talk — Dagstuhl Seminar (2010)
50. Wilson, R.: Social choice theory without the Pareto principle. J. Econ. Theory **5**, 478–486 (1972)
51. Young, H.P.: Optimal voting rules. J. Econ. Perspect. **9**(1), 51–64 (1995)

Introduction to Secret-Sharing

Andreas Blass

Abstract This is the written form of a talk that I gave at the Dagstuhl seminar "Dependence Logic: Theory and Applications". My purpose is to explain what the theory of secret-sharing is about; to point out its connections with the fundamental notions, dependence and independence, of dependence logic; and to indicate some of the results and open problems of this theory.

1 Basic Examples and Definitions

I shall begin with the simplest example of secret-sharing and gradually build up to more complex and interesting examples.

Suppose there is a secret code s, say a string of 100 binary digits, for opening a bank vault. Suppose the bank's rules say that the manager and assistant manager together should be able to open the vault, but neither one should be able to do so alone. An obvious way to achieve this is to give the manager the first 50 bits of s and give the assistant manager the remaining 50 bits (and give both of them the instructions for combining these parts of s). The problem studied in the theory of secret-sharing is more difficult than this example in three ways.

First, one requires not only that each individual is unable to recover s and open the vault by himself but rather that each individual has absolutely no information about s — not half of s as in the preceding paragraph.

At first sight, it may seem paradoxical that two people, neither of whom knows anything at all about s, could by cooperating recover s. Nevertheless, this goal is easily achieved as follows. Begin by (ignoring s and) flipping a fair coin 100 times, generating a uniformly random string r of 100 bits. Give this string r to the manager, and observe that he certainly learns nothing about s, since all he has is a random string independent of s. Give the assistant manager the bitwise XOR, $r \oplus s$, of the random string r and the secret s. The assistant manager also learns nothing about s. Indeed, he gets a random string (randomized over the probability space of the coin flips that generated r), whose probability distribution is the same no matter what

A. Blass (✉)
Mathematics Department, University of Michigan, Ann Arbor, MI 48109–1043, USA
e-mail: ablass@umich.edu

s is. So nothing can be inferred, even probabilistically, about s from the string $r \oplus s$. Nevertheless, the manager and assistant manager together can easily compute s, just by forming the bitwise XOR of their shares, $r \oplus (r \oplus s) = s$.

The second feature that makes secret-sharing more difficult than the example treated so far is that, in general, one has a larger number than 2 of people (usually called "participants") among whom the secret is to be shared. Fortunately, the idea in the preceding paragraph extends easily to any (finite) number n of participants. The goal is to give each participant a "share" of the secret s in such a way that all n participants jointly can recover s but no set of $n - 1$ participants has any information at all about s. To achieve this goal, form $n - 1$ independent random bit-strings $r_1, r_2, \ldots, r_{n-1}$ of the same length as s, and give these to the first $n - 1$ participants as their shares. Give the last participant the bitwise XOR $r_1 \oplus r_2 \oplus \cdots \oplus r_{n-1} \oplus s$. All n participants together can recover s as the bitwise XOR of all their shares. But any $n - 1$ participants together have only $n - 1$ independent, uniformly random bit-strings; in particular, the joint probability distribution of their shares is independent of s, so that their shares, even when combined, tell them nothing about s.

The third feature that makes secret-sharing more difficult is that, in general, the criterion for which sets of participants should be able to reconstruct the secret is more complicated than "everyone". For example, suppose the bank has eight tellers, any three of whom should be able to recover s and open the vault, but no two of whom should have any information about s.

This problem can also be solved, though at some cost, by the methods described above. For each set A of three of the eight tellers, give the three members of A shares as above, so that all three together can open the vault but no two learn anything about s. That is, give two of the members of A independent random strings r_1 and r_2, and give the third member $r_1 \oplus r_2 \oplus s$. Also, for different 3-element sets A, use independent coin flips to form the random strings r_1 and r_2 for A, so that the participants gain no information by comparing the shares they get as elements of different A's.

This scheme works, but it is inefficient in that the shares are quite large. Each participant is a member of $\binom{7}{2} = 21$ three-element sets A and gets a 100-bit share for each of these sets, so his total share is 2100 bits, much larger than the secret. Can one do better? If so, how much better can one do? These questions, asked for general criteria of who should be able to reconstruct s, are the central questions of secret-sharing theory. To formalize them, one makes the following definitions:

Definition 1. Let I be a nonempty finite set (whose elements are called *participants*). An *access structure* on I is a family Γ of subsets of I with the following three properties:

- If $X \in \Gamma$ and $X \subseteq Y \subseteq I$, then $Y \in \Gamma$.
- $\varnothing \notin \Gamma$.
- $I \in \Gamma$.

One calls the elements of Γ the *qualified subsets* of I. The access structure Γ is called *reduced* if

- For each $i \in I$ there is some $X \in \Gamma$ with $X - \{i\} \notin \Gamma$.

The intention behind this definition is that we want to distribute shares of a secret s among the participants in such a way that, if $X \in \Gamma$, then the shares of the members of X jointly determine s, whereas if $X \notin \Gamma$, then the shares of the members of X, even jointly, contain no information at all about s.

The three requirements in the definition of access structure follow immediately from this intention. The first requirement, monotonicity, means that, if the participants in X can determine s, then adding more participants to the set cannot prevent this determination; one could simply ignore the shares of the participants in $Y - X$. The second requirement expresses that, with no information at all, one cannot determine s, unless there is only one possible value for s (which would mean that s is not much of a secret). The third requirement means that all the participants together should have enough information to determine the secret. If it were violated, then we could just give all participants empty shares.

The additional requirement for a reduced access structure amounts to a normalization. If it failed for some i, then the share of i would never be relevant, so we could simply give i an empty share and reduce the problem to an access structure on $I - \{i\}$. So questions about secret-sharing for arbitrary access structures generally amount to the same questions for reduced access structures.

Definition 2. Let Γ be an access structure on I. A *(perfect) secret-sharing scheme*[1] for Γ is a randomized function $F : \{0, 1\}^l \times I \longrightarrow \{0, 1\}^t$, for some positive integers l and t such that:

1. If $X \in \Gamma$ and if $\langle F(s, i) : i \in X \rangle$ and $\langle F(s', i) : i \in X \rangle$ have a common possible value, then $s = s'$.
2. If $X \notin \Gamma$, then the probability distribution of $\langle F(s, i) : i \in X \rangle$ is independent of s.

The ratio l/t is called the *rate* of the scheme F.

Notice that in this definition, because F is a randomized function, $\langle F(s, i) : i \in X \rangle$ is a random variable. The phrases "possible value" and "probability distribution" are to be understood in the usual sense for random variables. Because I, l, and t are finite, we can take the relevant probability space to be finite as well.

The idea behind the definition is that the secret s to be shared is a bit-string of length l. The scheme randomly produces shares $F(s, i)$ of length t for each of the participants i. The first clause in the definition says that the members of a qualified set X can jointly recover s from their shares. Any possibility for the indexed family of their shares arises from only one s. So s is completely determined by the information available to these participants. The second clause says that the shares

---

[1]There are also imperfect schemes, in which unqualified subsets of I are allowed to have some limited information about s. We shall not consider such schemes in this paper, so we can and will safely omit "perfect". For even more brevity, we may sometimes omit "secret-sharing" and just say "scheme".

of an unqualified set give no information about s, because they are random with the same probability distribution regardless of s.

The rate measures how much larger the shares are than the secret. Thus, in our example above where any three of eight tellers form a qualified set, the rate was $1/21$. In the other schemes discussed above, the rate was 1. Higher rates are preferred because they represent more efficient sharing, i.e., shorter shares compared with the length of the secret.

Remark 1. Definition 2 incorporates the convenient simplifying assumption that the possible values of the secret s are all of the binary strings of some fixed length l. One can easily extend the whole theory to handle arbitrary finite sets S of possible secrets. In that case, the length l should be replaced by $\log |S|$. Here and throughout this paper, logarithms are to the base 2.

Definition 3. The *rate* of an access structure is the supremum of the rates of its secret-sharing schemes.

Definition 4. A secret-sharing scheme is called *ideal* if its rate is 1. An access structure is called *ideal* if it admits an ideal secret-sharing scheme; it is called *nearly ideal* if its rate is 1.

The terminology "ideal" is justified by the following result from [8]:

Proposition 1. *No secret-sharing scheme has rate greater than 1.*

Examples in [11, 13] show that an access structure can be nearly ideal without being ideal: The supremum in the definition of an access structure's rate need not be attained.

The central problem of secret-sharing theory is to find the rates of access structures, either for particular access structures or for families of them.

2 Dependence and Independence

Because this paper arose from a conference on dependence logic, it seems worthwhile to explicitly discuss the connection between these topics. In only this section, we presuppose familiarity with the basic notions of dependence logic.

The purpose of a secret-sharing scheme F for an access structure Γ is to make the secret s dependent on the combined shares of any qualified set of participants and independent of the combined shares of any unqualified set. To formalize this from the viewpoint of dependence logic, we proceed as follows.

We work with a two-sorted structure, in which one sort is the set $\{0, 1\}^l$ of possible values for the secret and the other sort is the set $\{0, 1\}^t$ of possible shares. We use the variable s to range over the first sort, and we use variables x_i, one for each participant $i \in I$, to range over the second sort. Let T be the team consisting of those assignments v to these variables that occur with positive probability in the scheme F. That is, the joint probability of the events defined by the equations $F(v(s), i) = v(x_i)$, for all $i \in I$, should be positive.

Then requirement (1) in Definition 2 says exactly that this structure and team satisfy the dependence relation $=((x_i)_{i \in X}; s)$ for every qualified X.

Requirement (2) implies that the structure and team satisfy the independence assertion $(x_i)_{i \in X} \perp s$ for every unqualified X, but requirement (2) is stronger than this assertion. To see the implication, suppose that our team T contains assignments v_1 and v_2. So $\langle F(v_1(s), i) : i \in X \rangle = \langle v_1(x_i) : i \in X \rangle$ has positive probability. Then, thanks to requirement (2), $\langle F(v_2(s), i) : i \in X \rangle = \langle v_1(x_i) : i \in X \rangle$ also has positive probability. Therefore, T contains the assignment v that agrees with v_1 on the x_i's and with v_2 on s. Thus, it satisfies $(x_i)_{i \in X} \perp s$.

The preceding argument did not use the full strength of requirement (2). The requirement says that certain probability distributions are equal, but we needed only that the same things have positive probability under these distributions; the actual probabilities could be different. Intuitively, $(x_i)_{i \in X} \perp s$ says merely that, given the shares of an unqualified set of participants, one cannot completely exclude any value of the secret. It leaves open the possibility that those shares might make certain possible values of the secret extremely unlikely and thus provide some probabilistic information about the secret. Indeed, they might even allow one to make a very good guess as to the actual value of the secret. Requirement (2), in contrast, prohibits getting any information about the secret, even probabilistic information, from the shares of an unqualified set.

3   Shamir's Scheme and Linear Schemes

In this section, we exhibit some clever secret-sharing schemes for certain sorts of access structures.

The first is due to Shamir [17]; we explain it with the three-of-eight tellers example considered above. Fix a prime number p slightly larger than 2^{100}, so that the secret s can be viewed as the binary expansion of an integer between 0 and $p - 1$ and thus as an element of the field $\mathbb{Z}/p$. Choose uniformly at random an (at most) quadratic polynomial $f(x) = ax^2 + bx + s$ over $\mathbb{Z}/p$, with constant term $f(0)$ equal to the secret s. So the coefficients a and b are chosen uniformly at random in $\mathbb{Z}/p$. Then give teller number i (for $1 \leq i \leq 8$) the share $f(i)$. Any three of the tellers have, jointly, the values of the quadratic polynomial f at three elements of the field $\mathbb{Z}/p$. This is enough to determine f by Lagrange interpolation and thus to determine its constant term s. Any two of the tellers, on the other hand, have, jointly, just two independent, uniformly random elements of $\mathbb{Z}/p$, independently of s, so they know nothing about s.

The rate of this scheme is nearly 1. Precisely, if p is chosen between 2^{100} and 2^{101}, which is always possible by a well-known theorem of Chebyshev, then the shares can be taken to be 101 bits long, so the rate is $100/101$. For an ideal scheme, instead of using a field of the form $\mathbb{Z}/p$, use the field of cardinality 2^{100}, and give teller i the share $f(a_i)$ for some fixed, distinct, and known (to all participants) elements a_i

of this field. (You can't use $f(i)$ as above, since the field now has characteristic 2 so the integers from 1 to 8 are not distinct in it.)

Shamir's scheme clearly works, with obvious modifications, for any access structure defined by a numerical threshold, i.e., the qualified sets are those of cardinality at least a certain, fixed r in the range from 1 to the number n of participants, provided the number of possible secrets 2^l exceeds n. Assign to each participant i, once and for all, a non-zero element a_i in the field K of size 2^l, using distinct elements a_i for distinct participants i. To share a secret s, regarded as an element of K in some standard way, choose uniformly at random a polynomial f over K of degree at most $r - 1$, with constant term s. Give each participant i the share $f(a_i)$. As before, any r of these shares suffice to determine by interpolation the polynomial f (since its degree is only $r - 1$) and thus its constant term s. But, as one can straightforwardly verify, any $r - 1$ shares are independent, uniformly random elements of K, regardless of s.

This construction proves that threshold access structures are ideal.

Another class of ideal access structures are the linear ones, obtained as follows. Let V be a finite-dimensional vector space over a finite field K (so V is also finite). Assume that K is large enough so that the desired secrets can be regarded as elements of K; so the cardinality $|K|$ should be 2^l or slightly larger. Let there be given a non-zero vector $v_i \in V$ for each participant $i \in I$ and one more non-zero vector w in the subspace spanned by the v_i's. Define

$$\Gamma = \{X \subseteq I : w \in \text{span}\{v_i : i \in X\}\}.$$

For any Γ of this form, we can obtain an almost ideal (and actually ideal if $|K| = 2^l$) secret-sharing scheme as follows. Given a secret $s \in K$ to be shared, choose a linear functional $f : V \longrightarrow K$ uniformly at random subject to the constraint that $f(w) = s$. Give each participant i the share $f(v_i)$.

If $X \subseteq I$ is qualified, then w is in the linear span of the v_i's for $i \in X$. That is, $w = \sum_{i \in X} \alpha_i v_i$ for some scalars $\alpha_i \in K$. It follows that $s = f(w) = \sum_{i \in X} \alpha_i f(v_i)$, so s is determined by the shares $f(v_i)$ of the participants $i \in X$.

If, on the other hand, $X \subseteq I$ is unqualified, then w is not in the linear span S of the v_i for $i \in X$. One can easily check that, when we chose f, the constraint $f(w) = s$ had no effect on the restriction $f \upharpoonright S$ to a subspace not containing w. The restriction is equally likely to be any linear functional on this subspace S. Thus, the participants $i \in X$ see only the values, at their v_i's, of a uniformly randomly chosen linear functional on S, independently of the secret s.

4 Matroids and Matroid-Related Schemes

It turns out that the existence of ideal schemes, and even the existence of schemes with rate near 1, depends on a connection between access structures and matroids. This section is devoted to explaining the connection, beginning with the definition of "matroid". Although infinite matroids have been considered in the literature, we confine our attention here to finite ones.

Definition 5. A *matroid* consists of a finite underlying set E together with a closure operator C on the subsets of E satisfying the exchange axiom. That is, $C : \mathscr{P}(E) \longrightarrow \mathscr{P}(E)$ must have the following four properties (of which the first three constitute the definition of "closure operator"): For all $X, Y \subseteq E$ and all $a, b \in E$,

Monotone   If $X \subseteq Y$, then $C(X) \subseteq C(Y)$,
Inflationary $X \subseteq C(X)$,
Idempotent $C(C(X)) = C(X)$,
Exchange   If $b \in C(X \cup \{a\})$ but $b \notin C(X)$, then $a \in C(X \cup \{b\})$.

We shall also impose the following convenient normalization on the matroids that we consider:

Loop-free $C(\varnothing) = \varnothing$.

If a matroid fails to satisfy this condition, then the interesting questions about it reduce to the same questions about the smaller matroid $E - C(\varnothing)$ with the induced closure operator.

Many authors impose the further requirement that all one-element subsets of E are fixed by C, but for our purposes it will be convenient to refrain from this additional normalization.

Although a matroid is technically pair (E, C), we sometimes refer to E as a matroid, when C is clear from the context.

There are a great many equivalent definitions of "matroid"; see [3] for many of these.

The original motivation for the concept of matroid was to generalize the key combinatorial properties of the following two examples:

Example 1. Let E be a finite subset of a vector space V over a field K. For any subset X of E, define $C(X)$ to be the set of those elements of E expressible as linear combinations (over K) of elements of X. That this is a closure operator is obvious. The exchange property is also easy to prove: The equation expressing b as a linear combination of $X \cup \{a\}$ must involve a with a non-zero coefficient; otherwise it would express b as a linear combination of X. But then this equation can be solved for a, expressing it as a linear combination of $X \cup \{b\}$. (Note that it is important here that we deal with a vector space over a field, not a module over a ring, because to solve for a we need to invert its coefficient.)

This matroid is loop-free if and only if E does not contain the zero vector.

A matroid of this form is said to be *representable* over K. To say that a matroid is *representable*, without specifying a field, means that it is representable over some field.

Given a matroid E, the statement that it is representable over a specific field K is a first-order, existential statement about K. It follows that, if E is representable over a field K, then it is representable over any extension of K, in particular over the algebraic closure of K. By the completeness of the theory of algebraically closed fields of any fixed characteristic, it follows that E is representable over the algebraic closure of the prime field of the same characteristic. If that characteristic is a prime p, then this field is the directed union of finite fields, and so E is also representable over all sufficiently large[2] finite fields of the same characteristic. If, on the other hand, the characteristic is 0, then a routine compactness argument shows that E is also representable over fields of any sufficiently large prime characteristic. In summary, any representable matroid is representable over some finite field, and in particular over all sufficiently large finite fields of at least one prime characteristic.

Example 2. Let G be an undirected graph, and E its set of edges. Define C : $\mathscr{P}(E) \longrightarrow \mathscr{P}(E)$ by letting $C(X)$, for any $X \subseteq E$, consist of those edges $uv \in E$ such that u and v are in the same connected component of the subgraph that has the same vertices as G but only the edges in X. That is, there is a path from u to v consisting entirely of edges in X. Equivalently, either $uv \in X$ or uv completes a circuit whose other edges are in X.

To see that this is a matroid, note that the monotone and inflationary properties are trivial, and that idempotence is easily proved by taking any path in $C(X)$ and replacing each of its edges by a path in X with the same endpoints (and eliminating detours, so as to get a path rather than a walk).

To verify the exchange property, let X, a, b be as there. So the endpoints of b are joined by a path P consisting of edges in $X \cup \{a\}$. That path must contain a, for otherwise it would show that $b \in C(X)$, contrary to hypothesis. So, by taking the parts of P before and after a, together with b, we obtain a path in $X \cup \{b\}$ joining the endpoints of a.

This matroid is loop-free if and only if G has no loops, i.e., no edges that join a vertex to itself. Of course, this is the origin of the terminology "loop-free" for matroids.

Matroids of this form are called *graphical*.

Every graphical matroid is representable over $\mathbb{Z}/2$ (and therefore over all fields of characteristic 2). Given a graph G, let V be the vector space over $\mathbb{Z}/2$ having the vertices of G as a basis. Associate to each edge $uv \in E$ the vector $u + v$ that is the sum of the two endpoints of the edge. It is not difficult to see that this association is an isomorphism from the graphical matroid given by G to the representable matroid given by this set of vectors.

Much of the terminology of matroid theory is inherited from either linear algebra (Example 1) or graph theory (Example 2). We already mentioned the graph origin of "loop-free"; here are some more examples that will be useful.

Definition 6. A subset X of a matroid E is said to be *dependent* if some $x \in X$ is in the closure of $X - \{x\}$; otherwise it is *independent*. A maximal independent set is called a *basis* of the matroid. A minimal dependent set is called a *circuit*.

---

[2]Here and below, "sufficiently large fields" is to be understood as all fields containing a specific subfield, not as all fields of sufficiently large cardinality.

The connection between matroids and access structures is as follows:

Definition 7. Let (E, C) be a matroid, and let d be an element of E such that $d \in C(E - \{d\})$. The access structure associated to (E, C) and d consists of the set $E - \{d\}$ of participants and the family

$$\Gamma = \{X \subseteq E - \{d\} : d \in C(X)\}$$

of qualified sets. An access structure associated in this way to a matroid and an element of it is said to be *matroid-related*. The matroid element d that is removed in forming the access structure is called the *dealer*.[3]

In this definition, the assumption that $d \in C(E - \{d\})$ ensures that, as required in our definition of access structures, the set of all participants is qualified. The requirement that $\varnothing$ is not qualified follows from our assumption that matroids are loop-free. That Γ is closed upward is immediate from the monotonicity of the closure operator C.

It is not difficult to show, using the exchange property of matroids, that the access structure associated to matroid E and an element d is reduced if and only if the circuits that contain d cover all of E.

The importance of matroids in the theory of secret-sharing arises from the fact [2] that every ideal access structure is necessarily matroid-related. In fact, the following stronger result was proved in [10]. Unfortunately, the proof of this result involves lengthy checking of cases, so we do not present it here.

Theorem 1. *If an access structure admits a secret-sharing scheme with rate $> \frac{2}{3}$, then it is matroid-related.*

The linear schemes described in Section 3 provide the following partial converse:

Theorem 2. *All access structures associated to representable matroids have ideal secret-sharing schemes.*

Questions that naturally arise here include: What are the rates of access structures that are not matroid-related? Is the bound $\frac{2}{3}$ in Theorem 1 optimal? What are the rates of matroid-related structures when the matroid is not representable? In the next section, we present some examples that partially answer these questions.

---

[3]The idea behind this terminology is to view d as a pseudo-participant who knows the secret s and distributes the shares to the others.

5 Examples

Every access structure with three or fewer participants is isomorphic to one of five examples, each of which admits an ideal scheme and is in fact associated to a representable matroid. These five examples can be described as follows:

1. The only qualified set consists of all three participants. The corresponding linear matroid consists of four vectors in general position in three-dimensional space. An ideal scheme was described in Section 1.
2. There are a manager and two tellers; the manager and at least one teller together can open the vault. The linear matroid consists of three vectors in general position in two-dimensional space, plus a fourth vector that is a multiple of one of the first three. The two linearly dependent vectors correspond to the two tellers. An easy way to share a secret s in this situation is to form a random bit-string r of the same length as the secret, give r to both of the tellers, and give $r \oplus s$ to the manager.
3. A threshold scheme where any two of the three participants form a qualified set. The matroid consists of four vectors in general position in a two-dimensional space. An ideal scheme, due to Shamir, was described in Section 3.
4. There is a manager, who should be able to open the vault by himself, and there are two tellers, who should, together, be able to open the vault. The matroid is the same as in (2) above, but now one of the two linearly dependent vectors corresponds to the manager and the other is the dealer d. To share a secret s, give one teller a random r, give the other teller $r \oplus s$, and give the manager s.
5. Each participant alone should be able to open the vault. The matroid consists of four multiples of a single vector. To share a secret, give every participant the secret.

With four participants, there is a non-matroid-related access structure, as follows:

Example 3. Let the participants be numbered $1, 2, 3$, and 4. Call a set of participants qualified if it contains some two participants whose numbers are consecutive. To see that this structure is not matroid-related, suppose it came from a matroid on the set $\{1, 2, 3, 4, d\}$ as in Definition 7. Then $d \in C(\{1, 2\})$ but $d \notin C(\{1\})$, because $\{1, 2\}$ is qualified and $\{1\}$ isn't. By the exchange property of matroids, $2 \in C(\{1, d\})$. Similarly, $3 \in C(\{2, d\})$. By monotonicity and idempotence of C, it follows that $3 \in C(\{1, d\})$. Since $d \notin C(\{1, 3\})$, the exchange property requires that $3 \in C(\{1\})$. Applying monotonicity and idempotence again, along with the fact that $d \in C(\{3, 4\})$ as $\{3, 4\}$ is qualified, we get that $d \in C(\{1, 4\})$, a contradiction because $\{1, 4\}$ is not qualified.

According to Theorem 1, the best rate that we could hope for in this situation is $\frac{2}{3}$. It turns out that this rate is achievable as follows. Let the secret be a string of length 2. (Analogous schemes are available for any even length.) To share a secret s whose two bits are s_1 and s_2, flip a fair coin four times to generate random bits r_1, r_2, r_3, r_4 and distribute shares as follows:

- Participant 1 gets r_1 and r_4.

- Participant 2 gets $r_1 \oplus s_1$, r_3, and $r_4 \oplus s_2$.
- Participant 3 gets r_1, $r_2 \oplus s_1$, and $r_3 \oplus s_2$.
- Participant 4 gets r_2 and r_3.

Participants 1 and 2 can reconstruct s_1 as $r_1 \oplus (r_1 \oplus s_1)$. So can 2 and 3. Participants 3 and 4 can reconstruct s_1 as $(r_2 \oplus s_1) \oplus r_2$. Similarly, each qualified pair can reconstruct s_2, using their shares involving r_3 or r_4. Each unqualified pair sees only four independent, uniformly random bits, regardless of s, so this is a secret-sharing scheme. Its rate is $\frac{2}{3}$ because the secret is two bits and the largest shares, those of participants 2 and 3, are three bits each. (Our formal definition of secret-sharing schemes requires all shares to be the same size, but this can be achieved trivially by giving participants 1 and 4 an uninformative extra bit, say 0.)

This example shows that the bound $\frac{2}{3}$ in Theorem 1 is optimal.

The rates of all the minimal, non-matroid-related access structures are known. Before stating the precise result, we need to clarify "minimal"; what ordering is used here?

Definition 8. A *minor* of an access structure (I, Γ) is an access structure of the form $(I - (P \cup Q), \Delta)$, where P and Q are disjoint subsets of I, and where

$$\Delta = \{X \subseteq I - (P \cup Q) : X \cup P \in \Gamma\}.$$

Intuitively, the transition from the access structure Γ to its minor Δ in this definition can be described by saying that the participants in P announce their shares publicly and leave, and the participants in Q leave quietly, not divulging their shares.

Any minor of a matroid-related access structure is matroid-related; indeed, the relevant construction is what is called a minor in matroid theory; see [4] for the definitions.

Returning to "minimal", we mean those non-matroid-related access structures all of whose minors are matroid-related. They are listed, with their rates, in the following result, in which the list of examples is due to Seymour [15], the rates in the first item and the $n = 3$ case of the second are due to Stinson [19], and the rates for larger n are due to Metcalf-Burton [14].

Theorem 3. *The minimal non-matroid-related access structures and their rates are as follows:*

- *Three access structures on four points, namely Example 3 above, the access structure obtained from that example by adding $\{1, 3\}$ as a qualified set, and the access structure obtained from that same example by making $\{3, 4\}$ unqualified but keeping $\{1, 3, 4\}$ qualified, all have rate $\frac{2}{3}$.*
- *For each $n \geq 3$, the access structure on $I = \{0, 1, 2, \ldots, n\}$, whose qualified sets are $\{1, 2, \ldots, n\}$ and all sets that contain 0 and at least one other participant, has rate $\frac{n-1}{2n-3}$.*

Apart from Theorem 1, little seems to be known about the rates of non-matroid-related access structures other than these minimal ones.

We turn next to access structures that are matroid-related but with non-representable matroids. All matroids on sets of cardinality seven or less are representable, but there is a famous example, the Vámos matroid [20], with eight points, that is not representable.

Definition 9. The Vámos matroid consists of eight elements, which we number from 1 to 8, and the closure operator C defined as follows:

- If $|X| \leq 2$, then $C(X) = X$.
- If $|X| \geq 5$, then $C(X) = \{1, 2, 3, 4, 5, 6, 7, 8\}$.
- If X is one of the five "special" 4-element sets

$$\{1, 2, 3, 4\}, \{1, 2, 5, 6\}, \{3, 4, 5, 6\}, \{3, 4, 7, 8\}, \{5, 6, 7, 8\},$$

then $C(X) = X$.
- If $|X| = 4$ and X is not one of the five special sets, then $C(X) = \{1, 2, 3, 4, 5, 6, 7, 8\}$.
- If $|X| = 3$ and X is included in one of the five special sets, then $C(X)$ is that special set.
- If $|X| = 3$ and X is not included in a special set, then $C(X) = X$.

Note that each special set is the union of two of the so-called Vámos pairs, the sets $\{1, 2\}, \{3, 4\}, \{5, 6\}, \{7, 8\}$; also, each union of two Vámos pairs *except* $\{1, 2, 7, 8\}$ is special.

We sketch briefly the proof that the Vámos matroid is not representable. If it were representable, then, because its bases have cardinality 4, we could identify its points with vectors in a 4-dimensional space in such a way that C corresponds to linear span. To simplify the picture, we can pass to the associated 3-dimensional projective space, where $y \in C(X)$ now means that y is in the projective subspace spanned by X. We consider the four lines spanned by the Vámos pairs; call these lines 12, 34, 56, and 78. Because $\{1, 2, 3, 4\}$ is special and thus dependent, the lines 12 and 34 are coplanar and thus meet, say at p; similarly, 12 and 56 meet, say at q; and 34 and 56 meet, say at r. If p, q, r were all distinct, they would determine a plane that contains all six of the points 1, 2, 3, 4, 5, and 6, which disagrees with C. So two of p, q, r are equal, so the three lines 12, 34, and 56 are concurrent. We have thus shown that the intersection point q of 34 and 56 also lies on 12; a symmetrical argument shows that the same q also lies on 78. But then the lines 12 and 78, meeting at q, are coplanar, which disagrees with C because $\{1, 2, 7, 8\}$ is not special.

By generalizing this argument, one can arrive at Ingleton's inequality [7], which asserts that, for any four subspaces A, B, C, D of a finite-dimensional vector space, the dimensions of these spaces and the subspaces spanned by two or three of them satisfy

$$\dim(A + B) + \dim(A + C) + \dim(B + C) + \dim(B + D) + \dim(C + D) \geq$$

$$\geq \dim(B) + \dim(C) + \dim(A + D) + \dim(A + B + C) + \dim(B + C + D).$$

Conversely, if one already knows Ingleton's inequality, then the non-representability of the Vámos matroid follows easily by taking A, B, C, D to be the subspaces spanned by the Vámos pairs in a hypothetical linear representation of the matroid.

The Vámos matroid gives rise to two associated access structures. A priori, there would be eight access structures, corresponding to the eight choices of dealer, but the matroid has automorphisms that make elements 1, 2, 7, and 8 equivalent and make 3, 4, 5, and 6 equivalent. So we need only consider the access structures V_1 and V_3 obtained by choosing 1 or 3 as the dealer. Perhaps surprisingly, the rates of these access structures are not known. Some upper and lower bounds are, however, known, beginning with the fact that neither of these access structures admits an ideal scheme [16]. Later, upper bounds of 10/11 and 9/10 were obtained in [1] for V_1 and V_3, respectively; these were improved to 8/9 and 17/19, respectively in [14]. The best lower bound (as far as I know) is 3/4 for both of these access structures [12].

6   Entropy and Inequalities

As we saw in Section 2, the concept of secret-sharing involves both a purely logical notion of dependence (of the secret on a qualified set of shares) and a probabilistic notion of independence (of the secret from an unqualified set of shares) that is stronger than a purely logical one. In fact, most of the literature uses a probabilistic formulation even for the logical part of the definition, and it makes heavy use of the concept of entropy. Because entropy considerations play a central role in computing or estimating rates, we give here a brief introduction to this point of view.

Definition 2 of secret-sharing schemes involved randomization in the function F that produces, from a secret s and a participant i, the share $F(s, i)$ of that participant. A common alternative definition introduces an additional randomization by making the secret a random variable also. From this point of view, which I'll call the "random-secret" point of view, a secret-sharing scheme is a family of random variables, all defined on the same probability space, such that one of these random variables S represents the secret and the rest, Q_i, indexed by the participants i, represent their shares. What we called $F(s, i)$ is then replaced by the variable Q_i conditioned on $S = s$. The only permitted values of the secret would be those for which $S = s$ has positive probability, so that this conditioning makes sense.

In the random-secrets approach, the required behavior for qualified and unqualified sets is expressed in terms of the notion of entropy introduced by Shannon in [18].

Definition 10.   The *entropy* of a random variable V is

$$H(V) = -\sum_v \mathbf{P}(V = v) \log \mathbf{P}(V = v),$$

where v ranges over all values for which the probability $\mathbf{P}(V = v)$ is positive. The *joint entropy* of several random variables, $H(V_1, \ldots, V_n)$, is the entropy of the tuple $\langle V_1, \ldots, V_n \rangle$ considered as a random variable in its own right.

The intuitive meaning behind the definition of $H(V)$ is that $-\log \mathbf{P}(V = v)$ is the amount of information one obtains by learning that, in a particular experiment, v was the value obtained by V. Thus, $H(V)$ is the average amount of information obtained by learning the value of V.

In the context of secret-sharing schemes, we abbreviate the joint entropy of the shares of several participants, $H(\langle Q_i : i \in X \rangle)$ as $H(X)$.

The requirements for a secret-sharing scheme, given an access structure Γ, are

- For each $X \in \Gamma$, $H(X, S) = H(X)$.
- For each $X \notin \Gamma$, $H(X, S) = H(X) + H(S)$.

The first of these says that, if X is qualified, then learning the shares of X and the secret involves no more information than just learning the shares of X. This is equivalent to item (1) in Definition 2 for the permitted secrets, the values that S takes with positive probability. The second is the entropy formulation of probabilistic independence and thus matches item (2) in Definition 2.

The importance of the random-secret point of view, and especially the use of entropy, is that the entropy function has various known properties that are useful in computing or at least bounding the possible rates of access structures. Among these properties are some basic inequalities, often called Shannon inequalities. With U, V, W denoting tuples of random variables, these inequalities are

Bounded   If U takes only n values with positive probability, then $H(U) \leq \log n$, with equality if and only if all n values are equally probable.

Monotone $H(U) \leq H(U, V)$.

Submodular $H(U, V, W) + H(W) \leq H(U, W) + H(V, W)$.

Remark 2. One defines *conditional entropy*, $H(U|V)$ as the average, over all values v of V, of the entropy of U restricted to the probability space defined by $V = v$ with the conditional probability distribution. Equivalently, $H(U|V) = H(U, V) - H(V)$. That is, it is the amount of information that must be added to knowledge of V to also determine U. In terms of conditional entropy, the requirements for a secret-sharing scheme say that $H(S|X)$ is 0 when X is qualified and $H(S)$ when X is unqualified. That is, if we know the shares of X, what is still needed in order to learn the secret is nothing if X is qualified (what we already know determines the secret) and all of S if X is unqualified (what we already know tells us nothing about the secret).

Among the Shannon inequalities, monotonicity says that conditional entropy is nonnegative. Submodularity says that $H(V|U, W) \leq H(V|W)$. That is, if we already know W, then learning U will not make it more difficult to learn V.

Remark 3. The monotonicity and submodularity inequalities echo elementary inequalities about dimensions of vector spaces. Specifically, suppose U, V, W are linear subspaces of some finite-dimensional vector space. Then we have

Monotone $\dim(U) \leq \dim(U + V)$ and

Submodular $\dim(U + V + W) + \dim(W) \leq \dim(U + W) + \dim(V + W)$.

This similarity is no accident; indeed, any linear inequality between entropy functions (of arbitrary random variables) implies the corresponding inequality between dimensions of finite-dimensional vector spaces.

To see this, notice first that, by a compactness argument and the completeness of the theory of algebraically closed fields of any fixed characteristic, it suffices to consider vector spaces over finite fields K. To any such vector space Z, associate the probability space Z^* consisting of all linear functionals $f : Z \longrightarrow K$, with the uniform probability distribution. Any linear subspace V of Z determines a random variable whose values are linear functionals $V \longrightarrow K$, namely the random variable $\hat{V}$ whose value at any $f \in Z^*$ is the restriction of f to V. Notice that, for the sum of two subspaces, the random variable $\widehat{U + V}$ is equivalent to the pair of random variables $\langle \hat{U}, \hat{V} \rangle$; that is, the restriction of any linear functional f to $U + V$ determines and is determined by its restrictions to U and to V. Notice also that, if V has dimension d, then the random variable $\hat{V}$ takes exactly $|K|^d$ values, all with the same probability. Therefore, $H(\hat{V}) = \log(|K|^d) = d \log |K|$. Therefore, given any linear inequality for entropies (of random variables and tuples of those), we can apply it to the random variables $\hat{V}$ associated to vector subspaces V of Z, and we obtain, after cancelling a factor $\log |K|$, the corresponding inequality for dimensions (of subspaces and sums of those).

Unfortunately, the converse does not hold. An inequality that is valid for dimensions of vector spaces need not be valid for entropies of arbitrary random variables. An easy counterexample is Ingleton's inequality, mentioned in Section 5 above. It is violated by four very simple random variables: Let B and C be two uniform, independent bits (`true` or `false`, each with probability $\frac{1}{2}$), let A be their conjunction and let D be their disjunction.

Ingleton's inequality, which dates from 1971, was the only known non-Shannon inequality for dimensions of vector spaces until Kinser [9] and Dougherty, Freiling, and Zeger [5] found new ones, involving more variables, in 2009. Since the analog of Ingleton's inequality for entropy fails, it was reasonable to suppose that the Shannon inequalities are the only linear inequalities satisfied by the entropies of arbitrary random variables and tuples of them. So it came as a surprise when Zhang and Yeung [22] exhibited a new linear inequality for entropy. For any random variables X, Y, Z, U, they obtained

$$2H(Z) + 2H(U) + 4H(XZU) + H(X) + H(YZU) + H(XY) \leq$$
$$\leq 3H(ZU) + 3H(ZX) + 3H(XU) + H(YZ) + H(YU).$$

More inequalities of this sort were found, partly with the aid of computer searches, by Dougherty, Freiling, and Zeger [6]. Metcalf-Burton [14] developed a fairly general procedure for using such inequalities to deduce bounds on the rates of access structures.

References

1. Beimel, A., Livne, N., Padró, C.: Matroids can be far from ideal secret sharing. In: Canetti, R. (ed.) Theory of Cryptography, pp. 194–212. Springer, Berlin (2008)
2. Brickell, E.F., Davenport, D.M.: On the classification of ideal secret sharing schemes. J. Cryptology **4**, 123–134 (1991)
3. Brylawski, T.: Appendix of matroid cryptomorphisms. In: [21], pp. 298–312
4. Brylawski, T.: Constructions. In: [21], Chapter 7, pp. 127–223
5. Dougherty, R., Freiling, C., Zeger, K.: Linear rank inequalities on five or more variables. http://arxiv.org/abs/0910.0284
6. Dougherty, R., Freiling, C., Zeger, K.: Six new non-Shannon information inequalities. In: IEEE International Symposium on Information Theory, Seattle, WA (July 2006)
7. Ingleton, A.W.: Representation of matroids. In: Welsh, D.J.A. (ed.) Combinatorial Mathematics and Its Applications, Academic Press, London, pp. 149–167 (1971)
8. Karnin, E.D., Greene, J.W., Hellman, M.E.: On secret sharing systems. IEEE Trans. Inf. Theory **29**, 35–41 (1983)
9. Kinser, R.: New inequalities for subspace arrangements. J. Comb. Theory A **118**, 152–161 (2011)
10. Kurosawa, K., Okada, K., Sakano, K., Ogota, W., Tsujii, S.: Nonperfect secret sharing schemes and matroids. In: EUROCRYPT '93, pp. 126–141 (1994)
11. Livne, N.: On matroids and non-ideal secret sharing. Master's thesis, Ben-Gurion University (2005)
12. Martí-Farré, J., Padró, C.: On secret sharing schemes, matroids, and polymatroids. Cryptology ePrint Archive, Report 2006/077 (2006). http://eprint.iacr.org/
13. Matuš, F.G.: Two constructions on limits of entropy functions. IEEE Trans. Inf. Theory **53**, 320–330 (2007)
14. Metcalf-Burton, J.R.: Information rates for secret sharing over various access structures. Ph.D. thesis, University of Michigan (2009)
15. Seymour, P.D.: A forbidden minor characterization of matroid ports. Q. J. Math. **27**, 407–41 (1976)
16. Seymour, P.D.: On secret-sharing matroids. J. Comb. Theory B **56**, 69–73 (1992)
17. Shamir, A.: How to share a secret. Commun. ACM **22**, 612–613 (1979)
18. Shannon, C.: A mathematical theory of communication. Bell Syst. Tech. J. **27**, 379–423 (1948)
19. Stinson, D.R.: An explication of secret sharing schemes. Des. Codes Crypt. **2**, 357–390 (1992)
20. Vámos, P.: On the In: representation of independence structures. Theory of Matroids, Encyclopedia of Mathematics, (1968)
21. White, N. (ed.): Theory of Matroids. In: Encyclopedia of Mathematics, vol. 26. Cambridge University Press, Cambridge (1986)
22. Zhang, Z., Yeung, R.W.: On characterization of entropy function via information inequalities. IEEE Trans. Inf. Theory **44**, 1440–452 (1998)

Printed in the United States
By Bookmasters